Rapid Assessment Program
Programa de Evaluación Rápida

RAP Bulletin
of Biological
Assessment

*Boletín RAP
de Evaluación
Biológica*

**A Biological Assessment of
Laguna del Tigre National Park,
Petén, Guatemala**

*Evaluación Biológica de los Sistemas
Acuáticos del Parque Nacional
Laguna del Tigre, Petén, Guatemala*

Brandon T. Bestelmeyer and Leeanne E. Alonso, Editors

CENTER FOR APPLIED BIODIVERSITY SCIENCE
(CABS)
CONSERVATION INTERNATIONAL (CI)
CI-PROPETEN
CONSEJO NACIONAL DE AREAS PROTEGIDAS
(CONAP)
CENTRO DE ESTUDIOS CONSERVACIONISTAS
(CECON)
ASOCIACION GUATEMALTECA PARA LA
CONSERVACION NATURAL (CĂNAN K'AAX)
COMISION NACIONAL DEL MEDIO AMBIENTE
(CONAMA)

RAP Bulletin of Biological Assessment is published by:
Conservation International
Center for Applied Biodiversity Science
Department of Conservation Biology
2501 M Street, NW, Suite 200
Washington, DC 20037
USA
202-429-5660 tel
202-887-0193 fax
www.conservation.org

Editors: Brandon T. Bestelmeyer and Leeanne E. Alonso
Assistant Editors: Debbie Gowensmith and Carlos Rodríguez Olivet
Design: Glenda P. Fábregas
Cover photographs: Leeanne E. Alonso
Translator: Monica Mack of Absolute Translations
Secondary translators: Francisco Castañeda Moya, Christian Barrientos, Carlos Rodríguez Olivet, Alejandro Queral-Regil and Glenda P. Fábregas

ISBN 1-881173-33-X

RAP Bulletin of Biological Assessment was formerly RAP Working Papers. Numbers 1-13 of this series were published under the previous series title.

Suggested citation:
Brandon T. Bestelmeyer and Leeanne E. Alonso (eds.). 2000. A Biological Assessment of Laguna del Tigre National Park, Petén, Guatemala. RAP Bulletin of Biological Assessment 16, Conservation International, Washington, DC.

Printed on recycled paper.

This study was funded by CI-USAID, the Global Environment Facility and The Rufford Foundation.

El Boletín RAP de Evaluación biológica es publicado por:
Conservation International
Centro para la ciencia de biodiversidad aplicada
Departamento de biología de conservación
2501 M Street, NW, Suite 200
Washington, DC 20037
USA
202-429-5660 tel
202-887-0193 fax
www.conservation.org

Editores: Brandon Bestelmeyer y Leeanne E. Alonso
Editores Asistentes: Debbie Gowensmith y Carlos Rodríguez Olivet
Diseño: Glenda P. Fábregas
Fotos de la carátula: Leeanne E. Alonso
Traductor: Monica Mack de Absolute Translations
Traductores secundarios: Francisco Castañeda Moya, Christian Barrientos, Carlos Rodríguez Olivet, Alejandro Queral-Regil y Glenda P. Fábregas

Conservation International es una organización privada, no lucrativa exenta del impuesto sobre la renta federal bajo la Sección 501 c(3) del Código de Rentas Internas.

ISBN 1-881173-33-X
©2000 por Conservation International.
Todos los derechos reservados.
Biblioteca de número de catálogo de tarjetas del Congreso 00-1002039

Las designaciones de entidades geográficas en esta publicación y la presentación de material no implican la expresión de ninguna opinión por parte de Conservation International o sus organizaciones patrocinadoras concerniente a la condición legal de cualquier país, territorio o área o sus autoridades, o concerniente a la delimitación de sus fronteras o límites.

Cualquiera de las opiniones expresadas en el Boletín RAP de las series de evaluación biológica pertenecen a los escritores y no necesariamente reflejan las de los miembros de Conservation International o sus co editores.

El Boletín RAP de Evaluación biológica inicialmente Documentación de trabajo de RAP. Los números 1-13 de esta serie fueron publicados bajo el título anterior de series.

Cita sugerida:
Brandon Bestelmeyer y Leeanne E. Alonso (eds.). 2000. Evaluación Biológica de los Sistemas Acuáticos del Parque Nacional Laguna del Tigre, Petén, Guatemala. Boletín RAP de Evaluación Biológica 16, Conservation International, Washington, DC.

Impreso en papel reciclado.

Este estudio fue financiado por CI-USAID y the Global Environment Facility y la Fundación Rufford.

TABLE OF CONTENTS

183 **Appendices**

PARTICIPANTS

Leeanne E. Alonso, Ph.D. (myrmecology [ants])
Rapid Assessment Program
Conservation International
2501 M Street, NW, Suite 200
Washington, DC 20037 USA
email: l.alonso@conservation.org

Ana Cristina Bailey (entomology)
Laboratorio de Ecología Sistemática
Universidad del Valle de Guatemala
18 Avenida 11-95 Zona 15, VH. III
Apdo. Postal 82
01901 Ciudad Guatemala, Guatemala
email: abailey@uvg.edu.gt

Christian Barrientos (ichthyology, RAP protocol)
CI-ProPetén
Calle Central
Ciudad Flores
El Petén, Guatemala
email: cbarrientos@conservation.org.gt

Brandon T. Bestelmeyer, Ph.D. (myrmecology, international team leader)
Department of Biology
MSC 3AF, Box 30001
New Mexico State University
Las Cruces, NM 80003 USA
email: bbestelm@jornada.nmsu.edu

Marcos Callisto, Ph.D. (limnology)
Universidade Federal de Minas Gerais ICB,
Depto. Biologia Geral
CP. 486, CEP. 30.161-970
Belo Horizonte, MG, Brasil
email: callisto@mono.icb.ufmg.br.gt

Francisco Castañeda Moya (herpetology, national team leader, RAP protocol)
CI-ProPetén
Calle Central
Ciudad Flores
El Petén, Guatemala
email: fmoya@conservation.org.gt

Miriam Lorena Castillo Villeda (ornithology)
CI-ProPetén
Calle Central
Ciudad Flores
El Petén, Guatemala
email: mcastillo@conservation.org

Barry Chernoff, Ph.D. (ichthyology)
Department of Zoology
Field Museum
Roosevelt Road at Lakeshore Drive
Chicago, IL 60605 USA
email: chernoff@fmnh.org

Javier Garcia Esquivel (RAP protocol)
CONAP
Villa 5 4-50, Zona 4
Edificio Maya
4° Nivel,
Ciudad Guatemala, Guatemala

Karin Herrera (limnology)
Universidad de San Carlos de Guatemala
2° Nivel, Edificio T-12
Ciudad Universitaria
Zona 12
Ciudad Guatemala, Guatemala
email: kherrera@usa.net

Herman A. Kihn, M.Sc. (ichthyology)
Centro de Estudios Conservacionistas (CECON)
Unidad de Investigación
Vida Silvestre
Avenida Reforma 0-78, Zona 10
Ciudad Guatemala, Guatemala

Oscar Lara, M.Sc. (herpetology, RAP protocol)
CONAP
4a. calle 6-17 z. 1
Ciudad Guatemala, Guatemala
email: seconap@guate.net
Escuela de Biología
Universidad de San Carlos de Guatemala
Edificio T-10
Ciudad Universitaria, zona 12
Ciudad Guatemala, Guatemala
email: oslara@usac.edu.gt

Blanca León, Ph.D. (botany)
Museo de Historia Natural
Universidad Nacional Mayor de San Marcos
Av. Arenales 1256, Apartado 14-0434
Lima-14 Peru
email: leon@umbc.edu

Mario Mancilla (RAP protocol)
CANANKAX
Barrio el Redentor
Colonia Itzá, Frente al Salón Social San Benito
Petén, Guatemala
e-mail: canankaxpet@guate.net

Julio Morales Can (botany)
Universidad de San Carlos de Guatemala
Edificio T-13, Ciudad Universitaria
Zona 12
Ciudad Guatemala, Guatemala
email: quinchobarrilete@yahoo.com

Jorge Ordoñez (entomology)
Proyecto Fauna
CI-Guatemala
7a-avenida 3-33
Zona 9, Edificio Torre Empresarial
4o nivel, oficina 406
Ciudad de Guatemala, Guatemala
email: jorge.e.ordonez@usa.net
        jorge.e.ordonez@citel.com.gt

Ismael Ponciano (RAP protocol)
Centro de Estudios Conservacionistas (CECON)
Unidad de Investigación
Vida Silvestre
Avenida Reforma 0-78, Zona 10
Ciudad Guatemala, Guatemala

Sergio G. Pérez (mammalogy)
Colecciones Zoologicas
Museo de Historia Natural
Universidad de San Carlos de Guatemala
Calle Mariscal Cruz 1-56
Zona 10
Ciudad Guatemala, Guatemala
email: museo@usac.edu.gt

Alejandro Queral-Regil, M.Sc. (herpetology)
11700 Old Columbia Pike, Suite 1902
Silver Spring, MD 20904 USA
fax: 202-547-6009
email: alejandro.queral@sierraclub.org

Carlos Rodriguez Olivet (RAP protocol)
CI-Guatemala
7a-avenida 3-33
Zona 9, Edificio Torre Empresarial
4o nivel, oficina 406
Ciudad de Guatemala, Guatemala
email: crodriguez@citel.com.gt

Edgar Selvin Pérez (ornithology)
Escuela de Biología
Universidad de San Carlos de Guatemala
Edificio T-10
Ciudad Universitaria, zona 12
Ciudad Guatemala, Guatemala
email: selvin_perez@latinmail.com

Christopher W. Theodorakis, Ph.D. (toxicology)
Texas Tech University
The Institute of Environmental and Human Health
P.O. Box 41163
Lubbock, TX 79409-1163 USA
fax: 806-885-4577
email: chris.theodorakis@tiehh.ttu.edu

Philip W. Willink, Ph.D. (ichthyology)
Fish Division
Field Museum
Roosevelt Road at Lakeshore Drive
Chicago, IL 60605 USA
email: pwillink@fmnh.org

Heliot Zarza (mammalogy)
Instituto de Ecología, UNAM
Circuito Ext. s/n junto al Jardín Botánico, CU
Ap. Postal 70-275, 04510
México D.F., México
fax: 56 22 98 95
email: hzarza@nosferatu.ecologia.unam.mx

ORGANIZATIONAL PROFILES

CONSERVATION INTERNATIONAL

Conservation International (CI) is an international, non-profit organization based in Washington, DC. CI acts on the belief that the Earth's natural heritage must be maintained if future generations are to thrive spiritually, culturally, and economically. Our mission is to conserve biological diversity and the ecological processes that support life on earth and to demonstrate that human societies are able to live harmoniously with nature.

Conservation International
2501 M Street, NW, Suite 200
Washington, DC 20037 USA
202-429-5660 (telephone)
202-887-0193 (fax)
http://www.conservation.org

CONSERVATION INTERNATIONAL - PROPETEN

In response to increasing threats to Guatemala's biodiversity in its nothernmost department of the Petén, CI established a Guatemalan branch, locally known as ProPetén, in 1991. CI-ProPetén works with local communities to conserve biological diversity by increasing environmental awareness and by developing economic alternatives for local communities. ProPetén strives to demonstrate that local communities can live within the Petén's remaining forests without destroying them. CI-ProPetén's efforts focus on the 1.6 million hectare (4 million acre) Maya Biosphere Reserve (MBR). The reserve contains the majority of the forests of the larger Maya Forest, which extends into Belize and Mexico.

Established in 1990, the MBR protects a number of ecosystems, most notably Laguna del Tigre National Park (LTNP), Central America's largest freshwater wetland. LTNP has been one of ProPetén's highest priorities in the last three years. Working closely with local, national, and international stakeholders, ProPetén has helped to create a viable park management infrastructure and has provided policy guidelines to national authorities to mitigate impacts from human activities in the region. Both ProPetén scientific staff and its Scarlet Macaw Biological Station, located on the southern frontier of Laguna del Tigre National Park, were utilized during the RAP expedition.

CI-ProPetén
Calle Central
Ciudad Flores
El Petén, Guatemala
502-926-1370 (telephone)
502-926-0495 (fax)

CECON: EL CENTRO DE ESTUDIOS CONSERVACIONISTAS (CENTER FOR CONSERVATION STUDIES)

The Center for Conservation Studies (CECON) is an institution of the University of San Carlos located in Guatemala City. CECON has the following as its principal purposes: 1) to undertake basic biological studies in Guatemala and use such studies as a basis for national conservation efforts, which include defining policies that the national government can use for better management of Guatemala's protected areas; 2) to rationally and technically manage a system of biotopes and other protected areas for the conservation of wildlife biodiversity; and 3) to lead a national effort to promote biological themes with an emphasis on conservation.

CECON manages four protected biotopes in the Maya Biosphere Reserve (MBR): Laguna el Tigre-Río Escondido, Cerro Cahuí, Naachtún-Dos Lagunas, and Zotz-San Miguel La Palotada. The Laguna del Tigre Biotope is located in the larger Laguna del Tigre National Park and is regarded as one of

Guatemala's most critical habitats. In an effort to gather more information on the regions and use such information as a basis for future protection efforts against human activity in the regions, CECON has played a critical role in the Laguna del Tigre RAP expedition. Personnel from the institute and an outpost in the park were utilized during the expedition.

Centro de Estudios Conservacionistas (CECON)
Avenida Reforma 0-63, Zona 10
Ciudad Guatemala, Guatemala
502-334-7662 (telephone)
502- 334-7664 (fax)

CONAP: CONSEJO NACIONAL DE AREAS PROTEGIDAS (NATIONAL COUNCIL FOR PROTECTED AREAS)

Founded in 1989 as an office of the presidency of Guatemala, the National Council for Protected Areas (CONAP) is in charge of administering all of Guatemala's protected areas. These areas, comprised of at least eighty-one units, contain just under two million hectares, or about eighteen percent of Guatemala's territory. The largest unit within this system of protected areas is the Maya Biosphere Reserve (MBR), located in the northern Petén. Under the Maya Biosphere Project and other major initiatives in the Petén, the government of Guatemala—via CONAP—has made the region the highest priority for national and international conservation investments. In 1996, CONAP declared Laguna del Tigre National Park (LTNP) and Sierra del Lacandon National Park (SLNP) its highest national priotrities. Since that time, CONAP has worked with local and international agencies—CI-ProPetén among them—to put into place management systems that reconcile human economic necessity with effective conservation.

Consejo Nacional de Areas
Protegidas (CONAP)
Villa 5 4-50, Zona 4
Edificio Maya
4º Nivel
Ciudad Guatemala, Guatemala
502-332-0465 (telephone)
502-332-0464 (fax)

CONAMA: LA COMISION NACIONAL DEL MEDIO AMBIENTE (NATIONAL COMMISSION OF THE ENVIRONMENT)

The National Commission of the Environment (CONAMA) is a governmental agency under the executive branch of the Guatemalan government. The agency acts as an assessor, coordinator, and facilitator for all necessary actions needed to protect and improve the environment. To achieve its goals, it works through the various government ministries, the Secretary General of Economic Planning, various autonomous and semi-autonomous agencies, municipalities, and the country's private sector.

La Comisión Nacional del Medio Ambiente (CONAMA)
7a. Avenida 7-09, Zona 9
Ciudad Guatemala, Guatemala 01013

CÄNAN K'AAX: ASOCIACION GUATEMALTECA PARA LA CONSERVACION NATURAL (GUATEMALAN ASSOCIATION FOR NATURAL CONSERVATION)

The Guatemalan Association for Natural Conservation (CÄNAN K'ÄAX) is a not-for-profit organization whose members are almost all from the Petén region. CÄNAN K'AAX was formed to respond through a local perspective to the conservation challenges and management of protected areas in the Petén and the larger Maya Forest. It derives its name from the Maya Itza word for "Guardian of the Forest."

The philosophy of the association is based in the principle of healthy co-existence between human beings and the environment. To realize this philosophy, its actions are based in the development and strengthening of the capacity and attitudes of civil society and the government in order to manage natural resources in a more sustainable manner, particularly with regards to protected areas. Presently, the association is dedicating itself to the co-management of Laguna del Tigre National Park in conjunction with CONAP.

Barrio el Redentor
Colonia Iztá, Frente al Salón San Benito
Petén, Guatemala
502-926-3732 (telephone)

ESTACION BIOLOGICA LAS GUACAMAYAS (THE SCARLET MACAW BIOLOGICAL RE-SEARCH STATION)

The Scarlet Macaw Biological Research Station, named after a threatened bird species within Laguna del Tigre National Park, was established in 1996 and is a base of operations for CI's broad conservation efforts in the most remote and threatened sections of this protected area. The field station addresses both the short- and long-term conservation needs of the park. By creating a permanent and well-equipped facility, high-quality personnel can be brought into the remote area for extended periods, enabling extensive ecological and species research to take place, which is ordinarily difficult in such a remote region. Presently, collaboration with the University of San Carlos in Guatemala provides field-based education opportunities for

Guatemalan students. Research partnerships with U.S.-based universities are also under negotiation.

Send inquiries to:
CI-ProPetén
Calle Central
Ciudad Flores
El Péten, Guatemala
502-926-1370 (telephone)
502-926-0495 (fax)

ACKNOWLEDGMENTS

We thank Juventino Galvez, Oscar Lara, and Javier García Esquivel at the regional office of the Consejo Nacional de Areas Protegidas for their participation, dedication, and cooperation in making this expedition possible and successful. We also thank Herman Kihn and the Centro de Estudios Conservacionistas of the University of San Carlos for coordinating our use of the Laguna del Tigre Biotope and for participating in the expedition. Thanks also to Mario Mancilla and CÄNAN K'AAX for their participation.

The staff at CI-ProPetén, especially Miriam Castillo, Rosita Contreras, Jorge Ordoñez, and Juan Mañuel Burgos, organized the logistics of the expedition. Imelda López Ascencio and Gloria Reyes Méndoza fed us extremely well, y siempre con muchas tortillas deliciosas! Manuel Alvarado provided us with boats and boatmen. We thank Francisco Bosoc, Baudilio Choc, Raquel Quixchán, Victor Cohuog, David Figueroa, Carlos Girón, Federico Godoy, Ramón Manzanero, Cesar Morales, Ramiro Lucero, Roman Santiago, and José Soza for their untiring efforts in the field, for their good humor and friendship, and for sharing their knowledge of the beautiful place in which they live. Without these friends our work would not have been possible.

We thank the following people for assistance with species identifications and other support: Mary Anne Rogers and Kevin Swagel (Chapter 4); Rodrigo Medellín, Fernando Cervantes, and Yolanda Hortelano (Chapter 8); and William P. MacKay, John T. Longino, and Phil S. Ward (Chapter 9).

Finally, we thank Jim Nations of CI-Washington and Carlos Rodriguez and Carlos Soza of CI-ProPetén for their guidance and support of this endeavor.

This research was supported by grants from the U.S. Agency for International Development and the Global Environment Facility to CI-ProPetén and by a grant from the Rufford Foundation to the Rapid Assessment Program (RAP) of Conservation International. We also thank Jay Fahn, Leslie Lee, and David and Janet Shores for supporting AquaRAP studies.

REPORT AT A GLANCE

A BIOLOGICAL ASSESSMENT OF LAGUNA DEL TIGRE NATIONAL PARK, PETEN, GUATEMALA

1) Dates of Studies
RAP Expedition: 8 – 30 April, 1999

2) Description of Location
Laguna del Tigre National Park (LTNP) is located within the Maya Forest, a largely dry forest system that extends from southern Mexico through northern Guatemala and Belize. LTNP occupies more than 289,000 hectares of the Maya Biosphere Reserve (MBR), which is the largest system of managed and protected areas of the Maya Forest. Within the MBR, the park is the largest protected core area. The western portion of the park is characterized by flooded savanna wetlands and lagoons. The wetland areas were included as "Wetlands of International Importance" under the Ramsar convention in 1990 and 1999. The eastern portion has more relief and possesses tall upland forests interspersed by low-lying forests of shorter stature and denser undergrowth. Tall riparian forests and sawgrass marshes line rivers within the park. The largest river within the park, the Río San Pedro, defines its southern boundary. This river is a tributary of the Río Usumacinta, which drains the Usumacinta Basin into the Gulf of Mexico. Other rivers within the park are tributaries of the Río San Pedro, including the Río Sacluc in the east and the Río Escondido to the west. The Río Candelaria is an independent affluent of the Laguna de Terminos.

3) Reason for RAP Studies
Despite the strategic importance of LTNP as a core conservation area within the MBR, the park is extremely threatened due to the development of core areas for oil exploitation. Human colonization of the park along roads created for oil operations is also pressuring the park's systems. Additional colonization is occurring along river courses, particularly the Río San Pedro and Río Escondido. Some consequences of this colonization include deforestation for slash-and-burn agriculture; the escape of fires into non-target areas such as sawgrass marshes; and unregulated hunting, fishing, and pollution. The effects of such activities on biological diversity are likely to become severe unless immediate action is taken to preserve the integrity of designated core areas of the MBR. Our studies highlight the regional importance of LTNP, ascertain the specific nature of the threats to its biological diversity, and suggest management options for areas within the park.

4) Major Results
A great variety of both aquatic and terrestrial habitats and taxa were observed in the course of the RAP sampling, including marshes, lagoons, rivers, streams, regenerating forests, and pristine forests. Each habitat was evaluated with respect to phytoplankton, aquatic insects, fish, amphibians, reptiles, ants, birds, and mammals. Environmental heterogeneity was recognized differently by each taxon. For some taxa, marshes were important; for others, lagoons, rivers, or complex forests were important.

In total, we observed 647 species. We recorded range extensions for one mammal, three birds, and two ant species. This RAP survey provided the first species lists of ants and phytoplankton for LTNP. Remarkably, the park harbors sixteen vertebrate species that are of international concern and that are listed by the International Union for the Conservation of Nature (IUCN) and/or the Convention on the International Trade of Endangered Species of Wild Fauna and Flora (CITES). Also, an extremely rare freshwater reef habitat, composed of living and dead bivalve molluscs, was discovered on the Río San Pedro. This reef harbored a unique assemblage of plants and animals.

Overall, the diversity within taxa and the representation of endemic, rare, or endangered species was great, underscoring the park's value for conservation in Central America and the park's potential to maintain longterm ecological integrity. Further, an ecotoxicological assessment of two fish species revealed an impact of mutagenic contaminants within some of the park's aquatic ecosystems. This result confirms the need for more intensive toxicological studies.

Numbers of species recorded:

Phytoplankton:	71 species
Aquatic insects associated with *Salvinia auriculata:*	44 species
Plants:	130 species
Fish:	41 species
Birds:	173 species
Reptiles and amphibians:	36 species
Mammals:	40 species
Ants:	112 species

New records for the region or Guatemala:

Birds:	3 species
Reptiles:	1 species
Mammals:	1 species
Ants:	2 species

5) Conservation Recommendations

LTNP as a whole needs immediate protection from deforestation and fires in order to preserve its current high levels of biodiversity. Several areas within the park are of particular importance and are threatened: flooded savannas in and around the Laguna del Tigre Biotope, sawgrass marshes along the ríos San Pedro and Escondido, and lagoons along the Xan-Flor de Luna road. The areas near the Guacamayas Biological Station, including the region near El Perú, enjoy relatively good protection; this should be maintained. Areas not currently inside the park boundaries but that are unique and should be protected include the Río Sacluc, varzea (inundated) forests near Paso Caballos, sawgrass marshes south of the Río San Pedro, mangroves along the Río San Pedro, and a freshwater reef system east of El Naranjo.

EXECUTIVE SUMMARY

INTRODUCTION

Laguna del Tigre National Park (LTNP) and the Laguna del Tigre Biotope within it are core conservation areas within the Maya Biosphere Reserve (MBR). These areas contain among the last remaining large tracts of Central American Maya Forest. This forest and the diverse habitats it contains are in danger of disappearing within this century given current rates of deforestation. The value of LTNP emerges from its relatively pristine state and its central location within the MBR. With the advent of road-building and other development in the area, the park's forests, wetlands, and rivers—and the biodiversity that these habitats harbor—are under increasing pressure from human colonization and the development of oil extraction facilities within the park. The primary objectives of this Conservation International (CI) Rapid Assessment Program (RAP) expedition were to augment the LTNP biodiversity database, highlight the park's regional importance, evaluate the nature of the threats to animals and plants within the park, and provide this information to natural resource managers and decision-makers. Other objectives included bringing international attention to an important scientific resource, the Guacamayas Biological Station (GBS), a research station located near the southern boundary the park, and training local students in species identification and ecological monitoring techniques. The ecological work focused on documenting the biodiversity value of several aquatic and terrestrial taxa in the park, investigating the relationships between taxa and different habitats and areas of the park in order to ascertain those areas' conservation value, and evaluating the potential effects of ongoing deforestation and petroleum development on LTNP's biodiversity. Throughout this project, CI has worked closely with scientists from CI-ProPetén (CI's Guatemala program), the Consejo Nacional de Areas Protegidas (CONAP), and the Centro de Estudios Conservacionistas (CECON) of the University of San Carlos. This project was made possible by generous grants from USAID to CI-ProPetén and from the Rufford Foundation to RAP.

This RAP expedition was unique in its emphasis on simultaneously providing data for both aquatic and terrestrial systems. In order to characterize patterns of species diversity, we sampled the following taxonomic groups: aquatic plants, fishes, phytoplankton and aquatic insects, aquatic reptiles, aquatic/riparian birds, ants, bats, and rodents. Each of these taxonomic groups represent important components of biodiversity in the region and reflect a wide array of ecological variation that affects biodiversity as a whole. In addition, we examined the toxicological effects of contaminants on individuals of two fish species using novel molecular techniques. This component allowed us to evaluate the threats by pollution on the park's biodiversity and on human populations connected with the park.

The RAP team examined taxa within four focal areas of the park that harbor varying kinds and amounts of terrestrial and aquatic habitats. These areas included sites near GBS and along the ríos San Pedro and Sacluc; within the Laguna del Tigre Biotope along the Río Escondido; near the Laguna Flor de Luna and the Xan Petroleum operation, where ecotoxicological studies were concentrated; and finally in the vicinity of the ríos Chocop and Candelaria (see Map 3).

CHAPTER SUMMARIES

Limnology and Water Quality

The limnologists, in conjunction with botanists, provided a first-ever classification of the diverse aquatic macrohabitats of LTNP. Eight distinct habitat types representing springs, deep rivers, marshes, and lagoons were classified. The varying physico-chemical characteristics of these habitats are the basis of the aquatic diversity of the park. The pH of the park's waters ranged from acid (6.78) to alkaline (8.30), from turbid and dark to aquamarine, and from eutrophic to oligotrophic. This diversity is based in the common karstic geology of the region. One unexpected discovery was the presence of a freshwater, mollusk-based reef in the Río San Pedro near the town of El Naranjo. This reef, which may serve to filter pollutants from the river, harbors a unique assemblage of invertebrates and is thus of great conservation value. The water quality of the park was generally good. Coliform bacteria were detected in 37% of the samples, but fecal coliform bacteria were detected in only 13% of the samples and were associated with the use of water bodies by settlers.

Phytoplankton composition also may be used to evaluate water quality and generally reflects variation in the physico-chemical characteristics of aquatic habitats. Fifty-nine genera and 71 species and morphospecies were recorded. Diatoms (Bacillariophycea), followed by chlorophytes (Chlorophyceae) and cyanophytes (Cyanophyceae) dominated the samples. Most taxa were cosmopolitan in their distributions. Several lagoons and sample points in the ríos Escondido and San Pedro harbored phytoplankton species that are indicative of pollution (e.g. *Microcystis aeruginosa* and *Synedra ulna*), but high densities of these species were observed only at two points in the Río San Pedro, one site in the Río Escondido, and in the Laguna la Pista. Multivariate analyses revealed that the single most important source of variation determining phytoplanktonic composition was pH, and the second was conductivity. Neither variable, however, was able to explain much of the variation.

In an intensive study of an insect-plant association, 44 morphospecies and 26 families of insects were found to inhabit a single aquatic plant species, *Salvinia auriculata*. This result highlights the great diversity of insects in the littoral aquatic systems of the park.

Aquatic Plants

RAP botanists recorded 67 families, 120 genera, and 130 species of aquatic macrophytes. Overall, the composition of this flora is similar to neighboring areas in the Yucatán Peninsula. Two general vegetation types, marshes (sibal, mainly herbaceous plants) and riparian forest (characterized by woody plants), form a mosaic within LTNP. Marshes, dominated by the coarse, grasslike herb *Cladium jamaicense*, were found to have the highest diversity of aquatic and terrestrial plants and appear to be in a constant state of succession.

The Cyperaceae and Fabaceae were the most diverse families in LTNP, and *Utricularia*, an aquatic floating plant that feeds on small invertebrates, was the most species-rich genus. The tree known as pucté (*Bucida buceras*) dominates riverine forest. Associations of pucté with other plant species define five riparian habitat types in the park. A few individuals of coastal mangrove (*Rhizophora mangle*) found along the Río San Pedro may be remnants of the Pleistocene, when this area was mostly under the ocean. This population is the most continental known in the Yucatán region. Since these trees are found outside the park, the mangroves may not be protected from destruction of their habitat. Among terrestrial habitats, the oak-dominated (*Quercus oleiodes*) encinal habitat is also rare. The molluscan reef habitat noted earlier harbors the rare aquatic plant *Vallisneria americana*.

Fishes

Fish diversity was found to have changed little from a previous survey performed more than sixty years ago. Forty-one species of fish were recorded, and three of these are new records for the Río San Pedro Basin. Two species were undescribed. Overall, many of the taxa found in the park are endemic to the Usumacinta drainage, and protection of the park would help to protect 25% of this fauna. The most species-rich area was the Río San Pedro, but most species were homogeneously distributed throughout the park. Lagoons, for instance, harbored 12 species, irrespective of the level of disturbance to the surrounding vegetation or the level of use by humans. Several habitats are particularly important to fish populations in LTNP. Sawgrass marshes composed of *Cladium jamaicense* (sibal) are breeding sites for cichlids and catfish and are abundant on the ríos San Pedro and Escondido. A varzea-type, or flooded-forest, habitat near Paso Caballos is an important nursery for *Brycon* and is generally a little-known habitat in Central America. Finally, small streams are unique fish habitats but are threatened by both pollution and siltation due to deforestation. Overfishing and habitat loss to fire are important threats to this economically and aesthetically valuable fauna.

Toxicology

In addition to assessing the diversity of the fish fauna, team members examined the effects of toxic compounds on two widespread fish species: *Cichlisoma synspilum* and *Thorychthys meeki*. Individual fish were examined in a pond adjacent to the Xan 3 oil well and at varying distances (from 5 km to 67 km) away from the source of contaminants under the hypothesis that the oil facility was the origin of toxic compounds. To characterize exposure to toxic compounds, two subcellular measures were used: DNA strand breakage and chromosomal breakage. These measures indicate exposure to compounds such as polycyclic aromatic hydrocarbons (PAHs), a contaminant associated with oil extraction. These techniques are unique in that they measure the effect of exposure to toxic compounds directly, thus indicating the degree to which populations of different kinds of organisms are affected by contamination.

Some fish at the Xan 3 site showed signs of fin rot, which indicates that these individuals were stressed, possibly due to contaminant exposure. Chromosomal breakage in *C. synspilum* was greater at Xan 3 than at the most distant site (in the Río San Pedro near LGBS) but was not greater at Xan 3 than at a nearer site. No variation in DNA breakage was apparent in *C. synspilum*, however. In contrast, *T. meeki* exhibited high values of DNA breakage at Xan 3, Laguna Buena Vista, and Laguna Guayacán and low values in the Río San Pedro and Laguna Flor de Luna, which was the third nearest site to Xan 3 and was nearer than Guayacán. Chromosomal breakage, however, did not vary in this species.

Assessments of the concentration of PAHs in the sediments where fish were collected revealed that concentrations at Xan 3 were similar to those at reference sites. PAH concentrations were higher at Laguna Buena Vista than at any other site. It is likely that a source of contamination is acting on fish populations, but the source or identity of the contaminant(s) will take more work to determine. It is possible that other kinds of contaminants which we did not measure are affecting fish populations, and it is unclear how these effects may be related to either oil extraction or other human influences. Nonetheless, the role of oil extraction in producing the toxic effects cannot be ruled out. For example, patterns of redistribution of contaminants and their effects may be related to temporally varying hydrological connections with respect to the timing of the release of contaminants as well as the behavior of the animal species examined. Our results support the notion that contamination is occurring in LTNP and that more intensive studies are warranted.

Birds

The bird fauna of LTNP is exceptionally rich and consists of several species of international concern. As a consequence, the wetlands of the western portion of LTNP were included in the Ramsar list of globally important wetlands in 1990 and 1999. We observed 173 species, including 11 new records for the park. A total of 256 species have been listed for the park. For the first time in 10 years, a jabiru (*Jabiru mycteria*), which is the largest bird in the Americas and is very rare, was observed. In addition, important distribution extensions for the fork-tailed flycatcher (*Tyrannus savanna monachus*), the gray-breasted crake (*Laterallus exilis*), and the vermilion flycatcher (*Pyrocephalus rubinus*) were documented. The western part of the park, with its flooded

savannas and shallow ponds (playones), contains unique habitats for wetland birds. Additionally, the Río Escondido and Laguna la Lampara are important sites. Forest specialists including the great curassow (*Crax rubra*), wrens (Trogloditidae), and woodcreepers (Dendrocolaptidae) were abundant in bosque alto on hills and in closed riparian (gallery) forests. Overall, patterns of topography and vegetation structure are closely related to patterns of variation in the composition of birds within LTNP. Preservation of the diversity of habitat types within LTNP, particularly the wetlands and bosques altos, is key to conserving its great bird diversity.

Aquatic Reptiles

Reptiles were surveyed near water bodies throughout the park, with a focus on populations of the Morelet's crocodile (*Crocodylus moreletii*). This animal is classified as "data deficient"—that is, understudied—by the International Union for the Conservation of Nature (IUCN) and may be in danger of extinction due to overhunting and habitat loss. In 87.14 km of shoreline surveyed, 130 crocodiles were observed. Lagoons had higher densities than did rivers; of the lagoons, the Flor de Luna area and La Pista had the highest densities. The combined densities in the lagoons along the Xan-Flor de Luna road are the highest yet recorded in Guatemala. The Laguna el Perú and Río Sacluc had a relatively high representation of adults, which are essential for the maintenance of breeding populations in the park. As do several fish species, crocodiles use sawgrass marsh (sibal) habitats for nesting, cover, and food; therefore, this habitat is particularly important. Overall, crocodile populations in the park are healthy, and monitoring programs should be initiated immediately.

In addition, 14 species of amphibians and 22 species of reptiles were observed. The most abundant amphibian was the fringe-toed foamfrog (*Leptodactylus melanotus*), and the most abundant reptile was the striped basilisk (*Basiliscus vittatus*), a Central American endemic. The beautiful and endangered Central American river turtle, or tortuga blanca (*Dermatemys mawii*), was observed during the expedition. The herpetology team also captured the striped spotbelly snake (*Coniophanes quinquevittatus*), a species that had never before been documented in this area. Additional records of note include the red-eyed leaf-frog (*Agalychnis callidryas*), Baudin's treefrog (*Smilisca baudinii*), Mesoamerican slider (*Trachemys scripta*), and Northern giant musk turtle (*Staurotypus triporcatus*).

Mammals

Despite the short period of time and the difficult circumstances for studies of mammals, 40 species were recorded, which is 30% of the potential number of species in LTNP.

Twenty-three species of bats, two species of primates, three species of carnivores, one species of perissodactyl, three species of artiodactyls, and eight species of rodents were observed. Among the bats, three species were common, including *Artibeus intermedius* (a large fruit-eating bat), *Carollia brevicauda*, and *A. lituratus*. Two Yucatán-endemic rodents were captured: the Yucatán deer mouse (*Peromyscus yucatanicus*)—a new record for Guatemala—and the spiny pocket mouse (*Heteromys gaumeri*). There were no apparent differences in mammal diversity among the four localities observed in the park. Populations of Mexican howler monkeys (*Alouatta pigra*) and Geoffroy's spider monkeys (*Ateles geoffroyi*) were abundant throughout these localities, suggesting that many areas of the park are relatively undisturbed from the mammalian point of view. Other rare or endangered species including Baird's tapir (*Tapirus bairdii*), collared peccary (*Tayassu tajacu*), and red brocket deer (*Mazama americana*) were also observed in the park. These were observed near the GBS station along the Río San Pedro, suggesting that this area may be especially important for mammal conservation in LTNP.

Ants

As for mammals, the time available and the time of year made it difficult to obtain a complete ant sample. Nevertheless, an impressive 112 species and 39 genera were recorded. These values are greater than those recorded in several other studies in the region. As in other Neotropical sites, the genus *Pheidole* contained the largest number of species, but several arboreal genera were also highly diverse. Three species endemic to the Maya Forest region were recorded, including *Sericomyrmex aztecus*, *Xenomyrmex skwarrae*, and *Odontomachus yucatecus*. One extremely rare ant genus, *Thaumatomyrmex,* was discovered for the first time in Guatemala. Species richness was highest in relatively open or disturbed habitats, contrary to our expectations. This was likely due to the high microhabitat diversity of these sites. A disturbed habitat was relatively similar to a nearby pristine forest site, which is consistent with the suggestion that the recovery of Neotropical ant faunas from disturbance is rapid. Overall, ant composition was very patchy and turnover rates between sites very high, precluding an assessment of species-habitat associations and suggesting that overall richness was much higher than that recorded. More intensive studies are needed to ascertain which habitats have the greatest representation of rare or unique species, but our preliminary results suggest that the bosques altos harbor a high number of canopy and litter ant species and thus may be of particularly high conservation value.

A COMPARISON OF RAP RESULTS WITH PREVIOUS STUDIES

A previous survey conducted by CI-ProPetén (Méndez et al., 1998) addressed some of the same objectives as our RAP but focused on several different sites and distinct taxa. Taxa observed by Méndez et al. (1998) but not observed in this RAP included butterflies and terrestrial plant communities. Birds and amphibians were also studied. Méndez et al. considered how varying topographic characteristics in different zones of LTNP corresponded to patterns of biological diversity in a west-east gradient of increasing topographic relief across the park. Their objective was to provide a prioritization of the park's habitats. They recognized three geographic components of the park at this broad scale: 1) the western area of the park, consisting of an open-canopy forest matrix with patches of tall, closed-canopy bosque alto forest; 2) a central area that is similar to the western area but with flooded savanna to the north; and 3) an eastern area with a greater cover of bosque alto and patches of open forest, marshland, and oak forest (encinales). Méndez et al. found that most plant species were canopy species, and this group contained a high number of species found in the eastern area but nowhere else in the park (i.e. "exclusive" species). Most of the bosques altos in which these trees reside were on hills of 100-170 m. Méndez et al. recorded 219 species of birds, 60 of which were habitat-restricted in some way. They found that the central area had the highest number of exclusive bird species. The greatest richness of amphibians was found in the central area, and this was attributed to the area's ecotonal nature within the park. Of the 97 species of butterflies that were recorded, species richness was greatest in the eastern and central areas, and the eastern area harbored the largest number of exclusive species.

Finally, cluster analyses based on species composition revealed that the eastern portion of the park clustered away from the western and central areas of the park for all taxa examined, indicating the uniqueness of the eastern area. Méndez et al. concluded that the topographic gradient was related to patterns of biodiversity in LTNP. Although the eastern, most topographically varied part of the park was richest overall in exclusive species, it was also afforded the greatest protection in the park. The western area, on the other hand, is least rich in exclusive species but is under-protected. Méndez et al. considered the western area to be extremely fragile due to the extreme vulnerability of the rich aquatic systems found there and the threat posed by petroleum development to those systems' species.

In many cases, the RAP results agree with those of Méndez et al. The bosque alto and several aquatic habitats present in the Río San Pedro focal area in the eastern part of LTNP were recognized for their importance to groups such as fish, mammals, and ants. Wetland and lagoon habitats found in central and western parts of the park (i.e. the Río Escondido and Flor de Luna focal areas), however, were also recognized as essential for the conservation of aquatic birds and the Morelet's crocodile. Further, some areas not currently located within the park's boundaries should be considered in the region's management. Overall, the value of particular habitats depends upon the taxa considered. Viewed in this way, and considering the status of LTNP as a core area within the MBR, all of the park's habitats warrant protection.

RESEARCH AND CONSERVATION RECOMMENDATIONS

Our observations lead us to believe that LTNP is a critical core area within the MBR. It is an important reservoir of tropical dry forest habitats and wetlands that harbor biotic communities and individual species of great regional and global conservation value. Our studies in LTNP bring to light specific information needs that can be provided only by more intensive studies in the park. Nonetheless, we can offer several specific recommendations for the park's management. The basis for the following recommendations can be found in the text of the executive summary and in the following chapters. Recommendations are not listed in order of priority. We first provide recommendations for further scientific research in the park, noting that the GBS is an ideal location from which to launch these investigations, then conservation recommendations.

Scientific research priorities:

- Additional research determining the hydrological linkages throughout the park's landscapes and their seasonal dynamics will be a necessary first step to building an understanding of the aquatic systems of the park, including the flow of contaminants and the dispersal and population dynamics of species.

- Considering the intense seasonal dynamics of tropical dry forest areas such as LTNP, studies of both terrestrial and aquatic systems should be conducted at different times of year, particularly during the wet season when aquatic systems are better connected and when many terrestrial organisms are reproducing.

- More detailed studies of the relationships between topography, soils (particularly soil depth), and plant and animal communities are needed to provide a more fine-scale view of patterns of biodiversity for purposes of prioritization. Considering the high degree of incursion of settlers into the park, such studies will aid in the optimal allocation of protection for habitats in invaded areas.

Specific conservation recommendations for LTNP:

- Reduce the rate of deforestation and the burning of habitats such as sawgrass marshes (sibales), with the ultimate goal of halting deforestation and deliberate burning altogether inside LTNP. The ecological integrity of LTNP, as a core area of the MBR, must be maintained if the MBR as a whole is to persist and flourish.

- Immediately conserve particular areas of relatively pristine bosque alto (high forest). Current levels of protection afforded to areas near the GBS and along the Río San Pedro should be maintained, and forested areas farther from this protected zone, such as El Perú or near the Laguna Guayacán, should be monitored carefully. Human disturbance of the forests in the San Pedro area should be minimized in order to favor populations of large mammals.

- Protect several unique aquatic habitats within the park, and, for management purposes, include within the park other aquatic and terrestrial habitats that are now outside the park. These include the varzea-like flooded forests near Paso Caballos and along the Río Sacluc, the sawgrass marsh (sibal) habitats along the southern margin of the Río San Pedro and along the Río Escondido, the molluscan reef system at the "rapids" area on the Río San Pedro near El Naranjo, the mangrove remnant along the Río San Pedro, and the encinal vegetation along the Río Sacluc. Of particular concern is the widening of canals into sawgrass marsh habitat along the Río Escondido, associated with boat passage, which should be controlled.

- Initiate a water-quality monitoring program within the park. Of particular interest are areas that are used by human populations, including tributary springs along the Río San Pedro and lagoons on the Xan-Flor de Luna road. Specifically, the levels and effects of fecal contamination and agricultural runoff should be examined.

- Investigate the determinants of stress and genotoxicity to fish—revealed in this assessment—at Xan 3 and in other lagoons. Several questions remain unanswered: What contaminants are involved? What are the sources of these contaminants? What is the role of hydrology in determining the redistribution of the contaminants? How does species behavior determine the toxic effects of the contaminants?

- Monitor species populations within the park in order to assess trends in their health. Of particular concern are those populations that are small and/or exploited by humans. Specifically, the dynamics of food fishes such as *Petenia splendida* and *Brycon guatemalensis* (especially in the heavily exploited Río San Pedro) should be monitored. A conservation and monitoring program for crocodiles (*Crocodylus moreletii*) should focus on sites where they are currently abundant, including the Río Sacluc, Laguna el Perú, and Laguna la Pista. Finally, bird populations should be monitored, especially in aquatic habitats in the central and western parts of the park due to this area's international significance and the threats posed to its habitats by oil extraction and human colonization.

The condition, threats, and conservation actions we recommend are summarized by habitat type in Table i.

We hope these recommendations provide some guidance in the management of LTNP, and we encourage interested parties to contact the authors of this report if further clarification is desired.

LITERATURE CITED

Méndez, C., C. Barrientos, F. Castañeda, and R. Rodas. 1998. *Programa de Monitoreo, Unidad de Manejo Laguna del Tigre: Los Estudios Base para su Establecimiento.* CI-ProPetén, Conservation International, Washington, DC, USA.

Table i. Summary of the conservation importance, threats, overall habitat condition, and management recommendations for the principal habitat types of Laguna del Tigre National Park and Biotope.

Habitat Types of Concern	Location(s)	Conservation Importance	Threats	Habitat Condition	Recommendations
Rivers, tributaries, and caños	Río San Pedro/ Sacluc, and Río Escondido	High richness of aquatic plants, fishes	Pollution, especially of tributaries, overfishing	Variable, poorer near settlements	Monitor, regulate human waste practices and fishing pressure.
Rapids/ freshwater reefs	Río San Pedro	Unique freshwater reef system of bivalve mollusks and associated diversity	Pollution, overharvest of mollusks	Still good	Include in LTNP, monitor and regulate mollusc harvest and river pollution.
Lagoons	Flor de Luna, Río Chocop	High densities of Morelet's crocodiles, unique aquatic birds	Pollution, petroleum contamination	Variable, one lagoon near an oil well may be polluted	Prohibit human use of certain lagoons, monitor contamination levels.
Oxbow lagoons	Río Escondido	High richness of aquatic plants	Erosion due to passage of boats in canals	Variable	Regulate use of boats and perhaps boat size or speed in park.
Sawgrass (sibal) marshes	All	High richness of aquatic birds, nurseries for fish species	Burning	Variable, many damaged by fire	Include more marshland in park, prevent spread of fires into marshes, regulate burning in park.
Varzea-type forest	Río San Pedro/Sacluc	Nurseries for fish species	Human settlements	Unknown	Include in LTNP, more studies needed.
Bosque alto	All	High richness of ants, mammals	Deforestation, burning	Variable, many damaged by fire	Control burning and protect relatively pristine patches.
Encinal	Río Sacluc	Unique plants, likely unique animals	Deforestation, fragmentation	Unknown	Include areas along Río Sacluc in LTNP.

CHAPTER 1

INTRODUCTION TO THE RAP EXPEDITION TO LAGUNA DEL TIGRE NATIONAL PARK, PETEN, GUATEMALA

Brandon T. Bestelmeyer

LAGUNA DEL TIGRE NATIONAL PARK, PETEN, GUATEMALA

Laguna del Tigre National Park (LTNP) is a core area of the Maya Biosphere Reserve (MBR), and the threats facing it underscore the need for both research and conservation efforts in the park. The biosphere reserve concept holds that conservation and development can be balanced by the careful design and management of reserve networks. The function of core areas, as opposed to extractive or buffer zones in which human economic activity may occur, is to preserve representatives of minimally disturbed ecosystems that serve as a benchmark for regional monitoring efforts and that protect wild populations. According to Batisse (1986), "The protection against any action that could endanger the conservation role assigned to the core area must...be fully ensured." The role of this RAP is to assess the value of LTNP as a core area of the MBR and to evaluate the threats to its role as a core area.

The park is part of the Maya Forest, which contains Central America's largest remaining tropical forest ecosystems. The forest expands from southern Mexico (especially Chiapas, Campeche, and Quintana Roo states) through northern Guatemala and Belize and forms a natural biogeographic and cultural unit (Nations et al., 1998). The 16,000-km² MBR in the Petén department of northern Guatemala contains the largest system of managed and protected areas in the Maya Forest. LTNP, with an area of 289,912 hectares, is the largest and most important protected core area within this reserve.

Despite its protected status, the park is increasingly threatened by colonization and development and is being deforested at a rate of 0.57% per year (1995-1997 data; Sader, 1999). As a consequence, 50% of the forest has already been lost, and one projection estimates that 98% of the original forest in the Petén may be lost by the year 2010 if current deforestation rates are not reduced (Canteo, 1996).

ECOLOGICAL CHARACTERISTICS OF LTNP

The park lies within a zone characterized as tropical dry forest (Murphy and Lugo, 1995; Campbell, 1998). While tropical moist forests have received a great deal of attention from ecologists and conservationists, tropical dry forest ecosystems—Neotropical dry forests in particular—are little studied, intensively used by humans, and extremely endangered by development throughout the world (Mooney et al., 1995; Gentry, 1995). For this reason, the Maya Forest has global significance.

The topography of the area is low and flat, with maximum elevations not exceeding 300 m. The area is underlain by limestone rocks of Miocene age, giving rise to the karstic landscape typified by limestone cliffs along large river courses, well-drained hills with thinner soils surrounded by lower-lying areas with deep soils (known as bajos), and few streams but numerous water holes that arise from depressions in the limestone (Wallace, 1997). Smaller limestone sinks are known as aguadas, and larger ones are akalchés or lagoons (Campbell, 1998). The wetlands of the western portion of the park are believed to be the largest that remain in Central America (Wallace, 1997). These wetlands are also the largest in Guatemala and are recognized under the Ramsar convention for their international importance.

As in all tropical dry forest ecosystems, the contrast between the wet and dry seasons is an ecological feature of overriding significance (Murphy and Lugo, 1995). The annual precipitation of the region averages 1600 mm per year, but little precipitation may fall in the January-to-April dry season, when temperatures may exceed 40° C. This period presents an important challenge to many of the moist tropical-derived organisms in the park. In addition, the seasonality in precipitation interacts with topography to produce extreme variations in environmental conditions. For example, bajos may be flooded during the wet season but drought-stressed during the dry season (Whitacre, 1998). Water persists in the large lagoons throughout the year (Campbell, 1998) and are thus critical features of the park's landscape and ecology.

Perhaps the most important axis of natural terrestrial environmental variation is between the low-lying, seasonally flooded forests and the well-drained uplands located on limestone bluffs. Whitacre (1998) reports that both forest tree and bird composition vary dramatically between these topographic conditions, such that areas with similar topography located miles apart are much more similar than areas of differing topography only one hundred miles apart. Upland forests (bosques altos) tend to be taller with abundant leaf litter and little undergrowth. Low-lying forests (bosques bajos) tend to be shorter (15 m to 20 m) and possess a dense undergrowth of shrubs, palms, and herbs. This variation is likely a strong influence on patterns of biodiversity in the region (Méndez et al., 1998).

The aquatic systems within the park are less well understood than the terrestrial systems. The major water-course in the park is the Río San Pedro, which is a major tributary of the Río Usumacinta. The Río Sacluc is a tributary of the San Pedro and drains areas to the southeast of the park. The Río Escondido is another tributary of the San Pedro that drains areas within and around the Laguna del Tigre Biotope in the western portion of LTNP. The Río Chocop drains the center of the park to the San Pedro, and the Río Candelaria is an independent affluent of the Laguna de Términos. The hydrological relationships among these rivers and the many lagoons—especially their subterranean linkages, which may be an important component in karstic landscapes—are unknown. Furthermore, there are likely seasonally varying linkages among water bodies with connections in the wet season that are not apparent in the dry season, during which our sampling took place. Such linkages should have important consequences both for the dispersal dynamics of plant and animal populations as well as for the spread of pollutants. These hydrological features should be an important priority for future research in the park.

Historical changes in climate and vegetation structure have left an important imprint on the ecology of the Petén forests as well. The region has been subjected to a great deal of climatic instability through the Holocene. For example, Leyden (1984) relates that vegetation in the region was dominated by pine-oak formations at the height of the last glacial period and that a period of increased aridity occurred from 2200 to 1140 years ago. More recent changes to vegetation are associated with the rise of the Mayan culture.

During the late Classic period (AD 600 to 850), the Maya reached peak population densities of 200 persons per square km (Culbert and Rice, 1990), and this suggests that extensive agricultural conversion of the landscape was needed to support them. As a consequence, the tropical forests we observe today are probably no older than 1000 years (Méndez, 1999). Understanding current ecological patterns within the Maya Forest requires us to be cognizant of the influence of these historical processes.

HISTORY OF LTNP

LTNP is administered by the Consejo Nacional de Areas Protegidas (CONAP), the Guatemalan government's natural resource agency. CONAP is responsible for protecting and conserving Guatemala's natural heritage, which it does through promoting sustainable use of natural resources and through conserving wildlife populations and ecosystems throughout the country. The Laguna del Tigre Biotope, a core area located within the park, is administered by Guatemala's University of San Carlos through its local conservation and research center, the Centro de Estudios Conservacionistas (CECON).

Ponciano (1998) relates the following history of the biotope and the park. CONAP was founded in 1989, and one of its first actions was to incorporate the pre-existing Tikal National Park, Laguna del Tigre Biotope, El Zotz Biotope, and Dos Lagunas Biotope into the MBR. Additionally, the Laguna del Tigre and Sierra del Lacandon National Parks were created. The creation of the MBR was catalyzed by two events: 1) the proposal of the "Ruta Maya" reserve network for conservation and tourism in a National Geographic article and 2) the threat posed by immigration and development due to a highway development project connecting Flores with Cadenas to the south. The German government was financing this project but, based on an environmental impact assessment, decided to suspend funds until the reserve was created. CONAP was able to create the reserve because it had strong political support and because a democratic government was in place at the time. The reserve was intended to harbor strictly protected core areas (national parks and biotopes), multiple-use (e.g. extractive) zones between these core areas, and a buffer zone in which private holdings may occur (Gretzinger, 1998). The park was registered as a Ramsar site of important wetlands in 1990 and 1999.

CONSERVATION CHALLENGES FOR LTNP

Population growth and immigration to the Petén have complicated sustainable management and biodiversity conservation. Population growth is estimated to be 7% to

10% per annum, largely due to the immigration of poverty-stricken farmers from the highlands to the south (Nations et al., 1998) as well as the repatriation of Guatemalan citizens from Mexico who fled the country during the 30-year civil war that ended in 1996 (Ponciano, 1998). As a consequence of these pressures, colonization has occurred within core areas and has proven difficult to manage throughout the reserve.

Recent immigrants have imported agricultural practices (broad-scale slash-and-burn techniques) from the highlands that are nonsustainable in the context of lowland forests. The unregulated and illegal harvest of valuable timber species and other products threatens to diminish the capacity of the forest to support human populations through nondestructive practices.

As in similar situations elsewhere, land tenure rights and natural resource concessions granted at the community level, enforcement of management policies, and expanded markets for forest products will be required in order to slow forest loss. Regulatory activities in the Petén, however, have been difficult to implement. Managed extraction of commercially valuable trees such as mahogany (*Swietenia macrophylla*) and Spanish cedar (*Cedrela odorata*), as well as nontimber forest products such as chicle (used in chewing gum), allspice, xate (a fern used in flower arrangements), and potpourri derived from forest products can provide income for Peteneros in the setting of a semi-intact forest. Whigham et al.'s (1998) studies of birds in Quintana Roo suggest that selective logging does not affect bird communities adversely (see also Whitacre, 1998). This may be due, in part, to the bird's adaptations to periodic natural disturbances in the region such as those caused by hurricanes. Furthermore, management of clearing and planting practices in agricultural landscapes (Warkentin et al., 1995) in conjunction with the preservation of intact old-growth forest areas kept out of farming cycles (Whitacre, 1998) may preserve a great deal of the Petén's biodiversity. These management alternatives should be explored further in order to develop the most appropriate strategy to preserve Petén's biodiversity.

An additional development challenge arises from the fact that 40% of the reserve overlaps with areas defined as having petroleum development potential (Ponciano, 1998). Basic Petroleum Inc. had rights to develop petroleum operations within the park before the biosphere's creation in 1990. Rosenfeld (1999) reports that Basic has been granted rights to explore for petroleum in approximately 55% of the area within LTNP. The direct effects of Basic's petroleum operations on the park are unknown, but it is clear that the creation of roads into previously unsettled areas has exacerbated forest clearing within the park (Rosenfeld, 1999). Agreements between conservation interests and the Ministry of Energy and Mines about how to proceed with development have yet to be reached.

In response to the threats posed by petroleum development and human colonization within the park, a major research/monitoring program was initiated in LTNP by CI-ProPetén in collaboration with CONAP and CECON (Mendez et al., 1998). This study documented for the first time the great diversity of plants, diurnal butterflies, reptiles and amphibians, and birds within the park. The vegetation survey performed in this study laid the groundwork for the terrestrial study effort presented here, and the data of Mendez et al. (1998) as a whole complement the information on aquatic fauna and flora and the terrestrial data on ants and mammals. It is hoped that our efforts, in combination with this previous survey, will help call attention to the value of LTNP's biodiversity, provide direction and impetus for more intensive studies in the region, and aid land managers in conservation and management of LTNP's biodiversity.

GENERAL STUDY DESIGN

We conducted this study at the end of the dry season in order to facilitate mobility through the park and to better sample certain groups such as fish and aquatic birds. Sampling was centered around four focal areas (see Map 3). The first was centered at Las Guacamayas Biological Station, which is located at the confluence between the Río San Pedro and the Río Sacluc. Sampling was conducted near the headwaters of and along the Río San Pedro and on nearby portions of the Río Sacluc. The second focal area was located along the Río Escondido to the west, which flows into the Río San Pedro from the north and passes through part of the Laguna del Tigre Biotope. The third focal area was centered on a system of lagoons near the northeast corner of the Laguna del Tigre Biotope. These lagoons are near the Xan petroleum fields currently operating inside the park. Toxicological studies were focused here in several lagoons located at varying distances from the Xan oil wells. The fourth and final area was centered at the Río Chocop, which includes several other lagoons as well as the Río Candelaria. Together, these focal areas represent a breadth of habitats found within the park, including areas that are relatively undisturbed and areas that are threatened by contamination and agricultural development.

Within each focal area, samples were collected at several sample points (or georeference points) representing different macrohabitats found in the area. Samples were collected by each group within a few hundred meters of a sample point. Sample points generally occur within a few kilometers of a point that defines the center of each focal area; this is the focal point (see asterisks in Appendix 1 and Appendix 2). This design will allow for hierarchical analyses of relationships among taxa at two spatial scales.

A classification of aquatic and terrestrial macrohabitats is used here as the foundation for comparisons among taxa and among regions within the park. These macrohabitat types are described below.

Aquatic Macrohabitats

There are few studies of aquatic habitats within the park. In this survey, we sampled aquatic habitats and organisms at three to twelve sample points in each focal area. Appendix 1 presents a preliminary classification of the aquatic macrohabitats by RAP team members. Discussions of the characteristics of these macrohabitats may be found in Chapters 2 and 3 and are summarized in Table 1.1.

Terrestrial Macrohabitats

Terrestrial macrohabitat types are based upon vegetation classifications recognized by Mendez et al. (1998). From one to three sample points were examined by terrestrial ecologists within each focal area, and these sample points include the focal point (Appendix 2). Detailed descriptions of terrestrial macrohabitat types may be found in Table 1.2.

Note that aquatic groups were able to sample many more points than terrestrial groups. This is due to differences in the effort needed to adequately sample different kinds of organisms at each point.

REPORTING OF RESULTS

In the chapters that follow, the results for each taxon or substudy of the RAP expedition are provided as self-contained manuscripts. Each study and the resulting presentation necessarily employ a slightly different organization, depending upon the difficulty in gathering and relating information about the focal taxon. Several analytical features, however, are shared in all taxon-based chapters in order to evaluate a group's diversity. Some or all of these measures may be used in a particular chapter, depending upon their utility given the data at hand. First, species richness is measured as the number of species. Second, species diversity (in the strict sense) is measured as the distribution of abundance among species. A more even distribution of individuals among species and a greater number of species generally lead to higher diversity values. Third, similarity (or conversely, complementarity) in species composition at a site is measured using similarity indices or using clustering or ordination techniques. Two sites that are dissimilar in the identity or abundances of the species occupying them have a greater combined diversity than two habitats that are relatively more similar. This is reflected in clustering and ordination diagrams by the distances between the samples in the diagram. And finally, we use the presence and abundance of globally or regionally rare or endemic species, such as those protected in international agreements including the Convention on the International Trade of Endangered Species of Wild Fauna and Flora (CITES) appendices and the International Union for the Conservation of Nature (IUCN) Red List. This requires information on the broad-scale distribution of taxa and may not be available for certain understudied groups such as plankton.

Table 1.1. Descriptions of aquatic macrohabitats in Laguna del Tigre National Park, classified for the first time in this study.

Macrohabitat Type	Characteristics
Riverine systems (ríos)	
River embayments (bahias)	Pockets of slower-moving water along the edges of rivers.
Permanent river (permanentes)	Deep portions of rivers that never dry, usually with a bed composed of limestone rocks.
River rapids (rápidos)	Fast-moving portions of rivers, often with rocky substrate, and in the case of the San Pedro, bivalve-based reefs.
Seasonal rivers (someras)	Small or shallow rivers or stretches of river that are seasonally dry. May have a variety of substrates, and deeper pockets may contain water year-round but may be largely stagnant in the dry season.
Non-riverine systems	
Tributaries (tributarios, riachuelos)	Small, often spring-fed, streams that feed rivers. Fast-moving, highly oxygenated waters.
Caños	Narrow, lotic environments connected to rivers from marshes. Waters are turbid, almost stagnant.
Oxbow lagoons (lagunas meándricas)	Lagoons formed by sharp river bends that become detached from the river course and exist as isolated lagoons alongside it.
Lagoons (lagunas)	Also known as alkachés or aguadas, lagoons formed by depressions in limestone. Waters may be clear or turbid.

Table 1.2. Descriptions of terrestrial macrohabitats examined during the RAP expedition to Laguna del Tigre National Park (based on Mendez et al., 1998).

Macrohabitat Type	Characteristics
Bosque alto	Characterized by a relatively closed canopy and sparse understory dominated by palms and *Piper* shrubs. Associations include the ramonal, dominated by the tree *Brosimum alicastrum*. Bosque alto occupies an area of 19,354 ha (5.72% of the park).
Bosque bajo	Short-stature forest with a denser understory due to periodic flooding. Dominated by trees such as the pucté and tinto, mixed with roblehicpo (*Coccoloba* sp.) and cojche (*Nectandra membranacea*). Palms including botan (*Sabal morisiana*) and escoba (*Crysophila argentea*) are also dominant.
Encinal	A relic of oak-dominated forest (*Quercus oleoides*) exists near to the confluence of the ríos San Pedro and Sacluc. Associated trees include pucté (*Bucida buceras*) and tinto (*Haematoxylum campechianum*) in flooded areas. This type occupies an area of 2,367 ha (0.70% of the park).
Grassland	Areas that recently have been cleared by slash-and-burn and that have been colonized by grass.
Guamil	Regenerating (secondary) forests with dense understory, sometimes lacking overstory trees altogether. Includes guarumo (*Cecropia pertata*), bullshorn acacia (*Acacia cornifera*), bayal (*Desmoncus* sp.), and palms as characteristic elements.
Riparian forest	Appears to be a mosaic of patches of bosque alto and bosque bajo in areas that are generally flooded in the wet season. The most abundant association is pucté and *Pachira* sp. See also the Aquatic Macrophytes section.
Tintal	A class of bosque bajo that is almost entirely composed of tinto (*Haematoxylum campechianum*). Often found near lagoon margins.

LITERATURE CITED

Batisse, M. 1986. Developing and focusing the biosphere reserve concept. Nature and Resources 22: 2-11.

Campbell, J. A. 1998. *Amphibians and Reptiles of Northern Guatemala, the Yucatán, and Belize.* University of Oklahoma Press, Norman, OK, USA.

Canteo, C. 1996. Destrucción de Biósfera Maya avanza año con año. Siglo Veintiuno, November 21. Guatemala City, Guatemala.

Culbert, T. P., and D. S. Rice, eds. 1990. *Precolumbian Population History in the Maya Lowlands.* University of New Mexico Press, Albuquerque, NM, USA.

Gentry, A. H. 1995. Diversity and floristic composition of Neotropical dry forests. Pp. 146-190 in *Seasonally Dry Tropical Forests* (S. H. Bullock, H. A. Mooney, and E. Medina, eds.). Cambridge University Press, New York, NY, USA.

Gretzinger, S. P. 1998. Community-forest concessions: an economic alternative for the Maya Biosphere Reserve in the Petén, Guatemala. Pp. 111-124 in *Timber, Tourists, and Temples: Conservation and Development in the Maya Forest of Belize, Guatemala, and Mexico* (R. B. Primack, D. Bray, H. A. Galletti, and I. Ponciano, eds.). Island Press, Washington, DC, USA.

Leyden, B. W. 1984. Guatemalan forest synthesis after Pleistocene aridity. Proceedings of the National Academy of Sciences 81:4856-4859.

Méndez, C. 1999. How old is the Petén tropical forest? Pp. 31-34 in *Thirteen Ways of Looking at a Tropical Forest: Guatemala's Maya Biosphere Reserve* (J. D. Nations, ed.). Conservation International, Washington, DC, USA.

Méndez, C., C. Barrientos, F. Castañeda, and R. Rodas. 1998. *Programa de Monitoreo, Unidad de Manejo Laguna del Tigre: Los Estudios Base para su Establecimiento.* CI-ProPetén, Conservation International, Washington, DC, USA.

Mooney, H. A., S. H. Bullock, and E. Medina. 1995. Introduction. Pp. 1-8 in *Seasonally Dry Tropical Forests* (S. H. Bullock, H. A. Mooney, and E. Medina, eds). Cambridge University Press, New York, NY, USA.

Murphy, P. G., and A. E. Lugo. 1995. Dry forests of Central America and the Caribbean. Pp. 9-34 in *Seasonally Dry Tropical Forests* (S. H. Bullock, H. A. Mooney, and E. Medina, eds.). Cambridge University Press, New York, NY, USA.

Nations, J. D., R. B. Primack, and D. Bray. 1998. Introduction: the Maya forest. Pp. xiii-xx in *Timber, Tourists, and Temples: Conservation and Development in the Maya Forest of Belize, Guatemala, and Mexico* (R. B. Primack, D. Bray, H. A. Galletti, and I. Ponciano, eds.). Island Press, Washington, DC, USA.

Ponciano, I. 1998. Forestry policy and protected areas in the Petén, Guatemala. Pp. 99-110 in *Timber, Tourists, and Temples: Conservation and Development in the Maya Forest of Belize, Guatemala, and Mexico* (R. B. Primack, D. Bray, H. A. Galletti, and I. Ponciano, eds.). Island Press, Washington, DC, USA.

Rosenfeld, A. 1999. Oil exploration in the forest. Pp. 68-76 in *Thirteen Ways of Looking at a Tropical Forest: Guatemala's Maya Biosphere Reserve* (J. D. Nations, ed.). Conservation International, Washington, DC, USA.

Sader, S. 1999. Deforestation trends in northern Guatemala: a view from space. Pp. 26-30 in *Thirteen Ways of Looking at a Tropical Forest: Guatemala's Maya Biosphere Reserve* (J. D. Nations, ed.). Conservation International, Washington, DC, USA.

Wallace, D. R. 1997. Central American landscapes. Pp. 72-96 in *Central America: A Cultural and Natural History* (A. G. Coates, ed.). Yale University Press, New Haven, CT, USA.

Warkentin, I. G., R. Greenberg, and J. Salgado Ortiz. 1995. Songbird use of gallery woodlands in recently cleared and older settled landscapes of the Selva Lacandona, Chiapas, Mexico. Conservation Biology 9: 1095-1106.

Whigham, D. F., J. F. Lynch, and M. B. Dickinson. 1998. Dynamics and ecology of natural and managed forests in Quintana Roo, Mexico. Pp. 267-282 in *Timber, Tourists, and Temples: Conservation and Development in the Maya Forest of Belize, Guatemala, and Mexico* (R. B. Primack, D. Bray, H. A. Galletti, and I. Ponciano, eds.). Island Press, Washington, DC, USA.

Whitacre, D. F. 1998. The Peregrine Fund's Maya Project: ecological research, habitat conservation, and development of human resources in the Maya forest. Pp. 241-266 in *Timber, Tourists, and Temples: Conservation and Development in the Maya Forest of Belize, Guatemala, and Mexico* (R. B. Primack, D. Bray, H. A. Galletti, and I. Ponciano, eds.). Island Press, Washington, DC, USA.

CHAPTER 2

THE AQUATIC HABITATS OF LAGUNA DEL TIGRE NATIONAL PARK, PETEN, GUATEMALA: WATER QUALITY, PHYTOPLANKTON POPULATIONS, AND INSECTS ASSOCIATED WITH THE PLANT *SALVINIA AURICULATA*

Karin Herrera, Ana Cristina Bailey, Marcos Callisto, and Jorge Ordoñez

ABSTRACT

- Water bodies in Laguna del Tigre National Park (LTNP) were classified according to eight macrohabitat types (refer to Table 1.1). The physico-chemical properties, water quality, phytoplanktonic flora, and insects associated with the aquatic plant *Salvinia auriculata* are reported for each water body examined.

- The pH of the park's waters ranged from acid (6.78) to alkaline (8.30), from turbid and dark to aquamarine, and from eutrophic to oligotrophic. The water quality of the park was generally good. Coliform bacteria were detected in 37% of the samples, but fecal coliform bacteria were detected in only 13% of the samples.

- A freshwater, mollusk-based reef was discovered in the Río San Pedro near the town of El Naranjo.

- Fifty-nine genera and 71 species and morphospecies of phytoplankton were recorded. Diatoms (Bacillariophycea), followed by chlorophytes (Chlorophyceae) and cyanophytes (Cyanophyceae), dominated the samples. Multivariate analyses revealed that the single most important source of variation determining phytoplanktonic composition was pH, and the second was conductivity.

- Forty-four morphospecies and 26 families of insects were found on *Salvinia auriculata*. This result highlights the great diversity of insects in the littoral aquatic systems of the park.

- The molluscan reef should be conserved, and a water-quality monitoring program should be initiated.

INTRODUCTION

The results of a limnological assessment of Laguna del Tigre National Park (LTNP) are presented, including the following: 1) descriptions of aquatic macrohabitats encountered during the study (refer to Table 1.1 and Appendix 1); 2) an evaluation of water quality using biotic measures such as the presence of coliform bacteria and *Escherichia coli* as well as abiotic measures; and 3) an evaluation of patterns in the diversity of phytoplankton within the park. In addition to this limnological component, we examined the composition of insects associated with floating plants, specifically *Salvinia auriculata*.

This limnological study is intended to serve as the basis for more intensive monitoring studies of aquatic systems in the park. The monitoring of aquatic organisms such as plankton and insects has certain advantages over chemical analysis in determinations of the health of aquatic ecosystems (UNESCO, 1971). Benthic insects, for example, maintain a relatively stable composition that is determined by responses to average conditions in their aquatic environment over time, whereas chemical analyses only reflect water quality at the time samples were taken, which may be quite variable depending upon the time of day or year (McCafferty, 1998). Monitoring using plankton and aquatic insects may represent a good option for recording environmental changes in LTNP, as they have in other freshwater systems in Guatemala (Basterrechea, 1986, 1988, 1991; Herrera, 1999). Furthermore, these organisms are important in their own right. Plankton are fundamental components of the aquatic food chain, and aquatic insects are among the principal groups of organisms that convert vegetable material into animal tissue in freshwater ecosystems. Thus, these organisms form the support for the existence of many organisms including fish, amphibians, and birds (Merritt and Cummins, 1984).

Although several studies have considered aquatic invertebrates in Guatemala, including studies of the Río Polochic by the biology department of the Universidad del Valle and by Basterrechea and Torres (1992) in LTNP, this is the first study in Guatemala to consider the relationships between insects and the floating plants *Salvinia auriculata*. *S. auriculata* is a relatively common and easily sampled aquatic plant in lotic aquatic systems in LTNP (see León and Morales Can, this volume) and thus serves as an ideal substrate from which to gather aquatic insects and to consider their relationships to the environment. The objective of this component of the limnological study is to examine the feasibility of more intensive studies of insects on aquatic plants as a monitoring tool for LTNP.

METHODS

Water Quality

Samples of water and sediment at 33 sample points in five focal areas were usually taken near the margins of water bodies (see sample depths in Table 2.1). Temperature, pH, electrical conductivity, and dissolved oxygen values were obtained using portable instruments. Depth at each point was measured using a handmade depth gauge. Transparency of water column was measured using a Secchi disk. Alkalinity was determined using the Gran method, modified by Carmouze.

Table 2.1. Limnological results at each sampling point.

Sample Point	Sample Depth (m)	Max. Depth (m)	Visibility (m)	Temp. (°C)	pH	Oxygen (mg/L - % saturation)	Conductivity MS/cm
SP1	2.10	2.20	1.95	27.5	7.15	5.80 - 84.5	0.81
SP2	3.30	3.90	2.55	25.9	7.03	6.10 - 78.0	1.02
SP3	1.75	5.40	1.75	24.8	6.89	4.72 - 58.0	1.22
SP4	2.12	6.60	2.12	27.5	7.33	7.36 - 95.0	1.55
SP5	1.75	7.10	1.75	28.3	7.33	7.67 - 110	1.32
SP6	1.65	6.10	1.65	29.5	7.20	7.77 - 103	1.98
SP7	0.80	4.10	0.80	28.1	7.40	*	1.51
SP8	2.60	2.60	1.45	29.5	7.41	*	1.48
SP9	2.33	9.00	1.45	29.5	7.43	*	1.42
SP10	0.95	7.85	0.95	29.6	7.36	8.09 - 103.7	1.66
SP11	3.00	5.10	1.30	29.9	7.47	*	1.44
SP12	0.50	3.10	2.50	29.3	7.67	*	1.37
E1	1.45	1.45	1.45	29.3	7.76	*	0.63
E2	0.85	0.85	0.85	29.8	7.27	*	1.57
E3	1.50	1.50	1.30	29.7	7.20	*	1.54
E5	0.40	0.40	0.40	29.6	7.02	*	1.62
E6	0.70	0.70	0.70	29.1	6.99	*	1.70
E7	2.80	2.80	1.30	29.4	7.46	*	1.90
E8	3.75	3.75	2.00	29.2	7.50	*	1.81
E9	0.20	0.20	0.20	27.3	7.57	*	0.73
E10	0.40	0.40	0.40	29.5	7.73	5.88 - 70.0	1.49
E11	0.50	0.50	0.50	25.4	8.30	4.24 - 53.5	1.95
F1	0.80	0.80	0.70	29.0	8.23	7.42 - 96.9	0.33
F2	1.05	1.05	1.0	29.3	7.96	6.36 - 46.5	1.98
F5 A	1.05	1.05	1.05	26.5	7.84	0.02 - 0.5	0.92
F5 B	0.9	0.9	0.9	26.8	7.84	0.95 - 9.5	0.91
F5 C	0.45	0.45	0.45	26.5	7.84	1.45 - 19.3	0.90
F5 D	0.30	0.30	0.30	26.3	7.84	1.40 - 17.4	0.89
C1	0.40	0.40	0.40	26.3	7.06	0.85 - 8.30	1.27
C2	1.85	1.85	1.85	30.0	7.16	8.51 - 106.5	1.06
C3	1.50	1.50	1.50	28.7	7.20	6.78 - 91.0	0.63
Ca1	1.45	1.45	1.45	25.3	6.84	2.80 - 40.7	1.41
Ca2	0.35	0.35	0.35	25.9	6.78	2.66 - 33.5	0.67

(* data not available).

Additionally, the presence or absence of micro-organisms that indicate water polluted by human waste, including total coliforms, fecal coliforms, and *Escherichia coli*, was determined. In the field only a qualitative test (Ready Cult Coliforms 100) was possible and is reported here. The addition of a chromogenic substrate that binds with coliform bacteria and an MUG fluorescent substrate that is highly specific to *E. coli* permits the simultaneous detection of total and fecal coliforms and *E. coli*. The presence of total coliforms was indicated if the solution turned blue-green. The presence of fecal coliforms was detected by a blue fluorescence and by a positive Indol reaction.

Phytoplankton

Plankton samples were taken at the water surface. Samples were obtained by filtering 100 L or 50 L of water through a 20-mm net; then samples were preserved in lugol. Sedgwick-Rafter chambers (100x and 450x magnification) were used to count and identify phytoplankton species and morphospecies. Patterns of species and generic richness were compared among sample points, and patterns of species composition and abundance among sample points were explored using multivariate techniques. Non-metric multidimensional scaling (NMMS; McCune and Mefford, 1999) was used to characterize patterns of change in phytoplankton composition and abundance across LTNP. Of the many multivariate techniques available, this technique is especially appropriate for characterizing differences in species composition among samples.

Aquatic Insects Associated with *Salvinia auriculata*

In order to sample aquatic insects inhabiting *Salvinia auriculata*, we used a 25x25-cm quadrat. At each sampling point where we observed *S. auriculata*, we placed the quadrat over a randomly determined portion of the plant and collected all parts of the plant that fell within the quadrat. We placed the collected *S. auriculata* into a plastic bag with some water and, in the lab, washed the plant material carefully through a 250-mm net. We discarded all the large plant parts and kept the smaller material, including parts of the roots and the insects caught in the net. This sample was fixed in a solution of 80% formol. Insects in the samples were separated and identified to genus and, where possible, to morphospecies.

RESULTS AND DISCUSSION

Below, we present general results for coliform sampling, patterns of phytoplankton diversity, and a description of the aquatic insects gathered from *Salvinia auriculata*. A description of macrohabitat characteristics, water quality, phytoplankton diversity, and aquatic insects for each focal area follows.

Coliforms: General Considerations

The results of qualitative tests for the presence or absence of coliform bacteria at the sampling points are presented in Table 2.2. It will be necessary to continue this investigation in greater detail in order to reach conclusions regarding the management of water resources in LTNP. These results, however, provide a cursory view of the impact of human activities on water bodies in the park. About 37% of the sample points revealed the presence of total coliforms, and 13.3% of the points revealed fecal coliforms. Patterns of water quality within the park is discussed below.

Table 2.2. Results of a qualitative determination of coliform presence at the sampling points.

Sample Point	Total Coliforms	Fecal Coliforms
SP1	–	–
SP2	–	–
SP3	–	–
SP4	–	–
SP5	–	–
SP6	–	–
SP7	–	–
SP8	–	–
SP9	–	–
SP10	–	–
SP11	–	–
SP12	–	–
E1	+	–
E2	+	–
E3	–	–
E5	+	–
E6	–	–
E7	–	–
E8	+	–
E9	–	–
E10	+	+
E11	+	+
F1	+	+
F2	+	+
F5	–	–
C1	+	–
C2	–	–
C3	–	–
CA1	+	–
CA2	+	–

Phytoplankton: General Considerations

Considering the rapid nature of the sampling regime, we can provide only very general statements regarding the distribution of phytoplankton in LTNP. The sampling strategy determines in large part the subset of species recorded in a particular locale, such that in this study most phytoplankton species recorded were characteristic of superficial waters. In some cases, however, we observed phytoplankton that are intimately associated with the substrate (benthic), such as *Pinnularia*.

Six classes, 59 genera, and a total of 71 species and morphospecies of phytoplankton were identified in this study (Appendix 3). Two algas (a chlorophyte and a diatom) were not identified. Diatoms (Bacillariophyceae), followed by chlorophytes (Clorophyceae) and cyanophytes (Cyanophyceae), predominated in the samples. This pattern of abundance followed our expectations based upon the conditions reported previously in studies carried out in the Petén. Three additional classes of phytoplankton were observed: Euglenophyceae, Dynophyceae, and Chrysophyceae. The majority of phytoplankton registered here are cosmopolitan, whereas others are more frequently observed in particular environments.

The presence of certain species of phytoplankton indicated particular environmental conditions at some of the sampling points. For example, cyanophytes often indicate human contamination. In the majority of sampling points located along the Río San Pedro; Río Escondido (except points E8 and E9); and lagunas Santa Amelia, Buena Vista, Flor de Luna, and La Pista, *Microcystis aeruginosa* was present and indicates the presence of wastewater-associated contaminants. Nevertheless, it is important to note that the densities of this cyanophyte surpassed 1000 organisms/L only at points SP2, SP7, E10, and C3 (refer to Appendix 1). In the rest of the sampling points, densities were less than 400 organisms/L. *M. aeruginosa* has been recorded in other bodies of water within other departments (counties) of Guatemala, such as Izabal Lake (Basterrechea, 1986), and it is the most common phytoplankton species in Amatitlán Lake, which is eutrophic (AMSA, 1998). In these cases, the densities of *M. aeruginosa* are much greater than those within LTNP, and it is important to mention that these bodies of water exist under different conditions and were sampled differently from those in LTNP. There are generally few studies available for other comparisons. *Synedra ulna* is another species that indicates contamination and would provide a good indicator species in future studies in LTNP. It will also be useful to combine phytoplankton sampling with information on hydrocarbon concentrations to provide more concrete linkages between contamination and phytoplankton indicators.

Several other phytoplankton genera were recorded in LTNP that have been reported in Amatitlán and Izabal lakes, including *Anabaena, Ankistradesmus, Ceratium, Cyclotella, Euglena, Fragilaria, Synedra, Lyngbya, Oscillatoria, Melosira, Merismopedia, Pediastrum, Staurastrum, Scenedesmus, Tabellaria,* and *Volvox. Coelastrum reticulatum* and *Golenkinia radiata* are species that were reported both in Izabal Lake and in this study. Most of the eight genera found in Petén Itzá Lake—*Cosmarium, Staurastrum, Cocconeis, Dinobryon, Navicula, Agmenellum, Microcystis aeruginosa,* and *Lyngbya*—were found in LTNP (Basterrechea, 1988).

We found a greater generic richness than that found by Basterrechea and Torres (1992) in the Río Escondido-Laguna del Tigre Biotope (22 genera), although the sampling season and methods were different. Basterrechea and Torres sampled areas only in lagoons within the biotope, as well as in Petén Itzá Lake; they reported 26 genera overall in this sampling: Petén Itzá (7 genera), La Carpa (3 genera), Pozo Xan (7 genera), El Toro (7 genera), and El Remate (12 genera). Sandoval (1997) reported 18 genera of phytoplankton in Petén Itzá Lake. The majority these genera were identified in LTNP. At some points, Sandoval reported dominance by Bacillariophytes, which were also common in this study.

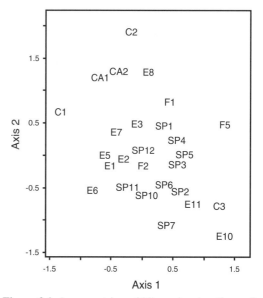

Figure 2.1. A non-metric multidimensional scaling ordination of sample points based upon phytoplankton densities.

The first two axes of the NMMS ordination (Figure 2.1) explained 58.8% of the variance in the original, unreduced distance matrix, and the axes were significantly different from those generated by randomized data (Monte Carlo permutation tests: P=0.035). The ordination indicates that the composition of phytoplankton from the ríos San Pedro and Escondido were broadly similar. All but one sample point (C3) in the Río Chocop/Río Candelaria focal area were distinct from the others. This difference in composition is related to the generally low richness in all of the Chocop/Candelaria samples except C3, which was relatively species rich (Appendix 3). Despite the ecological similarities between lagunas Guayacán (C2) and La Pista (C3) (see below), as well as their spatial proximity, these lagoons harbored distinct phytoplankton floras. NMMS axis scores were regressed upon environmental variables from Table 2.1 to explore which environmental variables are best correlated with patterns in plankton composition. Axis 1 scores positively correlated with only pH (R^2=0.14; F=4.13; df=1, 25; P=0.053), and axis 2 scores positively correlated with conductivity (R^2=0.15; F=4.46; df=1, 25; P=0.045). None of the environmental variables was able to explain a great deal of variation in species composition.

In most cases, the phytoplankton densities obtained were not particularly high (Appendix 4). It will be important to continue monitoring these and other sites within LTNP to evaluate changes in nutrient loads and other physico-chemical parameters and their effects on phytoplankton productivity.

Aquatic Insects Associated with *Salvinia auriculata*

S. auriculata was sufficiently common only at the Río San Pedro focal area for insect sampling. Insects were sampled from Arroyo Yalá (SP1), Paso Caballos (SP2), La Pista (SP3), El Sibalito (SP4), El Caracol (SP6), Río San Juan (SP7), and La Caleta (SP8) (see Appendix 5). In these samples, we identified 5 orders, 26 families, and 44 morphospecies. The order with the largest number of families represented was Coleoptera (10), and Ephemeroptera was represented by the least number (2; Figure 2.2).

It is important to mention that because the plants are mobile, there are occasionally connections between aquatic plants and the shoreline; therefore, many of the species inhabiting floating aquatic plants may also be semi-aquatic or terrestrial. Such processes may explain the presence of some Coleopterans such as the Scolitids. Despite the limited duration and extent of sampling, insect species richness on *S. auriculata* was considerable and indicates that more intensive descriptive studies and monitoring of this system are warranted.

Río San Pedro River Focal Area
Classification of macrohabitats
At the 12 sample stations along the Río San Pedro (with prefix S), five distinct macrohabitat types were encountered (Table 2.3; refer to Appendix 1).

Habitat characteristics, water quality, and phytoplankton diversity
The Río San Pedro is characterized by a relatively low current velocity. In the Río Sacluc (SP11), water levels were quite low. Sediment samples revealed considerable erosion of the river margin, probably due to the constant passing of boats.

Some general abiotic characteristics of samples from the ríos San Pedro and Sacluc are presented in Table 2.3 and include the following:

- The maximum depth of the central channel was found in front of the Scarlet Macaw Biological Station (7.10 m; Table 2.1), and the minimum was found in front of the Arroyo Yala (2.20 m).
- Samples were taken in the littoral zone of the rivers at depths ranging from 0.95 m to 3.30 m.
- Transparency values indicated that the euphotic zone reached the river bottom at 50% of the sample points.

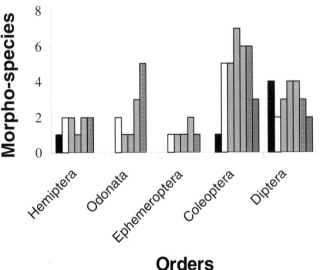

Figure 2.2. Patterns of richness of insect orders inhabiting *Salvinia auriculata* in samples in the Río San Pedro focal area of LTNP.

- Water temperature was high, varying between 24.8° C and 29.6° C.
- Dissolved oxygen concentrations were comparatively high at nearly all of the sample points.
- The waters of the Río San Pedro possess a comparatively high electrical conductivity. This characteristic is probably due to the karstic landscape through which the river flows. These waters may be classified as "hard" due to the high concentrations of calcium carbonate, which, in turn, may be related to the high conductivity values observed.
- Surprisingly, the pH of the water is neutral to slightly alkaline. Considering the geomorphological characteristics of this karstic region, we expected to observe a strongly alkaline pH (>8.5). The relatively large quantities of organic material in the water and high water temperatures may result in accelerated activity by decomposers (bacterias, fungi, and yeast). The production of organic acids in these alkaline waters may produce a nearly neutral pH (pH 8). The fast-running springs at arroyo La Pista (SP3) had colder water, a more acid pH, and greater electrical conductivity when compared to the San Pedro.
- The qualitative test for total coliforms, fecal coliforms, and *E. coli* was negative in all the samples that were tested. The number of phytoplankton species found at each sampling point was similar (Appendix 3), but the point with the lowest richness was SP 10. Genera such as *Tetraëdron* (SP2, SP3); *Kirchneriella* (SP2, SP3, and SP4); *Synura* (SP1, SP2, SP3, and SP6); and *Dinobryon* (SP3, SP4, and SP5) were recorded only within this focal area.

Table 2.3. Description of sampling points along the Río San Pedro.

Macrohabitat Type	Sample Points
Permanent river	SP 1, 4, 5, 9, 11
Tributaries	SP 2, 3, 6, 7
Caños	SP 8
River embayments	SP 10
River rapids	SP 12
Bottom Types	
Sediments with large particles of organic material	SP1, 3
Sediments with small particles of organic material and elevated amounts of interstitial water (typical areas of deposition)	SP7, 8, 9, 10
Current Velocity	
Slow waters, almost stagnant	SP1, 7a, 11
Waters with current	SP2, 3, 7b (springs)
Areas with a High Diversity of Aquatic Macrophytes	SP1, 2, 3, 6, 7
Areas without Aquatic Macrophytes	SP4, 5, 8, 9, 10, 11

Notes on other invertebrates

There appeared to be a high diversity of macroinvertebrates associated with *Salvinia auriculata*. About 44 distinct taxa representing herbivore, detritivore, predator, and rasping trophic groups were recorded (Appendix 3). Miners that use sediment as well as living foliar tissue were also present. Approximately 10 morphospecies of adult Odonates were also collected.

A molluscan reef

In a section of the Río San Pedro referred to locally as the "rapids," we noted the presence of freshwater reefs composed of bivalve molluscs. This type of aquatic system is unique in the region and is generally uncommon in the Neotropics. The river bottom is composed of a 35-cm-deep layer of bivalve shells. Organic sediments were absent. Living bivalves were found on the surface of the reef. The fish *Cichlasoma sunsphilium* was associated with the reef and fed upon the periphyton growing on the shells. Inside the shells various nymphs of Ephemeroptera (Baetidae) and Trichoptera as well as other bivalves, gastropods, and platyhelminths were found. We conclude that these shells are important microhabitats for aquatic invertebrates inhabiting the river bottom in this area.

In a stretch of river with low current, aquatic macrophytes were observed (*Nymphaea ampla* and *Valisneria americana*), forming another microhabitat for invertebrates. Within that stretch was an accumulation of fine organic material. The qualitative test for total and fecal coliforms and *E. coli* was positive, coinciding with human settlements near to the shoreline. Additionally, we identified the greatest richness of phytoplankton at this point (44 species), and densities were highest in the classes Bacillariophyceae, followed by Cianophyceae. This was one of the few points to harbor *Tetraëdron* (Appendix 3, 4).

Río Escondido Focal Area

Classification of macrohabitats

At the 10 sample stations along the Río Escondido (with prefix E), four macrohabitat types were encountered (Table 2.4).

Habitat characteristics, water quality, and phytoplankton diversity

The Río Escondido has relatively steep banks and a narrow river margin. The shores are composed mostly of forest, disturbed forest, and sibal. A large section of the sibal was being burned during our survey. The forest at the northeasternmost station, Punto Icaco (E8), had a relatively closed canopy forest of different character than the remaining parts of the Río Escondido.

The Río Escondido ends as a shallow channel and marsh system, above which are relatively deep sections that widen and narrow regularly. The bottom is principally soft mud, but at Punto Icaco there was much detritus, logs, and leaf litter. Oxbow lagoons and dead arms comprise many habitats to the sides of the Río Escondido, often through sibal habitats rather than in forest. These habitats are now shallow and have very soft mud and mud/marl bottoms. Characteristics of the Río Escondido area are summarized in Table 2.4 and include the following:

- The lowest depths were recorded in the lagoons and embayments, and the greatest depth in the river was at E8.
- Transparency reached the bottom at all sampling points.
- In general, the waters here had high temperatures (from 26° C to 29.8° C), and the pH values were neutral to slightly alkaline (6.99 to 7.76).
- Sediments had high levels of organic material and interstitial water.
- Electrical conductivity in the principal channel of the Río Escondido was low at 1 MS/cm. On the other hand, embayments had very high conductivity values.
- The presence of total coliforms was recorded at E1, E2, E5, and E8, and the presence of total and fecal coliforms and *E. coli* was recorded at E10 and E11. No contamination of fecal origin was detected at the other points.
- The greatest richness of phytoplankton was recorded at points E1 (38 spp.) and E5 (35 spp.). It rained during the sampling of this river, increasing its flow, so it is possible that the phytoplankton sampling was affected by dilution. This may have resulted in a general slight reduction in species richness when compared to samples taken from the San Pedro. Some genera of phytoplankton recorded here, such as *Cosmarium* (E1, E2, and E6) and *Mallomonas* (E1, E2, E5, E7, and E8), were not found in the San Pedro.

Table 2.4. Description of sampling points along the Río Escondido.

Macrohabitat Type	Sample Points
Permanent river	E3, 8, 11
River embayments	E1, 2
Ephemeral (shallow) river	E10
Oxbow lagoons	E5, 6, 7, 9
Bottom Types	
Sediments with large particles of organic material	none
Sediments with small particles of organic material and elevated amounts of interstitial water (typical areas of deposition)	all except E11
Rocky bottom	E11
Current Velocity	
Slow waters, almost stagnant	E1, 2, 7, 8, 10
Waters with current	E3, 9, 11
Areas with a High Diversity of Aquatic Macrophytes	E3, 8, 10
Areas without Aquatic Macrophytes	the remainder

Flor de Luna Focal Area

Classification of macrohabitats

Water bodies at this site consisted exclusively of lagoons, three of which were sampled (F1, F2, and F5). Habitat characteristics, water quality, and planktonic communities are discussed below for each lagoon individually.

Laguna Flor de Luna (F5)

- This lagoon is shallow (1.05 m) with very high surface-to-volume ratio. This favors the mixing of the entire water column due to the action of wind (principally northeast winds).
- The lagoon was homothermic (26.3° C to 26.8° C), electrical conductivity was less than 1 MS/cm, and the pH was neutral to slightly alkaline. The qualitative analysis for total and fecal coliforms and *E. coli* was negative.
- During the collection of sediment samples, the presence of hydrogen sulfide gas was detected. This gas, together with methane gas (CH_4), results in the chemical reduction of inorganic nutrients in the water because reduction is favored in the hypoxic or anoxic environment at the sediment-water interface. For this reason, benthic macroinvertebrates were not observed in the sediments. However, water striders (Gerridae) were abundant on the surface of leaves and in the flowers of *Nymphea ampla* throughout the lagoon.
- Thirty-four species of phytoplankton were recorded here. This was one of the few points at which *Tetraëdron*, *Micrasterias*, *Staurastrum*, and *Spirogyra* were recorded.

Laguna Santa Amelia (F2)

- This lagoon was also shallow (1 m), favoring wind-generated mixing.
- As in Flor de Luna, the lagoon was homothermic (27.7° C to 29.3° C).
- Electrical conductivity ranged from 1.85 and 2 MS/cm.
- Oxygen saturation was from 8.6% to 46.5% at the surface.
- The water exhibited a naturally dark color; nevertheless, transparency persisted to the lagoon bottom.
- Collection of the sediment revealed the presence of hydrogen sulfide gas, and no benthic macroinvertebrates were observed. *Nymphea ampla* was abundant across the lagoon.
- Tests for the presence of total and fecal coliforms and *E. coli* were positive, which indicated fecal contamination by humans and livestock. The waters of the lagoon appear to be used for human consumption, and contamination levels in the lagoon should be monitored. Several uncommon phytoplankton genera were found here, including *Staurastrum*, *Spirogyra*, and *Mallomonas*.

Laguna Buena Vista (F1)

- This lagoon was shallow as well (0.75 m) and homothermic (29° C to 29.1° C).
- The pH here was alkaline (8.23).
- The lagoon measured 200x600 m and was extremely homogeneous.
- Nearby human populations use this lagoon for consumption and for washing and bathing. Not surprisingly, the test for total and fecal coliforms and *E. coli* was positive.
- Several uncommon phytoplankton genera were identified here—although in some cases these were also recorded in F1 and F5—such as *Micrasterias*, *Staurastrum*, *Sorastrum*, *Spirogyra*, and *Mallomonas*.

Río Chocop Focal Area

Classification of macrohabitats

Five sampling points (with prefix C) were located here, representing three macrohabitat types (Table 2.5). Water quality and planktonic/invertebrate communities are summarized in Table 2.5 and are individually discussed below for each sample point.

Río Chocop (C1)

- The Río Chocop and the Río Candelaria are both classified as shallow rivers—that is, as semi-permanent streams—because during the dry season, the river levels drop dramatically and only a small amount of water flow is maintained. Both of these rivers had steep banks and overhanging forests with a closed canopy. The Chocop was almost dry with some areas of stagnant water and possibly a "biofilm" of bacteria on the surface of these pools.
- The test for the presence of total coliforms was positive, but for fecal coliforms and *E. coli*, the test was negative.
- There was little dissolved oxygen in the water column. The river bottom consisted of compacted sediments overlain by leaves from the adjacent terrestrial vegetation. Because of the lack of oxygen, there were few benthic invertebrates (the fragmenting trophic group was absent altogether). Large numbers of belostomatids (Hemiptera) were observed in shallower sections, however. This river is probably quite different during the wet season. For phytoplankton, this was one of the least species-rich points, with 28 species.

Table 2.5. Description of sampling points along the Río Chocop.

Macrohabitat Type	Sample Points
Ephemeral (shallow) rivers with muddy bottom	C1, Ca1
Permanent rivers with rocky bottom	Ca2
Lagoons	C2, 3

Lagunas la Pista (C3) and Guayacán (C2)

- These lagoons are ecologically similar and are thus treated together. Guayacán had some exposed limestone on its margin and aquamarine clear water (>2m visibility) over soft sediments. There was a margin with sawgrasses and margins of forest. There were many small pools in the forest associated with this lagoon that may represent springs.
- Wind probably mixes the entire water column in both lagoons. As a consequence, the lagoons were homothermic, dissolved oxygen was present to the bottom (>86% saturated), and conductivity was equal through the water column.
- The test for total coliforms, fecal coliforms, and *E. coli* was negative.
- In spite of the presence of dissolved oxygen in the water columns of both lagoons, we detected a strong odor of hydrogen sulfide gas in the sediment samples. These observations suggest a resistence from a microstratification of oxygen at the water-sediment interface. As a consequence, we expect an absence of macroinvertebrates in both lagoons.
- Guayacán harbored the fewest number of phytoplankton genera in the study (19), perhaps because of the oligotrophic status of this lagoon. Laguna la Pista exhibited a greater phytoplanktonic richness.

Río Candelaria (Ca1, 2)

- In this river the two sample points were quite distinct. At Ca1 the river was deep and had a slow current. Generally, the conditions were similar to those found in Río Chocop.
- Dissolved oxygen was present to the river bottom, although in low concentrations. Vegetable material was abundant on the bottom, and the presence of hydrogen sulfide gas was detected.
- The test for total coliforms was positive, and for fecal coliforms and *E. coli*, the test was negative. Twenty-seven species of phytoplankton were recorded here.
- The river at Ca2 was shallower (35 cm), had a more rapid current, and oxygen was present to the bottom.
- Electrical conductivity values were similar to those at Ca1.
- The test for total and fecal coliforms and *E. coli* was negative.
- Twenty-nine species of phytoplankton were recorded here, so both Candelaria sites were intermediate in richness when compared to the other sites. Several different trophic groups of benthic macroinvertebrates were observed at both points, including gastropod herbivores, filter-collecting Ephemeroptera, and odonate carnivores.

CONCLUSIONS AND RECOMMENDATIONS

With respect to the conservation and management of aquatic ecosystems in the park, two recommendations are offered:

- There is an urgent need to conserve the rapids of the Río San Pedro. These rapids possess a unique freshwater reef system that may also serve an important function by filtering suspended particles from the water.

- Many lagoons and streams within the park are little contaminated, and action should be taken to ensure their continued health.

With respect to future studies and monitoring the aquatic systems within the region, five recommendations are offered:

- Future studies should be executed at the basin scale and should be conducted over longer periods, encompassing the variation between the dry and wet seasons. Samples should be taken at different depths, and total and dissolved nutrients should be considered.

- The effects of human activities on the landscape, such as agriculture and the subsequent runoff of chemicals and waste, need to be evaluated at the basin scale.

- For phytoplankton monitoring, it would be useful to standardize sampling times among different sample points in order to avoid the effects of diel variation in the movement of plankton in the water column. This would reduce bias in comparisons among samples.

- Densities of coliform bacteria should be monitored and quantified in order to track accurately the degree of fecal contamination in the park and to identify those areas that are sources of contamination throughout the park.

- More intensive monitoring of aquatic insect populations needs to be initiated, and samples should be taken at different times of year in order to understand the effects of natural temporal variation and how this may interact with spatial variation caused by both natural and anthropogenic processes.

LITERATURE CITED

AMSA. 1998. *Listado de Géneros Determinados de Algas en el Lago de Amatitlan*. Secretaria de la República de Guatemala.

Basterrechea, M. 1986. Limnología del Lago de Amatitlán. Revista de Biologia Brasileira 46: 461-468.

Basterrechea, M. 1988. Limnología del Lago Petén Itzá, Guatemala. Revista de Biología Tropical 35: 123-127.

Basterrechea, M. 1991. *Evaluación del Impacto Ambiental de la Exploración Sísmica en la Cuenca del Lago de Izabal*. Guatemala.

Basterrechea, M., and M. Torres. 1992. *Hidrología y Limnología de los Humedales del Biotopo Río Escondido-Laguna del Tigre, Petén-Guatemala*. Centro de Estudios Conservacionistas y Fundación Mario Dary, Guatemala.

Herrera, K. 1999. Indicadores de la calidad del agua del Río Polochic y de la integridad biológica del Lago de Izabal. Master's thesis. Departamento de Biología, Universidad del Valle de Guatemala.

McCafferty, P. 1998. *Aquatic Entomology*. Jones and Bartlett Publishers, Sudbury, MA, USA.

McCune, B., and M. J. Mefford. 1999. PC-ORD. *Multivariate Analysis of Ecological Data*, version 4.0. MjM Software Design, Gleneden Beach, OR, USA.

Merritt, R. W., and K. W. Cummins. 1984. *An Introduction to the Aquatic Insects of North America*, 2nd ed. Kendall-Hunt Publishing Company, IA, USA.

Sandoval, K. 1997. Fitoplancton como indicadores de contaminación en tres regiones del Lago Petén Iztá. Informe final, Programa E.D.C. Escuela de Biología, Universidad de San Carlos de Guatemala.

UNESCO. 1971. *Memorias del Seminario sobre Indicadores Biológicos del Plankton*. Oficina Regional de Ciencia y Tecnología para América Latina y el Caribe, Montevideo, Uruguay.

CHAPTER 3

THE AQUATIC MACROPHYTE COMMUNITIES OF LAGUNA DEL TIGRE NATIONAL PARK, PETEN, GUATEMALA

Blanca León and Julio Morales Can

ABSTRACT

- A vascular flora of more than 130 species, 120 genera, and 67 families was recorded in and adjacent to aquatic habitats in Laguna del Tigre National Park (LTNP). This flora is similar in composition to that of neighboring areas in the Yucatán Peninsula, but it will be necessary to expand the study in order to provide more precise comparisons.

- LTNP's vegetation includes a variety of communities such as tall- or medium-statured riparian forests, short forests (bosques bajos), flooded palm savanna, and remnants of mangroves.

- A population of *Vallisneria americana* associated with bivalve reef rapids was unique. Another site with smaller bivalve deposits in the Río Escondido was especially rich in species. Natural disturbance such as seasonal flooding and erosion cause communities to be highly dynamic.

- Anthropogenic disturbances such as fire are altering the natural extent and composition of vegetation in LTNP.

INTRODUCTION

Guatemala is one of the few countries in Central America that has a well-studied flora and a long history of botanical collection (e.g. Standley, 1924-1964; Williams, 1966-1975). Nevertheless, few studies specifically address the vascular aquatic flora (e.g. de Poll, 1983).

For the Guatemalan Petén, the first botanical studies were performed in the first part of the 20th century and were associated with archeological explorations (Lundell, 1937). For this reason, the interest in understanding the cultural changes of the past has been tied to interpretations of changes in forest cover (Deevey et al., 1979; Islebe et al., 1996). These interpretive studies of historical, social, and environmental changes are based in paleolimnological studies; however, this work does little to broaden our knowledge of the current aquatic macroflora of the region.

The heterogeneity of the terrestrial forest vegetation in the Petén is unquestionable (Deevey et al., 1979). The seasonal flooding that occurs in marshes and the disturbances it produces also creates a great deal of heterogeneity in the expression of floristic elements of aquatic plant communities.

Laguna del Tigre National Park (LTNP) is associated with the presence of extensive areas of marshland, located among a broad range of types of aquatic habitats (Table 3.1) in which the macrophytes constitute the most conspicuous elements. Our objective is to describe the relationships between the aquatic flora and these macrohabitats and to describe the communities formed within them.

METHODS

Aquatic macrophytes were defined in this study as vascular plants that are permanently submerged in water (true aquatics) and plants with roots in supersaturated or flooded soils with the rest of the plant being emergent (amphibious).

Macrohabitat definitions were based on the hydrology of the sample points (see Herrera et al., this volume). With respect to aquatic macrophytes, six macrohabitats were recognized: rivers, streams, lagoons, oxbow lagoons, caños, and rapids (freshwater reefs). Of these, the river macrohabitat was subdivided into those exhibiting permanent (deep) water, ephemeral (shallow) water, and rapids depending upon the volume of water and substrate type. The rapids constitute a part of permanent river habitats.

Aquatic vegetation types were classified by physiognomy and were subdivided according to associations observed within transects. Two vegetation types were recognized in this study: riparian forest and marsh. Woody plants characterize riparian forests, while herbaceous plants characterize marshes. Names used for associations employed local terminology recognized in previous studies (Morales, unpublished).

We established temporary transects among the two vegetation types. Along each transect, we recorded the composition of macrophytes found within 50x50-cm quadrats. In lotic (i.e. running water) environments, quadrats were randomly placed along the riverbank at 10-m intervals. In lentic (i.e. standing water) environments, quadrats were placed across the maximum diameter of the water body at similar intervals. For each quadrat, we recorded the depth of the substrate, distance to shore, and the species of plants visible within two meters of the quadrat. To sample riparian forest, three transects (100 m in length) were established along the riverbank, each separated by 100 m. Along these transects we recorded all woody individuals that had more than 50% of their root mass in the water.

Voucher specimens of all fertile vascular plants that we recorded were collected. Plants were prepared in a plant press and later placed in alcohol in preparation for final oven drying. Identifications were performed using collections at the Field Museum of Chicago, USA.

RESULTS AND DISCUSSION

Floristic Composition of LTNP

A total of 67 families, 120 genera, and more than 130 species of vascular plants were recorded in marshes, riparian forests, and seasonally flooded forests (Appendix 6). Of these, 33 families include aquatic and/or amphibious species, and the other 34 are represented by terrestrial species, both woody and herbaceous, as well as epiphytes that can tolerate short periods of flooding.

The most diverse families were the Cyperaceae and Fabaceae, both of which contain terrestrial and amphibious species. The Cyperaceae and legumes also contain the most conspicuous elements of both marshes and riparian forests. Among the Cyperaceae, *Cladium jamaicense* (sibal) is the dominant element of marsh landscape. This plant forms extensive, dense colonies, and its basal parts form a substrate that favors the establishment of terrestrial plant seedlings such as those of the Asclepiadaceae, Cucurbitaceae, and Passifloraceae families. Among the legumes, *Mimosa pigra* and *Haemotoxylum campecheanum* (tinto) are dominant components of riparian forest. Other dominant families include the palms (Arecaceae) and grasses (Poaceae). Both are important in terrestrial and aquatic habitats.

The composition of the aquatic and amphibious plants reflects linkages to the Mexican flora (Lot and Novelo, 1990; Martínez and Novelo, 1993); there are no taxa that are endemic to the study area. A few species are known only from Central America, including *Cabomba palaeformis* and *Haemotoxylum campecheanum*. The most specioso genus of the aquatic plants was Utricularia, and the most speciose of the amphibious plants was Eleocharis.

Fifty-four of the species recorded in the park are aquatic or amphibious (Appendix 6). Nine species possess flowers that are strictly aquatic: *Brassenia schreberi*, *Cabomba palaeformis*, *Najas wrightiana*, *Nymphaea ampla*, *Pistia stratiotes*, *Utricularia foliosa*, *U. gibba*, *U.* sp., and *Vallisneria americana*. These species represent six families: Araceae, Cabombaceae, Hydrocharitaceae, Lentibulariaceae, Najadaceae, and Nymphaeaceae. Two were floating plants: *Salvinia auriculata* and *S. minima*. For the entire state of Campeche, and in the Yucatán Peninsula, Lot et al. (1993) mentioned 33 plants with strictly aquatic flowers. It is possible that we did not record several families that may exist in the park, including the Ceratophyllaceae, Lemnaceae, Menyanthaceae, and Sphenocleaceae.

Vegetation Types

The plains of the Yucatán Peninsula—particularly the "forested wetlands"—have rich marshlands (Lot and Novelo, 1990). These forested wetlands occur as a mosaic with marshes (herbaceous wetlands) in both the park and the biotope. The abundance of each of these vegetation types varies, in part, due to processes such as the seasonal variation in precipitation that modifies the saturation of the substrate as well as human disturbances. The hydrological properties of the area undoubtedly play an important role in maintaining the mosaic (Lugo et al., 1990).

Marsh
Marsh is without doubt the vegetation type most dominated by truly aquatic vegetation. Marsh occupies the lowest parts of the park and is probably associated with sediments deposited by the river or at the eroded margins of lentic environments. These plants also form small communities along banks which are dominated by riparian forest. The presence of truly aquatic plants is, in part, related to the depth and type of substrate, the level of a site's protection from strong currents, and the frequency of disturbance. Amphibious species dominate the marshes.

Marshes in LTNP are composed of various associations dominated by *Cladium jamaicense* (the association called sibal). *C. jamaicense* and *Phragmites australis* (sibal-carrizo) exist largely along the Río San Pedro to the mouth of the Río Escondido. The *C. jamaicense* and *Typha domingensis* association (sibal-tul) is apparently rare and occurs only at Arroyo la Pista (SP3). The *C. jamaicense* and *Acoelorraphe wrightii* association (sibal-tasiste) dominates the margins of the Río Escondido.

Marsh represents the most diverse vegetation type with respect to the numbers of aquatic and amphibious species as well as to terrestrial taxa such as the Asclepiadaceae, Convolvulaceae, Cucurbitaceae, and Passifloraceae. Terrestrial vegetation is favored in sibal that has recently burned, usually during the dry season. Fires burn heterogeneously in sibal, leaving a mosaic of heavily burned areas mixed with areas in which only the upper parts of plants have burned. The heterogeneity produced by fire is also reflected in the regrowth of woody species located on the banks of the marsh.

Another factor that affects the marsh and its biota is the change in drainage patterns due to the construction of roads. One can observe bodies of water along the access routes to the lagoons that are filled with grasses and sawgrass.

Forested Wetlands

This type of marsh is difficult to delimit during the dry season because only the lowest limit of its extent can be observed during this period. In Mexico this type of marsh is recognized in various classes, including tall and medium riparian forest, short riparian forests, wetland palm thickets, and mangroves (Lot and Novelo, 1990). All four of these types are present in LTNP.

The tall riparian forest is dominated by *Pachira aquatica, Inga vera, Lonchocarpus hondurensis,* and *Phitecellobium* sp. and can be seen near the Guacamayas Biological Station (LGBS) on the Río San Pedro. These are among the tallest forests in the area, but the trees rarely exceed 30 m due to the absence of deep soils. Short riparian forest comprised the majority of the forested areas we examined. They are represented by diverse associations which are dominated by *Haematoxylum campechianum* (tintal), which occurs at the Laguna Flor de Luna (F1). Additionally, the *Bucida buceras* and *Pachira aquatica* (pucte-zapote bobo) association occurs in parts of the ríos San Pedro and Sacluc; the *B. buceras* and *Metopium* (pucte-chechen negro) association occurs at Laguna Guayacán (F5); and *B. buceras* and *Diospyros* occurs in the upper part of the Río Escondido, where *Chrysobalanus icaco* and *Metopium* sp. were also found. The most common association was the *Bucida-Pachira,* especially along the Río San Pedro (SP6, SP9, and the rapids) and the Río Sacluc (Table 3.1). In general, the riparian forest is less species-rich in aquatic macrophytes than the marsh (Table 3.1). Nevertheless, other species—some of which have economic importance for local populations (e.g. *Calophyllum brasiliense*)— may appear in this forest in a cyclic fashion.

The flooded palm thicket is conspicuously developed in the middle and upper Río Escondido. The "tasistal" describes one of its principal associations—that in which

Acoelorraphe wrightii is dominant and is accompanied by herbs such as *Cladium jamaicense, Osmunda regalis* var. *spectabilis,* or trees of *Pachira aquatica.*

The mangrove is not developed in the study area. Only a few individuals of *Rhizophora mangle* occupy the margins of the Río San Pedro and are surrounded by an altered "sibal." This isolated population is the most continental known form of the peninsula (Nicholas Brokaw, pers. comm.). It is located outside the park and consequently receives no official protection.

All vegetation types are subjected to burning, either directly through the use of fire for deer hunting or indirectly through fires that escape adjacent slash-and-burn agricultural plots. The effects of fire in marshland and its impact on biodiversity was difficult to evaluate in this short-term study. SP7 was burned one year before the study, and SP13 was burned four years before the study (Table 3.1). In the former, we recorded 17 species; in the latter, we recorded 19 species. Both shared the majority of their species. It would be useful to record the impact of these disturbances over longer periods. Equally interesting would be a study documenting the survival probability of certain plant species when their aerial parts are lost during fires. We observed that in partially burned clusters of basal branches, the pseudobulbs of the orchids *Eulophia* and *Habenaria* can survive.

Aquatic Macrohabitats and Their Floral Characteristics

Based on the characteristics of the river bed and the seasonality of flow, riverine habitats can be subdivided into permanent river, seasonal river, and rapid macrohabitats (refer to Table 1.1).

Permanent (deep) river was sampled in six localities: four in the Río San Pedro, one in the Río Sacluc, and one in the Río Escondido (Table 3.1). These areas are characterized by beds of calcic rock, deep waters, and water flow throughout the year. This habitat sustains a mosaic of riparian forest types and marshes. The greatest variation in aquatic environments, vegetation types, and numbers of aquatic macrophyte species was found in the Río San Pedro. Our records suggest that the river macrohabitat is one of the most diverse even though the majority of its species are not unique to this macrohabitat.

The shallow river macrohabitat was sampled in three locations in three different focal areas (Table 3.1). Of these, only E10 was in an area that was extensively modified by human activity. This sample point was surrounded by a depauperate secondary forest, dominated by *Bucida buceras* with *Mimosa pigra* shrubs. Floating aquatic plants were limited to *Salvinia minima.* In the case of the ríos Chocop and Candelaria, each had riparian forests characterized by distinct associations that were apparently undisturbed. The riparian forest at Chocop harbored *Bucida buceras* and *Matayba,* while the forest at Candelaria consisted of the

Bucida buceras and *Diospyros* association. In these rivers, submerged aquatic plants are very scarce; only *Cabomba* was found in a bend of the Candelaria.

The rapids of the Río San Pedro are formed by deposits of bivalves and are floristically interesting because of the presence of a unique population of *Vallisneria americana*. Nonetheless, the species present in the riparian forest as well as the marsh that surrounds the rapids are found in other areas. Additionally, E9 (Canal Manfredo) in the Río Escondido also harbored a bivalve deposit and is comparatively rich in taxa. Bivalve deposits may contribute to microhabitat diversity for plants.

The tributary macrohabitat was found in three locations: two within marsh and one in riparian forest (Table 3.1). This macrohabitat is relatively species rich, although its species can be found elsewhere—with the exception of *Najas* cf. *wrightiana* (SP3).

The caño macrohabitat is characterized by turbid, nearly stagnant water. Our observations of this habitat are limited to marshland composed of a sibal on the Río San Pedro that was unburned for two years, although similar conditions likely occur at the seasonal tributaries of the Río Sacluc. At this sample point, we found that the edge of the sibal included terrestrial species of the genera *Ipomoea* and *Passiflora* and one plant of the Loganiaceae family.

The lagoon macrohabitat includes isolated water bodies and is represented in eight of our sample points (Table 3.1). In the majority of the lagoons, riparian forest is present in the surroundings, and aquatic macrophytes prosper, depending upon the type of substrate found to a one-m depth and the characteristics of the bank. The richness of this macrohabitat varies between three and nine species, the majority of which are also found in other macrohabitats. The flora at Pozo Xan (F6) contributed species known only from this macrohabitat, including *Brasenia schreberi* and *Xyris* sp. Laguna Flor de Luna (F5) appears to represent a lagoon in the process of eutrophication, with two vegetation types on opposite banks—a marsh dominated by *Cladium* on one and by *Eleocharis* on the other—and riparian forest dominated by *Haematoxylum campechianum*. The surface of the lagoon is almost completely covered by *Nymphaea ampla*. Laguna Guayacán is a large water body in which part of the bank is composed of exposed calcified rock covered by unflooded forest. Marsh vegetation in this lagoon is limited but can be found in protected inlets. These inlets appear to be streams feeding the lagoon.

The oxbow lagoons sampled had a very low (3) to very high (28) species richness. The variation in richness may be interpreted as a result of the difference in the ages of these lagoons, duration of isolation from the main river channel, and geological and hydrological characteristics. Sampling

Table 3.1. Macrohabitats, vegetation types, and numbers of plant species recorded at the 25 sampling points.

Sample Point	Macrohabitat	Vegetation Type	Number of Plant Taxa
SP3: Arroyo La Pista	tributary	marsh	14
SP4: El Sibalito	deep river	marsh	12
SP6: Caracol	tributary	riparian forest	9
SP7: San Juan	tributary	marsh	17
SP8: La Caleta	caño	marsh	10
SP9: Pato 1	deep river	riparian forest	11
SP11: Río Saclúc	deep river	riparian forest	12
SP12: Rápidos	deep river	reef, riparian forest	6
SP13: Est. Guacamayas	deep river	marsh	19
E5: Laguna Cocodrilo	oxbow lagoon	marsh	4
E6: Laguna Jabirú	oxbow lagoon	marsh	5
E7: Caleta Escondida	oxbow lagoon	marsh, riparian forest	14
E8: Punto Icaco	deep river	riparian forest	9
E9: Canal Manfredo	bivalve/rapids	marsh	28
E10: Nuevo Amanecer	shallow river	riparian forest	5
F5: Flor de Luna	lagoon	marsh, riparian forest	9
F6: Pozo Xan	lagoon	marsh	6
C1: Chocop	shallow river	riparian forest	4
C2: Guayacán	lagoon	marsh	7
C3: La Pista	lagoon	marsh	3
C4: Tintal	lagoon	riparian forest	5
C6: Laguna La Lámpara	lagoon	marsh, riparian forest	8
C7: Laguna Poza Azul	lagoon	riparian forest	7
Amelia 2	lagoon	marsh	3
CA1: Candelaria	shallow river	riparian forest	6

point E9 (Canal Manfredo) included species found nowhere else in this study, such as two species of *Utricularia*, *Sagittaria lanciforme*, *Hydrocotyle* cf. *bonariensis*, and an orchid, among others.

CONCLUSIONS AND RECOMMENDATIONS

• It should be a priority to understand the physical characteristics and hydrology of the areas within and around the park in order to better understand the processes that determine the heterogeneity of vegetation in marshes, as well as to manage and prevent contamination of the watersheds.

• The impact of oil development activities should be evaluated, and the changes generated by the construction of roads and their effects on water drainage should be considered.

• Future biological studies should evaluate the richness and composition in each watershed in the area and should relate these patterns to broad-scale patterns across the Yucatán Peninsula.

• It should be a priority to recognize and distinguish the natural and anthropogenic processes that affect the aquatic ecosystems of the park.

• Fire constitutes one the most important perturbations in the park and affects all types aquatic environments. Medium to long-term studies should be directed at determining the effects of fire on plant populations in marshes as well as riparian forests.

• Riparian vegetation that occurs along shallow rivers such as the Chocop and Candelaria differs from that occurring alongside larger, deeper rivers such as the San Pedro. Thus, each of these forest types should be conserved with similar effort. The presence of *Arthrostylidium* sp. along the Candelaria should be cause for concern because the species spreads aggressively in areas subjected to burning.

• The Río Sacluc should be incorporated within the LTNP so that the oak (*Quercus* sp.) forest (encinal) there may be protected, as well as the riparian forests that connect with the San Pedro.

• Areas with bivalve deposits in the Río Escondido are species rich and vulnerable to destruction due to erosion caused by the passage of motor boats, widening of the channel, and fire. The rapids of the Río San Pedro harbors unique populations, and this area should be conserved.

• There is a potential threat of petroleum contamination in Río Escondido because the pipeline passes near the headwaters of this river. Considering the devastating effect an oil spill could have throughout the watershed, a plan should be created to prevent such an event.

LITERATURE CITED

Deevey, E. S., D. S. Rice, P. M. Rice, H. H. Vaughan, M. Brenner, and M. Flannery. 1979. Mayan urbanism: impact on a tropical karst environment. Science 206: 298-306.

Islebe, G. A., H. Hooghiemstra, M. Brenner, D. A. Hodell, and J. Curtis. 1996. A Holocene vegetation history of lowland Guatemala (Lake Petén-Itza). Doctoral dissertation. Amsterdam, the Netherlands.

Lot, A., and A. Novelo. 1990. Forested wetlands of Mexico. Pp. 287-298 in *Ecosystems of the World 15: Forested Wetlands* (A. E. Lugo, M. Brinson, and S. Brown, eds.). Elsevier, Amsterdam, the Netherlands.

Lot, A., A. Novelo, and P. Ramirez-García. 1993. Diversity of Mexican aquatic vascular plant flora. Pp. 577-591 in *Biological Diversity of Mexico: Origins and Distribution* (T. P. Ramamoorthy, R. Bye, A. Lot, and J. Fa, eds.). Oxford University Press, New York, NY, USA.

Lugo, A. E., M. Brinson, and S. Brown. 1990. Synthesis and search for paradigms in wetland ecology. Pp. 447-460 in *Ecosystems of the World 15: Forested Wetlands* (A. E. Lugo, M. Brinson, and S. Brown, eds.). Elsevier, Amsterdam, the Netherlands.

Lundell, C. L. 1937. The vegetation of Petén. Publication 478. Carnegie Institution of Washington, Washington, DC, USA.

Martínez, M., and A. Novelo. 1993. La vegetación acuática del Estado de Tamaulipas, México. Anales Inst. Biol. Univ. Nac. Autón. México, Ser. Bot. 64: 59-86.

Morales Can, J. 1998. Tesis de bachillerato. Universidad Nacional de San Carlos, Guatemala.

de Poll, E. 1983. Plantas acuáticas de la region El Estor, Izabal. Universidad de San Carlos de Guatemala, Facultad de Ciencias Quimicas y Farmacia. Escuela de Biologia, Guatemala.

Standley, P. 1924-1964. Flora of Guatemala. Fieldiana Botany Vol. 24. Field Museum of Natural History, Chicago, IL, USA.

Williams, L. O. 1966-1975. Flora of Guatemala. Fieldiana Botany Vol. 24. Field Museum of Natural History, Chicago, IL, USA.

CHAPTER 4

AN ICHTHYOLOGICAL SURVEY OF LAGUNA DEL TIGRE NATIONAL PARK, PETEN, GUATEMALA

Philip W. Willink, Christian Barrientos, Herman A. Kihn, and Barry Chernoff

ABSTRACT

- Freshwater fishes were collected at 48 localities within five regional focal points in Laguna del Tigre National Park (LTNP) and the biotope.

- Forty-one fish species were captured out of 55 known or suspected to occur within the region. Several species captured represent new records for the Río San Pedro Basin in Guatemala (including *Batrachoides goldmani*, *Mugil curema*, and *Aplodinotus grunniens*). Two exotic species were found: the tilapia (*Oreochromis* sp.) and the grass carp (*Ctenopharyngodon idella*). Two undescribed species were captured, one ariid catfish and one belonging to the genus *Atherinella*.

- Our results show that most of the species known from the Río San Pedro 60 years ago are still present.

- The Río San Pedro focal point is the most diverse, but with a few exceptions, most species are distributed homogeneously among regions within the basin or among macrohabitats. The lagoon habitats all seemed to have about a dozen species regardless of the disturbance to marginal vegetation, human use, or proximity to roads.

INTRODUCTION

The fishes living in the fresh waters of the Petén of Guatemala and Mexico represent a relatively well-known and well-described ichthyofauna. Largely beginning with the publications of Alfreid Gúnther (1868) based upon the collections of Osberd Salvin in 1859, there has been a series of seminal papers dealing with the fishes of this interesting region. C. Tate Regan (1906-08) was the first to recognize and establish biogeographic provinces for fishes of Mexico and Central America. Carl L. Hubbs made critical explorations of the southern Petén, Lake Petén-Itza, and the Río San Pedro in 1935. His collections form an important basis of comparison for our survey. Since Hubbs, many species have been described, and debates about the nature of biogeographic provinces have ensued (Miller, 1976).

The Río San Pedro is a major tributary of the Río Usumacinta. The Río Candelaria is an independent affluent of the Laguna de Términos. The fishes of these basins are known to be comprised of 48 species, based upon documented museum specimens. In addition, knowledgeable sources and collections from the Río Pasión at Sayaxché suggest five additional species, to which we add two exotics. This brings the known and potential ichthyofauna up to 55 species. These species are listed in Appendix 7. This ichthyofauna is comprised of species that live primarily in fresh waters in addition to those that freely extend their ranges from marine or estuarine habitats into fresh waters, such as the mullets (*Mugil* spp.).

Most components of the ichthyofauna found in Laguna del Tigre National Park (LTNP) and the biotope belong to a broadly distributed group of fishes found between the Río Coatzacoalcos in the north and the Río Polochic/Río Sarstoon in the south. This general distribution has been referred to as the Río Usumacinta Province by Regan (1906-08) and Miller (1966, 1976). Although Bussing (1987) has included the Río Motagua in the Río Usumacinta Province, this has received little attention. However, Chernoff (1986) found that a monophyletic group of *Atherinella* includes the species found in the Motagua, Petén-Itza, and the Usumacinta-to-Coatzacoalcos drainages. Other presumably monophyletic groups (e.g. *Phallichthys* and *Xiphophorus*) also include the Motagua in their Usumacinta Province distributions. These sister-group relationships, including the

Río Motagua, hypothesize the presence of common ancestors once inhabiting the Motagua and Usumacinta regions, followed by some degree of local evolution after the two regions became isolated, probably during the Lower Tertiary (Rosen and Bailey, 1959; Rosen, 1979; Weyl, 1980).

With the exception of Lake Petén-Itza and related lakes, there is little or no local endemism of fishes within Petén proper. This includes LTNP and the biotope. But we note that there is a large degree of endemism at the level of the Usumacinta Province *sensu* Regan (1906-08). Thus, protected areas within the larger province must be established in order to conserve this interesting fauna. One of our objectives was to discover whether LTNP and the biotope can serve this purpose. This is particularly important because similar habitats have been totally or highly degraded in the neighboring Mexican states of Campeche and Tabasco due to cattle ranching and oil exploration.

The purpose of this survey was threefold: 1) to document the biodiversity of fishes in LTNP and the biotope; 2) to document, using rapid assessment protocols, the diversity of fishes associated with the focal points and macrohabitats; and 3) to identify as far as possible the state of health of the habitats, critical regions for maintenance of biodiversity, and potential or actual threats and their potential impacts.

METHODS

During the course of 20 days, 48 collections of fishes were made (Appendices 8-9). Fishes were captured using the following methods:
• seines (5 m x 2 m x 3.4 mm; 1.3 m x 1 m x 1.75 mm);
• cast nets (2 m, 1.5 m);
• experimental gill nets (48 m);
• bottom trawl (3 m); and
• dip nets.
Seining was very difficult because we did not find any beaches or clear areas to pull the seines.

All fish specimens collected were initially preserved in 10% formalin solution, then were transferred to 70% ethanol for archival storage. Identifications were made in the laboratory with the aid of original descriptions and more recent revisionary studies (Regan, 1905; Rosen and Bailey, 1959, 1963; Hubbs and Miller, 1960; Miller, 1960; Rivas, 1962; Suttkus, 1963; Deckert and Greenfield, 1987; Greenfield and Thomerson, 1997; and others). Specimens were deposited in the Field Museum in Chicago and the Museo de Historia Natural, Universidad de San Carlos de Guatemala. To locate the diversity of fishes, as many habitat types as possible were sampled (refer to Table 1.1). The areas sampled were treated hierarchically. At the top level of the hierarchy are the focal areas described in the "Expedition

Itinerary and Gazetteer." Within these focal areas, the entire aquatic sciences team coordinated studies at a finer scale (sample or georeference points). Our 48 samples mostly correspond to 40 georeference points (Appendix 1, 8). A few of the samples are outside of these coordinated sampling regimes. Simpson's Similarity Indices (number of taxa shared between two groups/number of taxa in the smaller group) were used to make comparisons among macrohabitats, focal areas, and gear types. We chose Simpson's Similarity Index because it emphasizes presences of taxa by using the number of species in the smaller group in the denominator (Chernoff et al., 1999). We consider absences of taxa to be ambiguous because a given taxa may actually be absent or it may have been present but avoided detection.

RESULTS AND DISCUSSION

Sampling Effectiveness

Although approximately 55 fish species are believed to inhabit the region surveyed (Appendix 7), we collected 41 (Appendix 8). Some of the fishes not collected are uncommon in the area. Examples include *Ictiobus meridionalis* and *Gobiomorus dormitor*. Others are large fishes which probably swim through the larger rivers but do not stay in any particular spot too long. These are difficult to collect during short surveys but are sometimes caught by local fishermen or those doing longterm studies. Examples include *Anguilla rostrata*, *Mugil cephalus*, and *Centropomus undecimalis*. The species accumulation curve demonstrates that we reached the asymptote after eight days, and we believe that the taxa collected are a good representation of which species were there at the time (Figure 4.1).

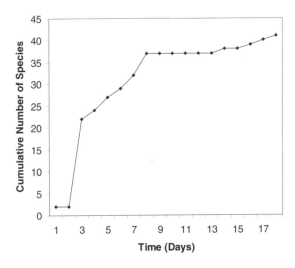

Figure 4.1. Species accumulation curve for fishes collected during the 1999 RAP expedition to Laguna del Tigre National Park, Petén, Guatemala.

Focal Points (see Expedition Itinerary and Gazetteer)

Río San Pedro

Fish Diversity

From the Río San Pedro, we collected 38 fish species during the RAP survey (Appendix 8) out of an estimated 55 species for the region. Fourteen of the species collected were cichlids. The second most dominant group was poeciliids, with 8 species. Ten species were collected throughout the focal point and in a variety of macrohabitats. They include *Astyanax aeneus, Cichlasoma pasionis, Belonesox belizanus, Cichlasoma synspilum, Cichlasoma helleri, Gambusia sexradiata, Cichlasoma heterospilum, Petenia splendida, Cichlasoma meeki,* and *Poecilia mexicana.*

Batrachoides goldmani, Mugil curema, and *Aplodinotus grunniens* are first official records for these species in this Guatemala basin. In comparison to Hubbs' collections from 1935, diversity and numbers do not appear to have changed significantly.

Several habitats here are especially noteworthy for the maintenance of fish diversity and are discussed below.

1) Lowland, varzea-like, inundated forests near Paso Caballos and along the Río Sacluc are critical as nursery areas for young fishes and probably function as areas in which fishes, such as *Brycon* spp., eat and disperse seeds. This type of habitat is unrecognized and undescribed in Mesoamerica.

2) Sibal (dominated by *Cladium jamaicense*) is probably an important microhabitat for cichlids and catfish. It probably also serves as a nursery ground for fishes which spawn concordantly with rising water levels and which feed on the seeds. Even during the dry season, water is an integral part of the sibal because the water penetrates all the way into the sibal. Leaf litter in the river bottom may provide an important food source for fishes, as microflora often grows on the leaves. Leaves may also provide cover for small fishes.

3) Tributaries entering the Río San Pedro from the north side of the river (Paso Caballos, SP2; Arroyo La Pista, SP3) were very clear and shallow and fast-running. Arroyo la Pista was over rocky substrate, almost forming riffles with downed logs. However, fish diversity was relatively low, with only 12 species total. In Arroyo la Pista, we observed huge aggregations of *Dorosoma petenense*, about 60 mm in length, schooling and flashing over the rocks. These stream habitats are possibly threatened due to habitat modification and pollution.

Observations

Gars and *Hyphessobrycon* were trawled up from the bottom. Gars were abundant and may be the dominant piscivore. Many fishes, including gars and *Petenia*, were full of eggs or had enlarged testes, indicating that they were preparing to reproduce. Specimens of *Atherinella alvarezi* were collected.

Although a few specimens were collected by Hubbs in 1935, the species has not been captured during more recent collections from the Río Pasión or near Sayaxché.

Threats

Human encroachment, in the form of burning of sibals, pollution of springs, and deforestation, is significant.

Fragility

Fragility is high for the sibal and varzea because they are fragmented.

Río Escondido

Fish Diversity

Only 27 fish species of the 55 were found among all the macrohabitats within the Río Escondido focal point (Appendix 8). The species of the Río Escondido contain the 10 common species (Table 4.1) plus *Atractosteus tropicus, Dorosoma petenense, Cichlasoma friedrichsthalii,* and *Cathorops aguadulce,* among others.

Observations

Many of the fishes were in reproductive condition.

Threats

The major threats include burning of the sibal and destruction of the forest. We witnessed burning sibal and active conversion of forest habitats for pasture and for agriculture. These activities tend to increase siltation of the river as well as eliminate important nursery areas and shadier cooler microhabitats for fish to live within.

Fragility

The sibal and forest habitats are fragile because their stands are becoming smaller and/or degraded. Almost all of the forested margin has been inhabited by people because this zone affords solid ground and is near the waterway. The fragility of the river from the perspective of providing adequate habitats for fishes to survive is unknown at this time.

Flor de Luna

Fish Diversity

Of the 55 species, 22 were found within the Flor de Luna focal point (Appendix 8). The fauna found within the lagoons was a subset of the Río San Pedro, including the 10 most common species.

Threats

At the Xan oil facilities, there is a potential threat of petroleum contamination from a functioning well adjacent to the lagoon and marsh. It was noted that fishes collected for toxicology had a high incidence of fin rot and that one cichlid had a fibrous mass on its liver.

It is hard to estimate the threats to the other lagoons because we found essentially the same species whether the lagoons had undisturbed margins, were used by local people for washing clothes, or were surrounded by disturbed forests and agricultural fields.

Fragility

It is difficult to estimate the fragility of the lagoon habitats because they appear robust to human impacts, forest modifications, and road construction. There are no data to compare our survey with samples prior to the modification of many of these lagoons.

Río Chocop

Fish Diversity

Only 15 species were found within the focal point (Appendix 8). However, *Xiphophorus hellerii* was found uniquely in this focal point.

Threats

The major threats include deforestation and burning. Deforestation and burning will increase siltation to these relatively clear water systems, which could ultimately affect the diversity of fishes. These threats are already occurring at a rapid rate even though these localities are within LTNP.

Fragility

Headwater streams such as we encountered in this focal point tend to be more fragile than downstream sections. Siltation due to deforestation and habitat conversion change the character of headwater streams dramatically.

Río Candelaria

Fish Diversity

Only one locality was sampled within this focal point, so estimates of diversity are difficult to make. We did collect 13 species (Appendix 8). *Atherinella* c.f. *schultzi* was only collected here. Since the Río Candelaria is a drainage independent from the Río San Pedro, there is a possibility of additional novel taxa being present.

Threats

As in the nearby Río Chocop focal point, deforestation and burning are major threats. Both were observed during our survey. Deforestation and burning will increase siltation to these relatively clear water systems, which could ultimately affect the diversity of fishes. These threats are already occurring at a rapid rate even though these localities are within LTNP.

Fragility

Headwater streams such as we encountered in this focal point tend to be more fragile than downstream sections. Siltation due to deforestation and habitat conversion change the character of headwater streams dramatically.

Comparisons among Macrohabitats, Focal Points, and Gear

Permanent river (e.g. the Río San Pedro) was the macrohabitat with the most species (N=33) (Appendix 9). Lagoons, tributaries, and seasonal rivers also had relatively high numbers of species (23, 22, and 24 respectively). Permanent river is the most common aquatic macrohabitat type in LTNP, although lagoons are abundant in the Flor de Luna and Chocop focal areas.

Table 4.1. Simpson's Similarity Indices among macrohabitat types. The average for all comparisons is 0.830, with a standard deviation of 0.142.

	Lagoons	Oxbow lagoons	Caños	Tributaries	Rivers, seasonal	Rivers, permanent	Rivers, bays	Rivers, rapids
Lagoons								
Oxbow Lagoons	1.000							
Caños	1.000	0.857						
Tributaries	0.818	0.882	1.000					
Rivers, seasonal	0.826	0.882	1.000	0.818				
Rivers, permanent	0.870	0.882	1.000	0.864	0.833			
Rivers, bays	0.923	0.692	0.571	0.846	0.846	1.000		
Rivers, rapids	0.857	0.714	0.571	0.642	0.714	0.857	0.462	

Overall, the similarity among macrohabitats is high (mean Simpson's Similarity Index = 0.830; standard deviation = 0.142; Table 4.1). This indicates that there are a large number of shared species among the macrohabitats. Certain macrohabitats differ more strongly in species composition—most notably rivers/bays versus oxbow lagoons, rivers/bays versus caños, and all macrohabitats versus rivers/rapids. The low similarity indices between the rivers/bays, oxbow lagoons, and caños was a surprise because they are often near each other and superficially appear to be very similar macrohabitats. All are slow water areas along the edges of rivers. We believe that the low indices are most likely an artefact of the low number of species collected in these macrohabitats when compared to the others (Appendix 9). Similarity indices tend to be low when comparing subsets with small sample sizes (Chernoff et al., 1999). A similar explanation may account for the low similarity between the rivers/rapids and other macrohabitats. An additional factor leading to the distinctness of the rapids is the presence of species that were collected only or largely in the rapids, including *Batrachoides goldmani*, *Cichlasoma lentiginosum*, and *Cichlasoma bifasciatum*. The rapids, with their swift current, rocky substrate, and bivalves, is a rare macrohabitat within the surveyed area.

The Río San Pedro focal area had the greatest richness of fishes. Only three species were found outside the San Pedro (Appendix 8). Based on Simpson's Similarity Indices,

the San Pedro, Escondido, and Flor de Luna were very similar (Table 4.2) and therefore represent the same fauna. The Río Escondido and Flor de Luna focal areas are essentially subsets of the San Pedro.

The mean similarity among all focal areas is 0.873 (standard deviation = 0.111). Most of the lower values are from comparisons of other areas with the Candelaria or Chocop. We attribute some of these strong differences to the fact that Candelaria is in a different drainage and that it harbors distinct macrohabitat types. The Candelaria's habitat is most similar to the Chocop's, but their similarity is low (Table 4.2). Again, this may be due to the low number of species collected in these areas (<16 species each) in comparison to the number of species collected in other areas (>21 species each; Appendix 8).

Using Simpson's Similarity Indices to compare gear types demonstrates that the same species were collected when using the seine, cast net, or trawl (Table 4.3). Fewer species were collected with the trawl (Figure 4.2), but these were a subset of those collected by the seine and cast net. Similarity was lower in contrasts with the gillnet (Table 4.3). This is due to the relatively large number of taxa collected exclusively with gillnets (e.g. *Mugil curema*, most of the catfishes, *Aplodinotus grunniens*; Figure 4.2). Only two species (*Brycon guatemalensis* and *Petenia splendida*) were caught with hook and line, and we used this method infrequently; there is little we can say about this gear type.

Table 4.2. Simpson's Similarity Indices among focal areas. The average for all comparisons is 0.873, with a standard deviation of 0.111.

	San Pedro	Escondido	Flor de Luna	Chocop	Candelaria
San Pedro					
Escondido	0.963				
Flor de Luna	1.000	1.000			
Chocop	0.867	0.933	0.733		
Candelaria	0.846	0.923	0.769	0.692	

Table 4.3. Simpson's Similarity Indices among gear types. The average for all comparisons is 0.694, with a standard deviation of 0.251.

	Seine	Cast Net	Trawl	Gillnet	Hook & Line
Seine					
Cast Net	0.935				
Trawl	1.000	1.000			
Gillnet	0.529	0.471	0.500		
Hook & Line	0.500	0.500	0.500	1.000	

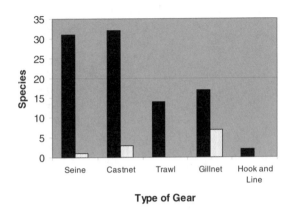

Figure 4.2. Number of fish species collected by each gear type used during the 1999 RAP expedition to Laguna del Tigre National Park, Petén, Guatemala. Solid is the total number of species collected by the gear type. Stippled is the number of species collected only by the gear type.

Overall, it is important to use a variety of collection techniques in order to inventory the complete fauna in a particular area.

In summary, we consider each of the focal regions to contain subsets of the ichthyofauna found within the Río San Pedro. Except for the fishes that prefer large river habitats—the ictalurid and ariid catfishes, *Mugil*, *Aplodinotus*, etc.—most of the fishes are distributed in smaller sets homogeneously among macrohabitats and focal regions. The exceptions to this pattern include *Xiphophorus hellerii*, *Heterandria bimaculata*, and *Atherinella* cf. *schultzi*, which were found in headwater habitats. The rapids/reef area also had two species, *Cichlasoma lentiginosum* and *Batrachoides goldmani*, that were not found elsewhere.

CONSERVATION CONSIDERATIONS

The homogeneity of fish species distribution among macrohabitats and among focal regions suggests that a general plan for conservation of fish biodiversity can easily be put into place in LTNP. We recommend that a sample of each of the types of macrohabitats be set aside as conservation areas—especially the reef and headwater habitats.

Introduced Taxa

Two non-native species were observed: *Oreochromis* sp., which is an African cichlid (no specimens kept), and *Ctenopharyngodon idella*, which is an Asian minnow. These species have been introduced throughout the world in order to control aquatic vegetation and to serve as a food source. They have not, as far as we know, been introduced into

northern Guatemala. But they have been introduced into some of the neighboring states of Mexico. It is probable that these fishes swam through the Usumacinta and up the San Pedro into LTNP. This is a graphic example of the interconnectedness of aquatic systems across national boundaries, and it highlights the need for ecoregion-based management strategies. Any conservation initiatives or introduction of exotics must take into consideration the mobility of the fauna.

Biogeographic Considerations

If we include *Oreochromis* sp., then we recorded 42 species during the LTNP ichthyological survey. Of these, 34 are endemic to the Usumacinta province. These endemic taxa include nine cichlids (*Cichlasoma bifasciatum*, *C. helleri*, *C. heterospilum*, *C. lentiginosum*, *C. meeki*, *C. pasionis*, *C. pearsei*, *C. synspilum*, and *Petenia splendida*), all of which are used for food. Another six species we recorded (*Astyanax aeneus*, *Brycon guatemalensis*, *Rhamdia guatemalensis*, *Belonesox belizanus*, *Poecilia mexicana*, and *Cichlasoma robertsoni*) are found in the Usumacinta, Motagua, and possibly other drainages. Two species (*Ctenopharyngodon idella* and *Oreochromis* sp.) were introduced. In short, many of the taxa we collected are found only in the Usumacinta drainage.

Two hundred or more species are estimated to exist in the Usumacinta-Grijalva drainage (Miller, 1976). We believe there are 58 species (56 native, 2 introduced) in the northern lowlands of Guatemala within the Usumacinta Basin. These 58 Guatemalan species represent 29% of the total Usumacinta fish fauna. LTNP occupies a large area of the Guatemalan northern lowlands, and it is relatively unmodified. Other portions of the Usumacinta drainage are in danger of being developed. Preservation of LTNP would help to protect a quarter of the Usumacinta fish fauna. This is significant because of the high degree of global endemism within the Usumacinta drainage and apparent environmental damage beyond the park boundaries, including Mexico.

SUMMARY OF FINDINGS AND CONCLUSIONS

The following are general conclusions that result from our interpretations of fish distributions among the five regional focal points.

- We collected 41 species of freshwater fishes out of a potential 55 species thought to inhabit this region. Fishes that we did not collect include *Centropomus undecimalis* (reported by fishermen and from Sayaxché), *Rivulus tenuis* (reported in previous studies), *Tarpon atlanticus* and *Hyporhampus mexicanus* (reported from Sayaxché), *Ictiobus meridionalis* (known from the Río Pasión), and *Rhamdia laticauda* (known from the Río Pasión and the Río Negro).

- We identified two introduced species of fishes, collected by fishermen: the tilapia *(Oreochromis* sp., no specimens kept) and the grass carp *(Ctenopharyngodon idella).* This is the first official record for the latter species in northern Guatemala.

- The mullet *(Mugil curema)* is the first record of the species from the San Pedro and the Usumacinta Basin of Guatemala.

- The freshwater drum (*Aplodinotus grunniens*) and the toadfish (*Batrachoides goldmani*) are first records from the San Pedro Basin.

- The fish fauna is more or less homogeneously distributed among the five regional focal points and among the eight macrohabitats. We could not unambiguously identify any special patterns among major components or communities of fishes.

- There was a group of 10 species found among most focal points and macrohabitats. This group includes *Astyanax aeneus, Belonesox belizanus, Cichlasoma helleri, C. heterospilum, C. meeki, C. pasionis, C. synspilum, Gambusia sexradiata, Petenia splendida,* and *Poecilia mexicana.*

- Certain small groups of species were only found in particular macrohabitats. The following are examples: rapids/reef—*Cichlasoma lentiginosum, Batrachoides goldmani,* and to a large extent *Cichlasoma bifasciatum*; headwaters with current—*Xiphophorus hellerii* and *Heterandria bimaculata*; main channel of the Río San Pedro—*Brycon guatemalensis,* ariid and ictalurid catfishes, *Mugil curema, Aplodinotus grunniens,* and *Eugerres mexicanus.*

- Lagoon faunas were remarkably similar despite differences in the proximity to human inhabitation, roads, or other major disturbances.

- The heterogeneity of macrohabitats is low. The river habitats and the lagoon habitats are very similar. The rivers and lagoons had very narrow margins.

- The fish fauna is basically the same as that found from brief surveys 60 years ago. We found no evidence of extinctions. But it should be noted that two exotic species have been introduced.

CONSERVATION RECOMMENDATIONS

These results suggest several specific recommendations for the conservation of fish populations and habitats within the park. Recommendations are not listed in order of priority.

- To develop an integrated conservation plan of the areas containing large numbers of lagoons, a hydrological study is necessary to provide critical information about the connection of lagoons, ponds, and sibals during the seasons of inundation. What are the roles or sources of the fish faunas in these relatively isolated habitats? Sets of lagoons and sibals should be identified as critical for the maintenance of biodiversity.

- Park managers should recognize the critical nature of the lowland varzea-like forest and the sibal habitats for the reproduction of fishes and for the service as nursery habitat for larval and juvenile fishes. These habitats must be considered fragile and threatened due to human modification, especially through burning. We recommend that the strip of unmodified sibal on the south side of the Río San Pedro and the unmodified varzea along the Río Sacluc and below Paso Caballos in the vicinity of the Las Guacamayas Biological Station be included within the park.

- The human impact on the Río San Pedro, both in Mexico and Guatemala, should be studied because there has been much deterioration of natural habitats along the river margins on the southern border of LTNP.

- According to preliminary studies by hydroecologists, the molluscan reef filters the water between the Las Guacamayas Biological Station and the town of El Naranjo. These types of reefs are rarely found in freshwater and have a unique fish community associated with them. This macrohabitat region must be protected in order to secure the health of the riverine ichthyofauna on the southwestern part of the park and also to secure good quality water for the people of El Naranjo.

- Studies on the population dynamics of many species of food fishes must be implemented immediately in order to protect the ichthyofauna from human-based pressures due to subsistence and some commercial fishing. The pressure mainly comes from unregulated gill netting, hook and line fishing, cast netting, and spearfishing. The principal targets are *Petenia splendida,* other cichlids, *Brycon guatemalensis,* and the large catfishes. We note that one of the main sources of fishing pressure occurs during the Easter Holy Week when fishermen exploit the resources very heavily throughout the basin.

- Data on the human use of fishes within the park and within the biotope are needed immediately in order to predict future demands on native populations of freshwater fishes as the human inhabitation of these regions increases.

- A sound conservation plan will require information about the ecology of the fish community during seasons other than the dry season.

- The lowland habitats of rivers and lagoons within LTNP and the biotope should be compared with similar habitats within the Usumacinta province of both Guatemala and Mexico, and elsewhere in the MBR. LTNP and the biotope may serve as a suitable conservation area for this broadly distributed fish community.

- Heavy deforestation and destruction of sibal should be avoided because it may adversely affect aquatic habitats within the basin due to increased siltation of the remaining waterways.

LITERATURE CITED

Bussing, W. L. 1987. *Peces de las Aguas Continentales de Costa Rica.* Editorial de la Universidad de Costa Rica, San Jose, Costa Rica.

Chernoff, B. 1986. Phylogenetic relationships and reclassification of menidiine silverside fishes, with emphasis on the tribe Membradini. Proceedings of the Academy of Natural Sciences of Philadelphia 138: 189-249.

Chernoff, B., P. W. Willink, J. Sarmiento, A. Machado-Allison, N. Menezes, and H. Ortega. 1999. Geographic and macrohabitat partitioning of fishes in Tahuamanu-Manuripi region, upper Río Orthon basin, Bolivia. Pp. 51-67 in *A Biological Assessment of the Aquatic Ecosystems of the Upper Río Orthon Basin, Pando, Bolivia* (B. Chernoff and P. W. Willink, eds.). Bulletin of Biological Assessment 15. Conservation International, Washington, DC, USA.

Collette, B. B., and J. L. Russo. 1981. A revision of the scaly toadfishes, genus *Batrachoides*, with descriptions of two new species from the eastern Pacific. Bulletin of Marine Science 31: 197-233.

Deckert, G. D., and D. W. Greenfield. 1987. A review of the western Atlantic species of the genera *Diapterus* and *Eugerres* (Pisces: Gerreidae). Copeia 1987: 182-194.

Greenfield, D. W., and J. E. Thomerson. 1997. *Fishes of the Continental Waters of Belize.* The University Press of Florida, Gainesville, FL, USA.

Günther, A. 1868. An account of the fishes of the states of Central America, based on collections made by Capt. J. M. Dow, F. Goodman, Esq., and O. Salvin, Esq. Transactions of the Zoological Society of London 6: 377-494.

Hubbs, C. L., and R. R. Miller. 1960. *Potamarius*, a new genus of ariid catfishes from the fresh waters of Middle America. Copeia 1960: 101-112.

Miller, R. R. 1960. Systematics and biology of the gizzard shad (*Dorosoma cepedianum*) and related fishes. Fishery Bulletin 173: 371-392.

Miller, R. R. 1966. Geographical distribution of Central American fishes. Copeia 1966: 773-802.

Miller, R. R. 1976. Geographical distribution of Central American fishes, with addendum. Pp. 125-156 in *Investigations of the Ichthyofauna of Nicaraguan Lakes* (T. B. Thorson, ed.). University of Nebraska Press, Lincoln, NE, USA.

Regan, C. T. 1905. A revision of the fishes of the American cichlid genus *Cichlasoma* and of the allied genera. Annals and Magazine of Natural History, Series 7, 16: 60-77, 225-243, 316-340, 433-445.

Regan, C. T. 1906-08. Pisces, in *Biologia Centrali-Americana* 8: 1-203.

Rivas, L. R. 1962. *Cichlasoma pasionis*, a new species of cichlid fish of the *Thorichthys* group, from the Río de la Pasión, Guatemala. Quarterly Journal of the Florida Academy of Sciences 25:147-156.

Rosen, D. E. 1979. Fishes from the uplands and intermontane basins of Guatemala: revisionary studies and comparative geography. Bulletin of the American Museum of Natural History 162: 267-375.

Rosen, D. E., and R. M. Bailey. 1959. Middle-American poeciliid fishes of the genera *Carlhubbsia* and *Phallichthys*, with descriptions of two new species. Zoologica 44: 1-44.

Rosen, D. E., and R. M. Bailey. 1963. The poeciliid fishes (Cyprinodontiformes), their structure, zoogeography, and systematics. Bulletin of the American Museum of Natural History 126: 1-176.

Suttkus, R. D. 1963. Order Lepisostei. Pp. 61-88 in *Fishes of the Western North Atlantic, Part 3* (H. B. Bigelow, ed.). Sears Foundation for Marine Research, Yale University, New Haven, CT, USA.

Weyl, R. 1980. *Geology of Central America.* Gebruder Borntraeger, Berlin, Germany.

CHAPTER 5

HYDROCARBON CONTAMINATION AND DNA DAMAGE IN FISH FROM LAGUNA DEL TIGRE NATIONAL PARK, PETEN, GUATEMALA

Christopher W. Theodorakis and John W. Bickham

ABSTRACT

- Sediment and tissue samples from two species of fish (*Thorichthys meeki* and *Cichlasoma synspilum*) were collected from a lagoon immediately adjacent to the Xan 3 oil facility, as well as from four reference sites (Laguna Flor de Luna, Laguna Buena Vista, Laguna Guayacán, and Río Escondido and Río San Pedro). Sediment was used to determine levels of polycyclic aromatic hydrocarbons (PAHs), which are chemicals indicative of petroleum pollution. Tissues were used to assay the level of DNA damage, which is indicative of exposure to PAHs and several other pollutants.

- DNA damage was analyzed using agarose gel electrophoresis (to determine DNA strand breakage) and flow cytometry (to determine chromosomal breakage).

- The sediment PAH concentrations from the lagoon at Xan 3 were not significantly higher than the reference sites, but those of Laguna Buena Vista were markedly higher than any other site.

- For *T. meeki*, the levels of DNA strand breakage were higher in the Xan 3 populations than in those from the Flor de Luna and Río San Pedro. For *C. synspilum*, the levels of chromosomal breakage were greater in Xan 3 than in the Río San Pedro site. In both cases the amount of DNA damage in Buena Vista was intermediate between Xan 3 and the other reference sites.

- Further research on the park's hydrological connections and their effects on contaminant redistribution, as well as on the levels of other contaminants, is warranted.

INTRODUCTION

One potential threat to the biota of Laguna del Tigre National Park (LTNP) that has not yet been addressed is the impact of oil extraction activities in the Xan oil fields. The primary concern is that such activities might result in accidental contamination of the surrounding environment. Environmental contamination may have adverse impacts on populations and communities (Newman and Jagoe, 1996). First, it could result in a decrease in reproductive success or an increase in mortality of affected organisms. This may lead to reductions in the population density of some or all species present. Some species may be more resistant to the effects of contamination than others. This could result in changes in community structure (species composition and relative abundance). Environments exposed to oil often possess fewer species than uncontaminated habitats because relatively few species are able to tolerate exposure to oil-associated contaminants (Newman and Jagoe, 1996).

Aquatic ecosystems are particularly susceptible to contamination for several reasons. First, terrestrial runoff and atmospheric deposition tend to remove contaminants from terrestrial to aquatic systems (Pritchard, 1993). Second, because of their highly permeable skin and gills, fish and other aquatic organisms are particularly susceptible to the effects of contamination (Pritchard, 1993). Many hydrocarbons are readily degraded in an aerobic environment, but the aquatic environment frequently contains little or no oxygen. Therefore, contaminants can persist in aquatic ecosystems for far longer than in terrestrial systems (Ashok and Saxena, 1995). Finally, contaminated aquatic organisms pose an important health risk for people consuming them or the water in which they live.

Here we evaluate the effects of hydrocarbon contamination on the physiological state of fishes as a consequence of oil extraction activities around the Xan oil wells in LTNP. We addressed this by examining fishes in lagoons located at different distances from the Xan oil wells. We hypothesized that the concentration of hydrocarbon contaminants would be highest near the Xan 3 oil facility and that this pattern would be paralleled by increasing physiological damage to fish individuals. There are many natural variables that can contribute to variation in biological systems across an area. Thus, if differences were found between Xan 3 and the reference areas, it would be difficult to determine if these differences were due to contaminant effects or to other sources of environmental variation. Parallel patterns of increasing damage to fish individuals alongside increasing concentrations of hydrocarbons among the sites (i.e. a toxicological assessment) would provide strong evidence that contamination is affecting fish populations.

To determine the levels of contaminants in the environment, levels of polycyclic aromatic hydrocarbons (PAHs) were measured. These compounds are characteristic components of petroleum contamination that accumulate in the sediment (Mille et al., 1998). The analysis includes a determination of PAH concentrations in both water and sediment.

The assessment of the effects of these compounds on fish focused on physiological consequences for a number of reasons. First, it is frequently difficult to infer the body tissue concentrations and toxic effects of contaminants from their environmental concentrations (McCarthy and Shugart, 1990). This is because the availability of certain chemicals for uptake into fish tissues ("bioavailability") is often determined by environmental variables (e.g. the amount of naturally occurring hydrocarbons in the water and sediment). Also, environmental contaminants are usually present as complex mixtures, and the toxicity of these mixtures may not be accurately predicted from measurements of concentrations of single chemicals (McCarthy and Shugart, 1990). Therefore, measurements of the physiological responses of fishes living in these environments are needed to rigorously evaluate the impacts of hydrocarbon contamination.
The particular physiological response used for this assessment was DNA (deoxyribonucleic acid) damage. This type of damage has been found to be characteristic of PAH and petroleum contaminant exposure (Bickham, 1990; Shugart, 1988). The two measures used to evaluate DNA damage were DNA strand breakage and chromosomal damage, measured as variation in cell-to-cell variation in DNA content. DNA is a double-stranded molecule, and exposure to PAHs can cause breaks in the DNA strands (Shugart et al., 1992). This is because PAHs chemically bind to DNA, and DNA repair enzymes remove portions of the strand that contains this modification. Subsequent enzymes fill in these gaps, but until they do, there remain temporary strand breaks (Hoeijmakers, 1993). If the breaks are on only one of the DNA strands (single-strand breaks), the DNA chromosome remains intact, but the integrity is compromised. If strand breaks occur on two adjacent strands, the chromo-

some may become fragmented. One way of measuring this fragmentation is to determine the cell-to-cell variation in DNA content (Bickham, 1990). When chromosomes are fragmented, cell division may result in an unequal distribution of DNA strands among daughter cells. This results in an increase in cell-to-cell variation in DNA content. Thus, exposure to PAH and petroleum contamination is indicated by both an increase in the amount of single-strand breakage and cell-to-cell variation in DNA content.

DNA damage can lead to tumor formation in fish (Bauman, 1998) or a shortened life span (Agarwal and Sohal, 1996). It may also have a detrimental effect on fertility and immune function (Theodorakis et al., 1996; O'Connor et al., 1996). This is because cells that are rapidly dividing are particularly prone to DNA damage, and organs, gonads, and developing embryos which form white blood cells have high rates of cell division. Thus, DNA damage may indicate increased mortality in species populations in addition to the presence of toxic contaminants.

METHODS

Sample Collection

Fish and sediment samples were collected in a lagoon adjacent to an oil-pumping facility which is hypothesized to be a source of contaminants in the study area, Xan 3, and at three lagoons occurring at varying distances from Xan 3 (Buena Vista, 5.63 km; Flor de Luna, 13.46 km; and Guayacán, 39.51 km). Additionally, a collection in the Río San Pedro (66.89 km from Xan 3) served as a distant, and likely uncontaminated, control site. Fish were collected using a cast net and were transported back to camp alive. The length of each fish was measured, and the fish were anesthetized. Liver, spleen, and blood samples were collected and frozen and preserved in liquid nitrogen. Water and sediment samples were also taken from the collection sites.

DNA Strand Breakage Analysis

DNA from liver tissues was extracted to determine the number of DNA single-strand breaks (Theodorakis and Shugart, 1993). DNA samples were analyzed by agarose gel electrophoresis in order to determine the average length of the DNA molecules. The length is inversely related to the number of single-strand breaks; if there are more breaks, then the average length of the DNA molecules will be lower.

Chromosomal Damage Analysis

Chromosomal damage was assessed by determining the amount of cell-to-cell variation in DNA content (Bickham,

1990). Nuclei were isolated from blood and spleen tissues, and the variation in DNA content was evaluated using a flow cytometer, which measures relative amounts of DNA in individual nuclei. A more detailed description of the methods for electrophoresis and flow cytometry can be found in Appendix 10.

Contaminant Analysis

Water and sediment samples were extracted with organic solvents (hexane/acetone or methylene chloride). Samples were concentrated by evaporation under nitrogen gas and analyzed by high-pressure liquid chromatography or gas chromatography.

RESULTS AND DISCUSSION

Physiological Condition and DNA Damage Analyses

The number of each species collected at each site is reported in Table 5.1. The only fish that showed external indication of disease (fin rot) were those collected from the Xan 3 lagoon. Fin rot is caused by a bacterial infection that leads to erosion of the fins. It is usually not seen in wild fish populations except under stressful conditions. Petroleum hydrocarbons and other pollutants may increase the occurrence of fin rot and other bacterial infections by suppressing the immune response in fish. This condition is cause for concern because fin erosion impairs a fish's ability to swim and maneuver and is usually a symptom of a more widespread bacterial infection. For captive fish, if left untreated, individuals with fin rot usually die from systemic infection.

Two common species, *Thorichthys meeki* and *Cichlasoma synspilum*, were used for analysis of DNA strand breakage and chromosomal damage analysis. Non-parametric statistical analyses (Kruskal-Wallis) were used to test for significant differences between sites. For *C. synspilum*, the variation in DNA content between cells (a reflection of the amount of chromosomal damage) was greater in fish from the Xan 3 than from the Río San Pedro, and this difference was statistically significant (Table 5.2; $P < 0.05$). The amount of chromosomal damage in fish from Laguna Buena Vista was not different from that of Xan 3. In contrast, there were no statistically significant differences in the amount of chromosomal damage between any *T. meeki* populations.

There were no statistically significant differences in the amount of DNA strand breakage between any *C. synspilum* populations. However, the amount of DNA strand breakage in the Xan 3 *T. meeki* population was higher than in the Flor de Luna and Río San Pedro populations, although this difference was not significantly different from populations inhabiting lagunas Buena Vista or Guayacán (Table 5.2). No fish from Río Escondido were collected for analysis.

The differing patterns of DNA damage between *C. synspilum* and *T. meeki* could stem from a number of causes. First, there could be differential exposure to different genotoxic compounds as a result of differences in habitat use or food preferences. Unfortunately, there is very little information of this type for Neotropical cichlids. Also, there may be differences in the molecular physiology of the two species. Again, such information is absent. It is clear that more research is needed in the natural history and physiology of Neotropical fish before definitive conclusions can be drawn as to differences between these two species. Additionally, the dissimilarity in genotoxic responses between species could be mediated by differences in the flow cytometry and DNA damage. This could be related to differences in sensitivities of the assays or the type of damage that they measure.

Table 5.1. Number of each species collected from Xan 3 and four reference sites.

Species	F6: Xan 3	F1: Laguna Buena Vista	F5: Laguna Flor de Luna	C2: Laguna Guayacán	SP2: Río San Pedro
Thorichthys meeki	12	14	20	17	18
Cichlasoma synspilum	6	8	0	0	14
C. uropthalmus	7	0	0	0	0
Astianyx fasctiatum	0	14	0	11	0
Poecilia mexicana	0	0	0	10	0
Gambusia sexradiata	0	0	23	0	0
G. yucatana	0	28	0	0	0

Table 5.2. Median variation in cell-to-cell DNA content (coefficient of variation between cells) and number of single-strand breaks/10^5 nucleotides (SSB) in two species of fish from Laguna del Tigre National Park, Petén, Guatemala.

Values with the same superscript indicate that they are not statistically significantly different.

Site	Number of Single-Strand Breaks		Variation in DNA Content	
	Median	Range	Median	Range
A. *Cichlasoma synspilum*				
F6: Xan 3	5.53[a]	4.01-17.80	2.57[a]	2.45-2.83
F1: Buena Vista	3.70[a]	1.38-10.28	2.41[a,b]	1.92-3.44
SP2: Río San Pedro	5.47[a]	2.85-8.56	2.13[b]	1.94-2.34
B. *Thorichthys meeki*				
F6: Xan 3	5.25[a]	1.65-17.40	2.36[a]	1.82-2.56
F1: Buena Vista	4.80[a]	1.24-28.27	2.07[a]	1.94-2.57
F5: Flor de Luna	2.53[b]	0.96-11.68	2.12[a]	1.68-2.48
C2: Guayacán	6.16[a]	0.48-25.12	2.06[a]	1.14-3.05
SP2: Río San Pedro	3.25[b]	0.02-11.52	2.01[a]	1.71-2.54

The electrophoretic assay measures single-strand DNA breaks, which could be due to direct nicking of the sugar-phosphate backbone or to conversion of certain DNA base modifications (also indicative of contaminant exposure) to single-strand breaks *in vitro* at alkaline pH (so-called "alkaline labile sites"). The flow cytometric assay, on the other hand, measures chromosomal damage, a result of double-strand breaks. Double-strand breaks are more difficult to repair (Ward 1988), so they may be a less transient effect than single-stand breaks or alkaline labile sites. These differences could also be due to the nature of the assays themselves since the flow cytometry assay used intact nuclei and the electrophoretic assay used extracted DNA.

DNA damage is a steady-state process; in other words, it is constantly being formed and repaired (Freidberg, 1985). Thus, any DNA damage that is apparent is that which remains unrepaired at the time the tissue was collected and frozen. This accounts for the fact that any differences between contaminated and reference sites are small. Another factor contributing to this pattern is that a large amount of DNA damage, especially chromosomal damage, is lethal to cells. Therefore, any cell that is afflicted by a large amount of DNA damage will die. This limits the range of DNA damage that can be detected in living tissue.

The fact that any damage is observable indicates that this steady state has been altered such that the equilibrium has been shifted toward accumulation of DNA damage, either through an increase in the number of DNA-damaging events or a decrease in DNA repair. This would be an argument in favor of viewing DNA damage as a biomarker of contaminant effect and not simply of exposure. Hence small differences in the amount of unrepaired DNA damage may indicate detrimental effects on the health of fish.

Contaminant Analysis

Sediment concentrations of PAHs were determined for the Xan 3 as well as for lagunas Flor de Luna and Buena Vista and ríos Escondido and San Pedro. Sediment samples for Guayacán were lost due to breakage of sample containers in transit to Texas A&M University. The total PAH sediment concentrations for Flor de Luna, Río Escondido, Río San Pedro, Buena Vista, and Xan 3 were, respectively, 202, 148, 612, 1190, and 225 ng PAH/g dry sediment (parts per billion). The concentrations of individual PAH isomers are listed in Table 5.3.

These data show that the amount of PAHs in the sediment sample from the Xan 3 are less than or equal to those of the reference sites. However, the data from the DNA and chromosomal damage analyses indicates that fish from Xan 3 are being exposed to a DNA-damaging agent. This is in accordance with the observation that the fish from Xan 3—and no fish from any other site—showed signs of disease. This suggests that the fish are being stressed by a contaminant other than PAHs. There are many other compounds besides PAHs that can cause DNA damage—for example, heavy metals, chlorinated hydrocarbons, arsenic, chromium, pesticides, and nitroaromatic compounds. Detailed risk assessment procedures could be implemented which would be able to identify the source of this genotoxicity. An unexpected finding was that the amount of PAHs in the sediment was much higher in Laguna Buena Vista than in any other site. This may explain the elevated amounts of DNA strand breakage in the fish from this site. There is also evidence that fish at Guayacán are exposed to DNA-damaging agents.

Table 5.3. Sediment concentrations of polycyclic aromatic hydrocarbons.

Compound	Surrogate-Corrected Concentration (ng PAH/g dry sediment)[a]				
	F6: Xan 3	F1: Buena Vista	F5: Flor de Luna	C2: Guayacán	SP2: Río San Pedro
Naphthalene	6.1	15.4	2.8	56.0	2.4
C1-Naphthalenes	9.7	25.2	3.9	44.1	3.0
C2-Naphthalenes	13.6	90.1	21.5	44.0	6.7
C3-Naphthalenes	17.1	54.4	11.2	23.4	7.2
C4-Naphthalenes	13.9	85.4	11.6	34.1	8.8
Biphenyl	2.7	5.3	0.7	11.9	0.8
Acenaphthylene	BMDL[b]	0.7	BMDL[b]	BMDL[b]	BMDL[b]
Acenaphthene	2.3	9.7	0.9	6.0	0.8
Fluorene	5.5	29.2	1.8	17.0	1.8
C1-Fluorenes	11.3	37.6	3.6	9.5	3.2
C2-Fluorenes	27.5	190.0	17.0	75.7	15.2
C3-Fluorenes	13.0	53.5	9.3	18.6	13.9
Phenanthrene	15.4	69.3	12.2	23.7	12.7
Anthracene	1.4	5.8	0.5	4.6	0.8
C1-Phenanthrenes/Anthracenes	18.3	94.8	29.0	34.8	16.8
C2-Phenanthrenes/Anthracenes	15.1	72.7	13.0	17.7	14.4
C3-Phenanthrenes/Anthracenes	4.6	29.8	6.9	7.1	4.7
C4-Phenanthrenes/Anthracenes	3.0	17.8	5.4	3.8	1.9
Dibenzothiophene	1.1	12.8	1.7	5.5	1.3
C1-Dibenzothiophenes	4.5	31.0	4.6	4.2	4.7
C2-Dibenzothiophenes	6.9	52.4	8.5	7.5	7.5
C3-Dibenzothiophenes	4.5	42.5	4.4	3.8	3.5
Fluoranthene	1.8	17.6	2.3	10.0	2.4
Pyrene	1.5	13.5	2.0	8.4	2.4
C1-Fluoranthenes/Pyrenes	1.5	4.2	BMDL[b]	6.0	0.7
Benz(a)anthracene	0.3	1.8	0.2	3.3	0.2
Chrysene	0.3	2.1	0.3	3.1	0.3
C1-Chrysenes	4.9	16.4	2.6	6.1	1.2
C2-Chrysenes	10.9	37.4	16.8	14.9	1.7
C3-Chrysenes	2.8	49.3	1.1	24.5	2.0
C4-Chrysenes	0.7	11.0	1.3	3.4	2.8
Benzo(b)fluoranthene	0.6	1.8	0.7	2.8	0.5
Benzo(k)fluoranthene	BMDL[b]	2.1	0.8	2.7	0.3
Benzo(e)pyrene	0.4	1.6	0.4	1.3	0.4
Benzo(a)pyrene	0.3	0.8	0.7	1.5	0.4
Perylene	BMDL[b]	4.4	BMDL[b]	64.2	BMDL[b]
Indeno(1,2,3-c,d)pyrene	BMDL[b]	1.5	0.7	BMDL[b]	BMDL[b]
Dibenzo(a,h)anthracene	BMDL[b]	BMDL[b]	0.5	BMDL[b]	BMDL[b]
Benzo(g,h,i)perylene	BMDL[b]	1.1	BMDL[b]	BMDL[b]	BMDL[b]
Total PAHs	225	1190	202	612	148
Spiked Surrogate % Recovery					
Surrogate	% Recovery				
Naphthalene-d8	88	75	74	72	83
Acenaphthene-d10	95	102	97	78	109
Phenanthrene-d10	108	71	103	96	117
Chrysene-d12	70	65	72	62	77
Perylene-d12	81	51	92	62	87

[a] Each sediment sample was spiked with a sediment PAH, and concentrations were normalized for % recovery of surrogate.
[b] Below minimum detection limit.

CONCLUSIONS AND RECOMMENDATIONS

The hypothesis that the Xan 3 oil facility has an impact on the surrounding environment cannot be rejected because there was some evidence of stress and DNA damage in fish from the adjacent lagoon. We conclude that the amount of DNA damage in the fish at Xan 3 was greater than at Río San Pedro, with intermediate results for fish occupying the lagoons in the vicinity of Xan 3. The finding of higher PAH concentrations in Laguna Buena Vista than in any other site was unexpected. This result may be due to the complex patterns of water interchange among lagoons during the wet season or to other contaminant sources. The concentrations of individual PAHs in the sediment, however, were much lower than the US EPA's criteria for protection of benthic invertebrates, which are based on short-term toxicity tests of single chemicals (US EPA, 1991a, 1991b, 1991c). However, the longterm effects of exposure to multiple chemicals on environmental and human health are unknown. Of particular concern is the fact that the local human population heavily uses Buena Vista for fishing, bathing, laundering, and recreation. Although the data do not indicate an extreme threat to environmental and human health, they do suggest that improved management should be initiated.

We recommend that an environmental monitoring program be initiated and a detailed ecological and human-health risk assessment be carried out in order to address several issues, which include the following:

- a determination of the source of stress, fin disease, and genotoxicity in the fish from the lagoons adjacent to the Xan oil facility;

- a more detailed screening of the possible contaminants at the Xan oil facility;

- a determination of the origin of the PAHs in Laguna Buena Vista and, more generally, studies of the seasonally varying hydrological connections around the Xan oil facility so the redistribution patterns of potential contaminants can be understood; and

- an assessment of any possible health effects from PAH exposure in the local population inhabiting the areas surrounding the Xan facility.

LITERATURE CITED

Agarwal, S., and R. S. Sohal. 1994. DNA oxidative damage and life expectancy in houseflies. Proceedings of the National Academy of Sciences 91: 12,332-12,335.

Ashok, B. T., and S. Saxena. 1995. Biodegradation of polycyclic aromatic-hydrocarbons—a review. Journal of Scientific & Industrial Research 54: 443-451.

Bauman, P. C. 1998. Epizootics of cancer in fish associated with genotoxins in sediment and water. Mutation Research—Reviews in Mutation Research 411: 227-233.

Bickham, J. W. 1990. Flow cytometry as a technique to monitor the effects of environmental genotoxins on wildlife populations. Pp 97-108 in *In Situ Evaluation of Biological Hazard of Environmental Pollutants* (S. Sandhu, W. R. Lower, F. J. DeSerres, W. A. Suk, and R. R. Tice, eds.). Environmental Research Series Vol. 38. Plenum Press, New York, NY, USA.

Bickham, J. W., J. A. Mazet, J. Blake, M. J. Smolen, Y. Lou, and B. E. Ballachey. 1998. Flow cytometric determination of genotoxic effects of exposure to petroleum in mink and sea otters. Ecotoxicology 7: 191-199.

Freeman, S. E., and B. D. Thompson. 1990. Quantitation of ultraviolet radiation-induced cyclobutyl pyrimidine dimers in DNA by video and photographic densitometry. Analytical Biochemistry 186: 222-228.

Freidberg, E. C. 1985. *DNA Repair*. Plenum Press, New York, NY, USA.

Hoeijmakers, J. H. J. 1993. Nucleotide excision repair II, from yeast to mammals. Trends in Genetics 9: 211-217.

McCarthy, J. F., and L. R. Shugart. 1990. Biomarkers of environmental contamination. Pp. 3-16 in *Biomarkers of Environmental Contamination* (J. F. McCarthy and L. R. Shugart, eds.). Lewis Publishers, Boca Raton, FL, USA.

Mille, G., D. Munoz, F. Jacquot, L. Rivet, and J. C. Bertrand. 1998. The Amoco Cadiz oil spill: evolution of petroleum hydrocarbons in the Iie Grande salt marshes (Brittany) after a 13-year period. Estuarine Coastal and Shelf Science 47: 547-559.

Newman, M. C., and C. H. Jagoe (eds.). 1996. *Ecotoxicology: A Hierarchical Treatment.* Lewis Publishers, Boca Raton, FL, USA.

O'Connor, A., C. Nishigori, D. Yarosh, L. Alas, J. Kibitel, L. Burley, P. Cox, C. Bucana, S. Ullrich, and M. 1996. DNA double strand breaks in epidermal cells cause immune suppression in vivo and cytokine production in vitro. Journal of Immunology 157: 271-278.

Pritchard, J. B. 1993. Aquatic toxicology—past, present, and prospects. Environmental Health Perspectives 100: 249-257.

Shugart, L. R. 1988. An alkaline unwinding assay for the detection of DNA damage in aquatic organisms. Marine Environmental Research 24: 321-325.

Shugart, L. R., J. Bickham, G. Jackim, G. McMahon, W. Ridley, J. Stein, and S. Steinert. 1992. DNA alterations. Pp. 125-154 in *Biomarkers: Biochemical, Physiological, and Histological Markers of Anthropogenic Stress* (R. J. Huggett, R. A. Kimerle, P. M. Mehrle Jr., H. L. Bergman, eds.). Lewis Publishers, Boca Raton, FL, USA.

Theodorakis, C. W., B. G. Blaylock, and L. R. Shugart. 1996. Genetic ecotoxicology I: DNA integrity and reproduction in mosquitofish exposed *in situ* to radionuclides. Ecotoxicology 5: 1-14.

Theodorakis, C. W., and L. R. Shugart. 1993. Detection of genotoxic insult as DNA strand breaks in fish blood cells by agarose gel electrophoresis. Environmental Toxicology and Chemistry 13: 1023-1031.

Theodorakis, C. W., S. J. D'Surney, J. W. Bickham, T. B. Lyne, B. P. Bradley, W. E. Hawkins, W. L. Farkas, J. F. McCarthy, and L. R. Shugart. 1992. Sequential expression of biomarkers in bluegill sunfish exposed to contaminated sediment. Ecotoxicology 1:45-73.

U.S. Environmental Protection Agency. 1991a. Proposed sediment quality criteria for the protection of benthic organisms: Acenapthene. U.S. EPA, Washington, DC, USA.

U.S. Environmental Protection Agency. 1991b. Proposed sediment quality criteria for the protection of benthic organisms: Fluoranthene. U.S. EPA, Washington, DC, USA.

U.S. Environmental Protection Agency. 1991c. Proposed sediment quality criteria for the protection of benthic organisms: Nanthrene. U.S. EPA, Washington, DC, USA.

Vindelov, L. L., I. J. Christianson, N. Keiding, M. Prang-Thomsen, and N. I. Nissen. 1983. Long-term storage of samples for flow cytometric DNA analysis. Cytometry 3: 317-320.

Vindelov, L. L., and I. J. Christensen. 1990. A review of techniques and results obtained in one laboratory by an integrated system of methods designed for routine clinical flow cytometric DNA analysis. Cytometry 11: 753-770.

Ward, J.F. 1988. DNA damage produced by ionizing-radiation in mammalian-cells-identities, mechanisms of formation, and repairability. Prog. Nucl. Acid. Res. Mol. Biol. 35: 95-125.

CHAPTER 6

A RAPID ASSESSMENT OF AVIFAUNAL DIVERSITY IN AQUATIC HABITATS OF LAGUNA DEL TIGRE NATIONAL PARK, PETEN, GUATEMALA

Edgar Selvin Pérez and Miriam Lorena Castillo Villeda

ABSTRACT

- A total of 173 bird species were recorded in Laguna del Tigre National Park (LTNP), approximately 60.7% of the total number of species (N=285) reported from the park. Of these, 157 were reported in point counts; the remainder from nonsystematic counts.

- Eleven new records of birds for the park's official list were recorded, for a total of 296 bird species. The new records include *Aramus guarauma, Calidris minutilla, C. melanotus, C. fuscicollis, Cyanocorax yucatanicus, Cardinalis cardinalis, Cyanocorax yucatanicus,* and *Himantopus mexicanus.* Distribution extensions were also recorded for three of the new records: *Tyrannus savana monachus, Laterallus exilis,* and *Pyrocephalus rubinus.*

- Overall, ciconiiforms (large wading birds) were distributed widely across the park; they were most abundant in the marshes of the Río Escondido and the open canopy forest of the Río Sacluc and Río San Pedro. Forest specialists were especially numerous on hills in closed forests of the eastern San Pedro area; these include *Crax rubra, Agriocharis ocellata, Automolus ochrolaemus,* dendrocolaptids, and trogloditids.

- The principle source of variation in species composition in the park is the contrast between the western portions, including the biotope and the surrounding wetlands, and the forested areas of the central (Chocop/Candelaria) and eastern (Río San Pedro/Sacluc) portions of the park. The western area was unique and contained many species found only there, including *Calidris minutilla, C. fusciola, C.melanotus, Himantopus mexicanus, Laterallus exillis, Phorphyrula martinica, Jacana spinosa* (associated with playones, or shallow stagnant waters), *Tyrannus savana monachus,* and *Pyrocephalus rubinus* (associated with flooded savannas in La Lámpara).

INTRODUCTION

Laguna del Tigre National Park represents a very important international wetland area for Guatemala. The park was registered as a Ramsar site in 1990 and is also included in the Montreaux 1993 register. The former listing was necessary because of the threat posed by petroleum exploitation in the park (Carbonell et al., 1998). The goal of this study was to document and compare the diversity of wetland birds in different areas within LTNP and to contribute to the development of monitoring studies of birds in this important wetland area.

Three important avian habitat types can be distinguished in the LTNP wetlands: 1) riparian forests; 2) flooded savannas, swamps, or wetlands; and 3) zones of disturbance (petroleum operations, agriculture, houses, etc). These habitats and the microhabitats they contain tend to support distinct bird guilds, including species that use low riparian and wetland trees and bushes; shallows with weedy vegetation; tall and solid emergent vegetation; low entanglements in vegetation, usually in open areas; and tall trees in riparian areas. Thus, vegetation structure is expected to be a major determinant of species composition patterns for wetland birds (Zigûenza, 1994).

A previous baseline survey of birds for a monitoring program in LTNP (Mendez et al., 1998) recorded 41 novel records for a total of 285 species, including a new record for the Petén (*Zenaida asiatica*) and the range extension of the greater potoo (*Nyctibius grandis*). Our objective was to complement this effort by broadening the number of vegetation associations and areas sampled within LTNP. We also provide a focus on the relationships between birds and aquatic habitats that has not been seen in previous studies.

METHODS

Birds were sampled in three broad habitat types. Riparian areas were sampled along the ríos San Pedro, Sacluc, Candelaria, Chocop, and parts of the Río Escondido. Three macrohabitat types within riparian areas were considered. We studied open and closed riparian forest, marshes in combination with riparian forest, flooded savanna composed of sawgrass (sibal), trees (e.g. *Crescentia* sp.), tasiste (*Acoelorraphe wrightii*), and grasses. This latter macrohabitat was most common in the Río Escondido area and in certain lagoons such as La Lámpara. Second, marsh/ wetlands were surveyed, exemplified by the vegetation of the Río Escondido within the biotope and at its confluence with the Río San Pedro. This category of habitat was the least common of the three habitat types. Finally, areas of disturbed and regenerating vegetation (guamiles) such as those near settlements, oil operations, and abandoned agricultural fields were examined.

Sampling was centered at the four focal points used in this RAP survey (refer to "Expedition Intinerary and Gazetteer"). As for the reptile and amphibian sampling (Castañeda Moya et al. this volume), the use of discrete sampling points was inappropriate because birds use habitats over relatively broad scales. Thus, we sampled birds along transects in bodies of water using 20 point counts per location. The individual sample units treated here include the Río San Pedro, Río Sacluc, Río Escondido, Laguna Flor de Luna, Río Chocop, Río Candelaria, and Laguna la Lámpara (Table 6.1, Appendix 1). At the Río Chocop we were able to sample only 10 points. For surveys on rivers, the point counts were distributed along the length of the river. For lagoon surveys, point counts were taken while walking along the water's edge.

In order to choose the most effective and convenient methods for sampling aquatic bird species in the park, we initially compared three sampling strategies. Audiovisual point counts were conducted by establishing 10 points equally spread over a distance of three km (i.e. one point count every 300 m.) that were surveyed in a single session. Species seen or heard were recorded during 10 min at each point. Birds were also sampled by traveling at a constant speed (10 km/h) on a motorboat and recording every bird seen or heard over a three-km swath. Finally, we sampled birds at night along three-km transects in order to record nocturnal perching sites, especially for communal species. For each sampling scheme, we recorded the habitat type in which each species was found and their behavior in that habitat (e.g. feeding, perching).

After an initial evaluation of these methods, we decided to use audiovisual point counts alone. This method proved to be the most effective for recording the greatest number of species in different kinds of habitats as well as for identifying very active birds such as wrens, woodcreepers, parrots, and cracids. Our census was aided by the use of a directional microphone to record calls that were not immediately recognizable. Nonsystematic observations were also recorded but not included in the data analysis.

Table 6.1. Distribution of sampling intensity per focal point, Laguna del Tigre National Park, Petén.

Focal Areas	Extent of Sampling	Types of Habitat Sampled
Río San Pedro	From Paso Caballos (SP2) to La Caleta (SP8)	Gallery forest connected with high forest on hills, marshland
Río Sacluc	Almost the entire Río Sacluc	Open and closed gallery forests
Río Escondido	Almost two-thirds of the river (when navigable)	Open and closed gallery forests, sawgrass marsh, flooded grassland, caños, burned areas
Laguna Flor de Luna	Laguna Flor de Luna	Open gallery forests, flooded savanna, "guamiles," and sawgrass marshes
Río Chocop	Parts of the ríos Chocop and Candelaria; separately, Laguna la Lámpara	Open and closed gallery forests, marshes, disturbed areas (oil operations, guamiles, burned areas), vegetation rafts with *Nymphea*

RESULTS AND DISCUSSION

We recorded a total of 2103 individuals in 173 species, 42 families, and 17 orders (Appendix 11). Most individuals were passeriformes. The number of species tallied corresponds to 60.7% of the avifauna reportedly found in the park (Méndez et al., 1998).

New species records for LTNP include *Aramus guarauma, Calidris minutilla, C. melanotus, C. fuscicollis, Cyanocorax yucatanicus, Cardinalis cardinalis, Cyanocorax yucatanicus,* and *Himantopus mexicanus.* The ranges of distribution were expanded for *Tyrannus savana monachus, Laterallus exilis,* and *Pyrocephalus rubinus* (based on maps in Howell and Webb 1995). *Jabiru mycteria* was observed and had not been recorded in this area for 10 years (Herman A. Kihn, pers. com.).

Avifaunal Comparisons Among Focal Areas

Overall, bird species were homogeneously distributed across the park. The greatest richness was found along the San Pedro, Sacluc, and Escondido rivers, while relatively fewer species were observed on the Chocop and Candelaria rivers (Figure 6.1). The latter may be due to the small area of closed gallery forest there. The high species richness found on the San Pedro and Sacluc rivers may be due to favorable weather conditions during sampling and the proximity of the large bosques altos. Many forest specialist bird species (c.f. Whitacre, 1996) were recorded in these areas, including woodcreepers, trogons, motmots, guans, chachalacas, curassows, and certain flycatchers. Some of these species were found only at these sites, such as *Xiphorynchus flavigaster, Agriocharis ocellata, Crax rubra,* and *Trogon massena.*

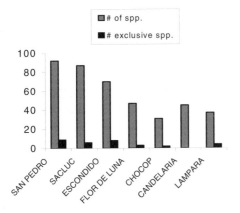

Figure 6.1. Patterns of bird species richness and the numbers of exclusive species (found in no other sample) among the sample areas in Laguna del Tigre National Park.

The Río Escondido area harbored several riparian macrohabitats, including sawgrass marshes, open gallery forest, and closed gallery forest (i.e. Punto Icaco, E8). The adjacent zones on the west side of the park, together with the Laguna la Lámpara, represent a unique habitat composed of swamps and flooded savannas. These habitats harbor bird species such as *Rostramus sociabilis, Laterallus exillis, Aramus guarauma,* and *Jabiru mycteria.*

The central area of the park, represented here by the Chocop/Candelaria focal area, presented unique bird species such as *Dendrocygna anabatina* (the black-bellied whistling-duck), which was observed in a lagoon near Pozo Xan 1. This zone had closed gallery forests and harbored primary forest species such *Micrastur semitorquatus* and *Xenops minutus,* which were also recorded in the ríos Sacluc and San Pedro during prior surveys (Pérez, 1998).

The jacana (*Jacana spinosa*) is strongly associated with floating vegetation and was recorded only in lagoons possessing the waterlily *Nymphea ampla* and in the shallow wetlands of the Río Escondido, where it occurred with *Porphirula martinica, Laterallus ruber,* and *L. exillis. Busarellus nigricollis* and *Rostramus sociabilis* were recorded in isolated lagoons and were associated with emergent vegetation in sawgrass marshes.

Analyses of similarity revealed that there was generally a high degree of similarity in species composition among sample areas (Table 6.2). The San Pedro and Sacluc areas shared 60 species due to their proximity. The majority of the avifauna of the Río Escondido and Laguna la Lámpara was found in the San Pedro, and most of these species were associated with the open gallery forests and forested wetlands (see Léon and Morales Can, this volume). Species inhabiting the flooded grasslands, however, were not found in the San Pedro area; these include *Tyrannus savana monachus, Pyrocephalus rubinus,* and *Aramus guarauma.* Of the sites visited, only La Lámpara exhibited flooded

Table 6.2. Numbers and percentages of shared species among the study sites. NS is the number of species shared, % is the percentage of species of the smaller fauna shared with the larger fauna in each comparison (Simpson's Similarity Index).

	Río Candelaria		Río Chocop		Río Escondido		Flor de Luna		Laguna la Lámpara		Río San Pedro	
	NS	%	NS	%	NS	%	NS	%	NS	%	NS	%
Chocop	20	64.51										
Escondido	22	48.88	15	48.38								
Flor de Luna	20	44.44	11	35.48	31	66						
Lámpara	13	35.13	5	16.13	29	78.38	19	51.35				
San Pedro	28	62.22	16	51.61	45	64.28	26	55.31	28	75.67		
Sacluc	30	66.67	26	83.87	38	54.28	30	63.83	25	67.57	60	69

grasslands, composed primarily of *Cladium jamaicense* and other species of Cyperaceae, grasses, trees (*Crescentia alata*), and tasiste (*Aceolorapphe wrightii*) in certain areas. This type of habitat is also called "campería" by local inhabitants. A small number of families had settled on the edge of this lagoon. A combination of stresses due to variation in climate (Kushlan et al., 1995) and local anthropogenic alteration may threaten the existence of critical habitats here.

The closed gallery forest type is shared among the ríos Chocop, Candelaria, and Sacluc; as a consequence, the avifauna of these sites is similar (Table 6.2) The aquatic bird species of this type of gallery forest are typified by the presence of kingfishers such as *Chloroceryle americana* and *C. aenea*, which frequent low perches located near bodies of water.

Avifaunal Distribution Among Macrohabitats

Riparian forest
The largest number of bird species (117) was recorded in this type of habitat. There are two types of riparian forest: closed and open. In closed gallery forest, a large number of flycatcher species is associated with the undergrowth. Other species that are typical of this habitat include *Thryothorus maculipectus,* several woodcreepers (*Dendrocinchla homochroa, D. autumnalis, Automolus ochrolaemus, Xiphorynchus flavigaster, Xenops minutus*), *Micrastur ruficollis, Platyrhynchus cancrominus,* two kingfishers (*Chloroceryle americana, C. aenea*), *Geotrygon montana,* and *Columba nigrirostris.* Studies on the use of birds as biological indicators indicate that bird species that forage in the middle and undergrowth strata of mature forests (i.e. woodcreepers, wrens, and tyrant flycatchers) are sensitive to habitat disturbance (Whitacre, 1996). This is due to the fact that these sedentary species are exposed to the effects of disturbance to a greater degree than those species inhabiting the upper canopy (Levey, 1990).

Open gallery forest (including forested wetlands) is the most common forest type in this area, found along the ríos San Pedro, Sacluc, and Escondido, and near all the lagoons we observed. This habitat harbored the majority of species we recorded. The complex forest structure presents opportunities for a variety of foraging niches, including canopy insectivores, hawkers (flycatchers and swallows), large wading birds (Ciconiiformes) that hunt along the river edge, and predators such as hawks and kingfishers—most of which require tall vegetation. Inhabitants of this type of forest include raptors (*Busarellus nigricollis, Pandion haliaetus,* and *Rostramus sociabilis*); ciconiiforms (*Ardea herodias, Egretta thula, E. caerulescens, E. alba, Butorides striatus, Tigrisoma mexicanum, Cochlearius cochlearius, Nycticorax violaceus, Anhinga anhinga,* and *Phalacrocorax olivaeus*); flycatchers (*Miozetetes imilis, Pitangus*

sulphuratus, Megarhynchus pitanga, and *Legatus leucophaius*); and doves (*Columba speciosa* and *C. cayenensis*).

Threatened species such as the scarlet macaw (*Ara macao*) were recorded in open forest near a hillside near the Chocop focal area. Previous studies (Pérez, 1998) suggest that many of the macaw's preferred food plants occupy this habitat, including ramón (*Brosimium alicastrum*), members of the Zapotaceae (*Pouteria spp.*), vines of the family Bignoniaceae (malerio bayo, malerio blanco, guaya [*Talisia olivaeriformes*] and jocote jobo [*Spondias spp.*]), as well as the tree species in which the macaws make their nests, such as *Acacia glomerosa*. This tree is associated with flooded lowlands, often near lagoons. Slash-and-burn agriculture is the primary threat to riparian forests of the park.

Flooded savanna
This type of habitat was recorded only in an area near Laguna la Lámpara. Species exclusive to this site include *Tyrannus savana monachus, Pyrocephalus rubinus,* and *Aramus guarauna*. Other species common in the area are *Tyrannus melancholicus, Myiozetetes similis, Megarynchus pitanga, Agelaius phoeniceus, Columba speciosa, C. flavirostris,* and *C. cayenensis*. This unique habitat type may be threatened by human colonization around the lagoon—especially by fires started during the dry season to clear land for agriculture.

Flooded grasslands
This habitat is also known as wet savanna, salt marsh, flood plains, or sibal. It is one of the most extensive tropical ecosystem types (Dugan 1992). Flooded grassland is found exclusively on the Río Escondido. It was also one of the most distinct in terms of bird species composition. There is a high diversity of microhabitats in this habitat due to its varying hydrological characteristics. For example, sandy banks are suitable habitat for species such as *Calidris melanotus, C. minutilla, C. fuscicollis, Laterallus exilis, L. ruber,* and *Tringa flavipes*. We recorded most of the egret species observed in this survey in this habitat, as well as *Columba* spp. and the snail kite (*Rostramus sociabilis*), which depends principally on aquatic snails (*Pomacea* sp.). The jabiru (*Jabiru mycteria*) was observed here as well. This habitat is threatened by development due to the proximity of roads and the oil facilities; in particular, the widening of the river canal for navigation would be detrimental.

CONCLUSIONS AND RECOMMENDATIONS

We judge the overall health of avian biodiversity in the park's gallery forests to be intermediate. The abundance of generalist species that are indicative of disturbance was not as great as in similar studies of guamil sites in the Lacandona jungle of nearby Chiapas, Mexico (Warkentin et al., 1995). Continued forest disturbance and loss, however, is imminent if the agricultural activity of colonists continues unabated. In this regard, the central area of LTNP (i.e. Flor de Luna and Chocop focal areas) is likely to be most affected by colonization due to the presence of roads in this zone and their proximity to wetland habitats. Due to the uniqueness of the bird fauna in this zone, the impacts of habitat destruction here will have disasterous consequences for the park's avifauna (Méndez et al., 1998).

Our observations suggest that critical habitats for aquatic birds in LTNP, including both wetlands and bosque alto/riparian forests, are rare in the core areas of the Maya Biosphere Reserve (MBR; Carbonell et al., 1998). The strict conservation of such habitats in core areas should be an important consideration in the design and management in the MBR (Méndez et al., 1998). To this end, it is imperative that future studies in the park provide a more detailed understanding of what constitutes different avian habitats and their distributions (rather than relying on simple zonations). We offer several specific recommendations for studies and conservation of birds within LTNP in no particular order of importance:

- Investigate the effects on the bird populations of channel widening in the Río Escondido.

- Investigate the association between *Tyrannus savana monachus* and *Pyrocephalus rubinus* with flooded savanna habitats of La Lámpara.

- Implement a monitoring program to document changes in bird diversity along disturbance gradients in different habitat types and use these data to establish tolerance limits for particular bird species.

- Provide a unified, regional perspective for management of the wetland avifauna of the western MBR.

- Provide a detailed classification of avian habitats for areas throughout the MBR.

LITERATURE CITED

Carbonell, M., O. Lara, J. Fernández-Porto, and others. 1998. Convention on Wetlands. Proceedings of the Orientation of Management. Ramsar Site: Laguna del Tigre, Guatemala. Gland, Switzerland.

Dugan, P. J. (ed.) 1992. Conservation of Wetlands: An Analysis of Actual Issues and the Necessary Actions. IUCN, Switzerland.

Howell, S., and S. Webb. 1995. *A Guide to Birds of México and Northern Central America.* Oxford University Press, New York, NY, USA.

Kushlan J. A., G. Morales, and P. Frohing. 1985. Foraging niche relations of wading birds in tropical west savannas. Pp. 663-682 in *Neotropical Ornithology.* Ornithological Monographs 36.

Levey, D. 1990. Habitat-dependent fruiting behaviour of an understorey tree, *Miconia centrodesma*, and tropical treefall gaps as keystone habitats for frugivores in Costa Rica. Journal of Tropical Ecology 6: 409-420.

Méndez, C., C. Barrientos, F. Casteneda, and R. Rodas. 1998. Programa de Monitoreo, Unidad de Manejo Laguna del Tigre. Los Estudios Base para su Establecimiento. CI/ProPetén, Conservation International, Guatemala.

Pérez E. S. 1998. Evaluation of available habitat for the scarlet macaw (*Ara macao*) in Petén, Guatemala. Tésis de Licenciatura. Universidad de San Carlos de Guatemala, Facultad de Ciencias, Químicas y Farmacia, Escuela de Biología.

Warkentin I. G., T. Russell, and J. Salgado. 1995. Songbird use of gallery woodlands in recently cleared and older settled landscapes of the Selva Lacandona, Chiapas, México. Conservation Biology 9: 1095-1106.

Whitacre, D. 1996. *Ecological Monitoring Program for the Mayan Biosphere Reserve*. Draft. The Peregrine Foundation.

Zigüenza R. 1994. Evaluación de las fluctuaciones poblacionales de 31 especeis de aves en el área de protección especial Manchon-huamuchal. Tésis de Licenciatura. Universidad de San Carlos de Guatemala, Facultad de Ciencias, Químicas y Farmacia, Escuela de Biología.

CHAPTER 7

THE HERPETOFAUNA OF LAGUNA DEL TIGRE NATIONAL PARK, PETEN, GUATEMALA, WITH AN EMPHASIS ON POPULATIONS OF THE MORELET'S CROCODILE (*CROCODYLUS MORELETII*)

Francisco Castañeda Moya, Oscar Lara, and
Alejandro Queral-Regil

ABSTRACT

Population Study of *Crocodylus moreletii*

- The densities of *C. moreletii* recorded along the Xan-Flor de Luna road and in Laguna la Pista are the highest yet recorded in Guatemala.

- Laguna la Pista, Laguna el Perú, and the Río Sacluc are ideal sites in which to implement a management program for *Crocodylus moreletii* in Laguna del Tigre National Park (LTNP) due to the high percentages of adults at these sites as well as the infrastructure available to researchers.

- The sibal (stands of *Cladium jamaicense*) habitat was most often used by *C. moreletii* in LTNP. This habitat likely offers suitable insolation, food availability, and protection for this species.

- Transects of 500 m sampled in 150 min appears to be a good sampling strategy for *C. moreletti* in areas within LTNP.

Survey of Herpetofauna in Riparian Vegetation

- There is a gradient of decreasing herpetofaunal abundance from the shorelines of water bodies to the forest interior during the dry season in LTNP.

- The species of reptiles and amphibians recorded in LTNP have broad distributions within the Yucatán Peninsula and exhibit little specificity to the Petén area. *Dermatemys mawii*, however, is a regional endemic and should be a focus of park managers.

INTRODUCTION

The Guatemalan herpetofauna is extremely rich. Campbell and Vannini (1989) listed 326 species for Guatemala, and 160 of these occur in the Petén region (Campbell, 1998). However, during the last ten years, there have been several taxonomic revisions and phylogenetic analyses that have revealed a richer herpetofauna than was previously thought (Campbell and Mendelson, 1998). Currently we know of 145 species of amphibians and 242 species of reptiles for a total of 387 species in Guatemala, although recent estimates suggest that there may be more than 400 species (Campbell and Mendelson, 1998).

Despite these studies, the herpetofauna of the Petén and of Laguna del Tigre National Park specifically is poorly known. Recently, Mendez et al. (1998) conducted a survey of the plants, birds, and amphibians and reptiles of the park. This study provided baseline data on herpetofauna, although the geographic scope of this study was very limited.

An important component of the herpetofauna of the Petén is Morelet's crocodile (*Crocodylus moreletii*). *C. moreletii* is an endemic species of the Yucatán Peninsula and is listed in the IUCN Red Book (1996) as data deficient and in Appendix I of CITES. Previous population studies of *C. moreletii* in Guatemala have shown that the persistence of the species in the area is threatened by illegal hunting and by increased destruction of habitat due to human encroachment (Lara, 1990; Castañeda, 1997).

METHODS

Population Study of *Crocodylus moreletii*

Sampling of crocodiles took place in the following sites:
　　1) San Pedro focal area: Río San Pedro (ca. 38 km along the river, from the community of Paso Caballos to the

community of El Buen Samaritano) and Río Sacluc (ca. 7 km from the confluence with the Río San Pedro to the end of the Río Sacluc).

2) Río Escondido focal area (ca. 29 km covering most of the navigable section of the river).

3) Flor de Luna focal area (ca. 1.3 km of shoreline of Laguna Flor de Luna and ca. 1 km of shoreline of Laguna Vista Hermosa).

4) Río Chocop focal area: Laguna Tintal (C4, 1.4 km of shoreline); La Pista (C3; 2 km of shoreline); Laguna Guayacán (C2; 1.3 km of shoreline); and several other aguadas (small ponds) and lagoons along the Xan-Flor de Luna road (refer to "Expedition Itinerary and Gazetteer").

5) In addition, a sample was taken at Laguna el Perú (5 km of shoreline), which is near the El Perú archeological site (a trail to the site begins near SP10).

We conducted the population census of *Crocodylus moreletii* by shining spotlights along the shorelines of rivers and lagoons from an inflatable boat with an outboard motor. We counted any visible crocodile. We attempted to capture every animal we saw, although they were often out of reach from the boat. We categorized both observed and captured animals as neonates, juveniles, sub-adults, or adults based on approximate body size (Castañeda, 1998). We estimated the size of every individual using the relationship between eye-nostril distance and total length (following Lara, 1990).

In order to determine types of habitat used, we also recorded the habitat in which an animal was first observed. The habitat types considered were 1) water: those records of floating individuals not on the shoreline or in vegetation; 2) riparian forest: those records in arborescent vegetation on the shorelines of water bodies; 3) guamil: secondary-growth vegetation lacking trees; 4) sibal: stands of sawgrass (*Cladium jamaicense*) in marsh vegetation; and 5) emergent vegetation.

Survey of Herpetofauna in Riparian Vegetation

The herpetofauna occupying the banks of rivers was examined at terrestrial sample points along the Río San Pedro (SPT1), Río Sacluc (SPT5), and Río Escondido (ET2, ET3; refer to "Expedition Itinerary and Gazetteer").

We laid out belt transects 500 m in length and 5 m wide at the shoreline of these three rivers. Each transect was divided into five smaller 100-m transects. We standardized sampling efforts by limiting the search to 30 min for every 100 m. In addition, opportunistic observations in areas sampled for crocodiles were recorded.

RESULTS

Population Study of *Crocodylus moreletii*

Overall, 130 individuals of *C. moreletii* were recorded in 87.14 km of surveys in water bodies of LTNP. The average density at sample points was 4.3 crocodiles/km. Closed aquatic systems had a greater average density when compared to open systems (rivers; Table 7.1). Among open aquatic systems, the Río Sacluc possessed a greater density than the Río San Pedro or Río Escondido. The lagoons of the Flor de Luna focal area and the Laguna la Pista (SP3) had the greatest densities among the closed aquatic systems. The population structure of *C. moreletii* in LTNP is biased toward younger age classes (Figure 7.1).

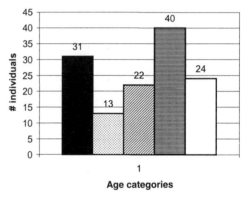

Figure 7.1. The number of individuals of *Crocodylus moreletii*, by age class, observed during surveys within Laguna del Tigre National Park.

The Laguna el Perú had the highest percentage of adults, constituting 32.26% of all adults in the study, followed by the Río Sacluc (Table 7.2). Laguna la Pista also contained a large number of adults and exhibited the largest percentage of sub-adults in the study (ca. 46% of all sub-adults). In the Río Escondido, as in the San Pedro, we observed a high percentage of neonates as well as juveniles. Samples from many lagoons in the Flor de Luna and Río Chocop focal areas exhibited low numbers; in these areas, most individuals were juveniles.

The largest percentage of habitat records of *C. moreletii* were in sibal (39%), followed by water away from shore (31%), riparian forest (25%), emergent vegetation (4%), and guamil (1%; Table 7.3). Adults had similar patterns of abundance among habitats, with relatively high values in water, riparian forest, and sibal (Table 7.2). On the other hand, neonates and juveniles were most strongly associated with sibal.

Survey of Herpetofauna in Riparian Vegetation

We found a total of 14 amphibian species and 22 species of reptiles (including *Crocodylus moreletii*; Appendix 12) along the sample transects. We recorded 240 individual reptiles and amphibians, excluding *C. moreletii*. The most abundant

Table 7.1. Estimated densities for *Crocodylus moreletii* in Laguna del Tigre National Park, Petén, Guatemala.

Open Aquatic Systems			
Sample point	**Length (km)**	**# individuals**	**Individuals/km**
Río San Pedro	38	23	0.61
Río Sacluc	7	14	2.00
Río Escondido	29	22	0.76
TOTAL	**74**	**59**	**Average= 1.12**
Closed Aquatic Systems			
Laguna Flor de Luna	1.3	5	3.85
Laguna Vista Hermosa	1	3	3.00
Laguna Tintal	1.4	5	3.57
Laguna la Pista	2	22	11.00
Laguna Guayacán	1.3	2	1.54
Laguna el Perú	5	22	4.40
Other lagoons along Flor de Luna road	1.14	14	12.28
TOTAL	**13.14**	**73**	**Average= 5.66**
GRAND TOTAL	**87.14**	**130**	**Average= 4.30**

amphibian species were the fringe-toed foamfrog (*Leptodactylus melanonotus*; 119 individuals), *Hyla microcephala* (25), and *Ranaberlandieri* (24). These species were found along all the transects sampled. The most abundant species of reptiles were the striped basilisk (*Basiliscus vittatus*; 12) and *Norops bourgeai* (7). Among turtles, *Kinosternon leucostomum* was the most abundant, although most individuals of this species were recorded as anecdotal observations. We found the highest species richness in the Río San Pedro (17 species), followed by the Río Sacluc (13) and the Río Escondido (11; Appendix 12). With respect to abundance, however, the highest values were observed in the Río Sacluc (123), followed by the Río San Pedro (68) and Río Escondido (40).

DISCUSSION

Population Study of *Crocodylus moreletii*

C. moreletii is commonly called the "marsh lizard" (lagarto pantanero) due to its preference for these habitats (Alvarez del Toro, 1974). This explains the relatively high densities of this species in lagoons and other closed aquatic systems when compared to open aquatic systems such as rivers.

Within the zone of surrounding the Las Guacamayas Biological Station, the Río Sacluc is one of the most important sites for the reproduction of this species (Castañeda, 1998). The densities recorded in this study (two individuals/km), when compared to values in other rivers such as the San Pedro (0.60) and Escondido (0.76), confirm the importance of the Río Sacluc in the region. The densities recorded in the combined lagoons along the Xan-Flor de Luna road (12.28) and Laguna la Pista (11) are the highest densities recorded in Guatemala. The only other studies available for comparison—

those of Lara (1990) and Castañeda (1998)—reported a range of densities from 0 - 5.9.

The lagunas El Perú and La Pista and the Río Sacluc are ideal sites from which to initiate management programs for *C. moreletii* within LTNP for the following reasons: 1) La Pista is easy to access and is relatively near the CONAP control post; 2) El Perú is adjacent to a monitoring post of CI-ProPetén, has facilities, and has a large number of adult individuals; and 3) Río Sacluc is near the biological station. The fact that each of these sites has a large proportion of adult individuals is an important factor to consider in managing the reproductive success of *C. moreletii*. The lagoons and ponds along the Xan-Flor de Luna road are of great interest not only because of the number of animals, but also because these are excellent sites in which to study migration patterns of this species. The majority of the recorded animals were juveniles that possibly migrated from other sites, which is consistent with the animals' ages.

By comparing density values obtained by Castañeda (1998) in the ríos Sacluc (4.35 individuals/km) and San Pedro (2.10) with those recorded in this study in the same locations (2.00 and 0.61 respectively), we see that the values may differ but that the Sacluc maintains a greater density in both samples. This variation may be due to the use of a more intensive sampling design in a smaller area by Castañeda (1998; two transects and eight surveys), while in this study ten transects were sampled only once, allowing for greater coverage of the park.

The population structure recorded in this study is atypical in that sub-adults were the smallest class. Stable populations typically have adults as the smallest age class (see references in Castañeda, 1998). This might be explained by a more secretive behavior of sub-adults or by the possibility that adults were migrating from natal areas to other sites in search of territories.

Table 7.2. Population structure of *Crocodylus moreletii* by sample.

Data given are: Total # of individuals
        % of sample
        % of individuals

Age Class → Sampling Site ↓	Adult	Neonate	Unknown	Juvenile	Subadult	Total Indiv./ Sample
Aguada 1	0 0.00 0.00	0 0.00 0.00	0 0.00 0.00	1 100.00 2.50	0 0.00 0.00	1
Aguada 2	0 0.00 0.00	0 0.00 0.00	0 0.00 0.00	1 100.00 2.50	0 0.00 0.00	1
El Perú	10 45.45 32.26	0 0.00 0.00	5 22.73 20.83	5 22.73 12.50	2 9.09 15.38	22
Escondido	2 9.09 6.45	10 45.45 45.45	1 4.55 4.17	8 36.36 20.00	1 4.55 7.69	22
Establo	0 0.00 0.00	0 0.00 0.00	0 0.00 0.00	6 100.00 15.00	0 0.00 0.00	6
Flor de Luna	1 20.00 3.23	3 60.00 13.64	0 0.00 0.00	0 0.00 0.00	1 20.00 7.69	5
Garza 1	0 0.00 0.00	0 0.00 0.00	1 33.33 4.17	1 33.33 2.50	1 33.33 7.69	3
Guayacán	0 0.00 0.00	1 50.00 4.55	0 0.00 0.00	1 50.00 2.50	0 0.00 0.00	2
La Pista	6 27.27 19.35	4 18.18 18.18	5 22.73 20.83	1 4.55 2.50	6 27.27 46.15	22
Román	0 0.00 0.00	0 0.00 0.00	0 0.00 0.00	3 100.00 7.50	0 0.00 0.00	3
Sacluc	5 35.71 16.13	0 0.00 0.00	7 50.00 29.17	2 14.29 5.00	0 0.00 0.00	14
San Pedro	4 19.05 12.90	4 19.05 18.18	3 14.29 12.50	8 38.10 20.00	2 9.52 15.38	21
Tintal	2 40.00 6.45	0 0.00 0.00	1 20.00 4.17	2 40.00 5.00	0 0.00 0.00	5
Vista Hermosa	1 33.33 3.23	0 0.00 0.00	1 33.33 4.17	1 33.33 2.50	0 0.00 0.00	3
TOTAL # of Individuals	31	22	24	40	13	130

CONSERVATION INTERNATIONAL

Rapid Assessment Program

Table 7.3. Habitat use of *Crocodylus moreletii* by age class.

Habitat	Body of Water	Adults	Neonates	Unknown	Juvenile	Subadults	Habitat Totals
				Age class			
Open water	Lagoon	13	0	7	11	3	34
	River	4	1	1	0	0	6
Total		17	1	8	11	3	40
Riparian forest	Lagoon	1	1	3	4	1	10
	River	5	3	7	7	1	23
Total		6	4	10	11	2	33
Emergent vegetation	Lagoon	0	0	0	4	0	4
	River	0	0	0	1	0	1
Total		0	0	0	5	0	5
Guamil	Lagoon	0	0	0	1	0	1
	River	0	0	0	0	0	0
Total		0	0	0	1	0	1
Sibal	Lagoon	6	7	3	3	6	25
	River	2	10	3	9	2	26
Total		8	17	6	12	8	51
Overall total		31	22	24	40	13	130

The similarity of the abundance patterns between the adults and the "unknown" class (Table 7.2, 7.3) with respect to sample points and habitats suggests that in many cases the unknowns were likely adults. It is possible that these individuals have had negative experiences due to human presence in the area, which includes hunting (Castañeda, 1998), and thus tend to hide before they can be tallied. The lagunas El Perú and La Pista and the Río Sacluc may be considered important conservation areas for *C. moreletii* within LTNP due to the high proportion of adults that they possess—especially because the reproductive success of crocodile populations depends on a large number of breeding adults. In this regard, it is also important to note that Laguna la Pista also had a large number of sub-adults that will contribute to the breeding population if they survive to maturity. The relatively high proportions of young found in the Río Escondido, Río San Pedro, and Laguna la Pista may be due to earlier eclosion from eggs at these sites or to varying reproductive success among crocodile populations. The influence of human activity on crocodile reproduction is an important topic for study in LTNP.

The sibal (marsh) vegetation type is relatively common in LTNP and is an important habitat for animal species such as fish and turtles, as well as for crocodiles which nest in this vegetation. The importance of this habitat may be reflected in the relatively high use of sibal by crocodiles recorded in this study (Table 7.3). It is possible that crocodiles use sibal for other reasons, including its relatively high levels of insolation throughout much of the day, the increased food availability due to the use of sibal by fish, the reduced accessibility for humans, and its reduced level of disturbance.

Despite the importance of sibals within the park, this habitat is extensively burned each year, and this process constitutes a major threat to *C. moreletii* populations in the region (Castañeda, 1998).

Adults were often associated with open water, possibly because our study corresponded to the crocodiles' mating season that begins in March in Mexico (Alvarez del Toro, 1982). In the case of LTNP, this period may begin in March or April (Castañeda, 1998). Adults were often found in lagoons, perhaps due to the reduced accessibility of lagoons to humans when compared to rivers.

Survey of Herpetofauna in Riparian Vegetation

The overall high abundance of specimens recorded along the Río Sacluc is due to the high abundance of the fringe-toed foamfrog (*Leptodactylus melanonotus*), which is a common species with a broad distribution in LTNP. The majority of animals were detected on the shorelines of water bodies, and few were found at greater than five meters inland from the shoreline. This is likely due to the dependence of the herpetofauna on areas near water during the dry season (Mendez et al., 1998).

It is apparent that modifying the transect design employed by Méndez et al. (1998) of 200 m sampled in 120 min to a strategy of sampling 500 m in 150 min was more efficient. However, more intensive survey efforts are needed in LTNP. New records for the area, such as the striped spotbelly snake (*Coniophanes quinquevittatus*) found in this survey, are likely to emerge from such efforts.

Other threatened reptiles of the Petén that were observed during these surveys include the Central American river turtle (*Dermatemys mawii*), whose populations are declining due to hunting (Campbell, 1998), and the boa (*Boa constrictor*). *D. mawii* is listed by the IUCN as endangered, and both reptiles are CITES Appendix II listees. Along with Morelet's crocodile, these species should be a focus of future surveys and species-based management plans within the park.

RECOMMENDATIONS

- Monitor *Crocodylus moreletii* in LTNP, emphasizing the Río Sacluc and lagunas El Perú and La Pista.

- Initiate a captive management program for *C. moreletti* in which wild-reared eggs are collected and incubated under controlled conditions. This will help to reduce the impact of natural losses, and the surplus may be used as an economic alternative for local human populations.

- Use the 500 m/150 min transect sampling strategy, employing both visual and auditory signs, for future herpetofaunal inventory and monitoring efforts in LTNP.

- Include the Central American river turtle (*Dermatemys mawii)* and the boa (*Boa constrictor)* in monitoring programs.

LITERATURE CITED

Alvarez del Toro, M. 1974. *Estudio Comparativo de los Crocodylia de México.* Ed. Inst. Instituto Mexicano de Recursos Naturales Renovables, A.C. Mexico.

Alvarez del Toro, M. 1982. *Los Reptiles de Chiapas.* Instituto de Historia Natural del Estado, Departamento de Zoología.

Campbell, J. A. 1998. *Amphibians and Reptiles of Northern Guatemala, the Yucatán, and Belize.* The University of Oklahoma Press, Norman, OK, USA.

Campbell, J. A., and J. R. Mendelson III. 1998. Documentación de los anfibios y reptiles de Guatemala. Mesoamericana 3:21.

Campbell, J. A., and J. P. Vannini. 1989. Distribution of amphibians and reptiles in Guatemala and Belize. Proceedings of the Western Foundation of Vertebrate Zoology 4:1-21.

Castañeda, F. 1997. Estatus y manejo propuesto de *Crocodylus moreletii* en el Departamento de El Petén, Guatemala. Pp. 52-57 in *Memorias de la 4ta. Reunión Regional del Grupo de Especialistas de Cocodrilos de América Latina y el Caribe.* Centro Regional de Innovación Agroindustrial, S.C. Villahermosa, Tabasco.

Castañeda, F. 1998. *Situación actual y plan de manejo para Crocodylus moreletii en el área de influencia de la Estación Biológica Las Guacamayas, Departamento de Petén, Guatemala.* Tésis de Licenciatura en Biología, Facultad de Ciencias Químicas y Farmacia, Universidad de San Carlos, Guatemala.

Lara, O. 1990. *Estimación del tamaño y estructura de la población de Crocodylus moreletii en los lagos Petén-Itza, Sal-Petén, Peténchel y Yaxhá, El Petén, Guatemala.* Tésis de Maestría, Universidad Nacional, Heredia, Costa Rica.

Méndez, C., C. Barrientos, F. Casteñeda, and R. Rodas. 1998. *Programa de Monitoreo, Unidad de Manejo Laguna del Tigre. Los Estudios Base para su Establecimiento.* CI-ProPetén, Conservation International, Washington, DC, USA.

CHAPTER 8

THE MAMMAL FAUNA OF LAGUNA DEL TIGRE NATIONAL PARK, PETEN, GUATEMALA, WITH AN EMPHASIS ON SMALL MAMMALS

Heliot Zarza and Sergio G. Perez

ABSTRACT

- The mammal fauna was surveyed in four focal areas of Laguna del Tigre National Park (LTNP). The study focused on small mammals.

- Bats were captured with mist nets and rodents with snap traps. Other mammals were recorded by tracks and sign.

- A total of 40 species of mammals were identified from a potential list of 130 for the park. There were no significant differences in bat diversity among the four focal areas. Rodent species showed specific habitat associations as has been reported previously in other studies.

- We report a new species record for Guatemala: the deer mouse *Peromyscus yucatanicus*. Another regional endemic, the spiny pocket mouse *Heteromys guameri*, was also observed. Several rare and/or threatened large mammals, including tapir and the red brocket deer were observed near the Las Guacamayas Biological Station. This area may be of great conservation value for mammals.

INTRODUCTION

Deforestation produces loss and fragmentation of mammal habitats and is one of the worst threats to mammalian biodiversity worldwide. It produces a significant change in the composition and relative abundance of mammal species, reflected in the local extinction of rare species (Williams and Marsh, 1998) and in increases in the abundance of generalist or opportunist species (Malcolm, 1997). For example, disturbances often lead to the invasion of species that generally benefit from human activities, such as rabbits (*Sylvilagus* spp.) and the coyote (*Canis latrans*; Garrott et al., 1993).

Among mammals, bats are important components of biodiversity in tropical forests and may represent up to 60% of the local mammal fauna (Eisenberg, 1989). They are trophically diverse, including species that feed on invertebrates and vertebrates, pollen, nectar, fruit and blood (Wilson, 1989). The Neotropical forests have the greatest diversity of bats in the world. Of the 75 genera present in the Neotropics, 48 belong to the Family Phyllostomidae (Wilson, 1989). The diversity of phyllostomid bats is reflected in the numbers of species and their diversity of diets (Gardner, 1977), and these characteristics make them good biological indicators (Fenton et al., 1992).

The composition and relative abundance of bats are determined by the complexity and heterogeneity of the habitat (LaVal and Fitch, 1977). If the habitat maintains a high complexity and heterogeneity, it can provide a great number of niches per unit of space (August, 1983). Thus, patterns of bat composition and diversity have great potential as detectors of habitat disturbance in tropical forests (Amín and Medellín, *in press*). A study in Lacandona, Mexico (Medellín et al., *in press*), found significantly higher values of bat diversity in places with little perturbation and higher abundance of the subfamily Phyllostominae in undisturbed habitats. Similar patterns were reported for eastern Quintana Roo, Mexico (Fenton et al., 1992). It has been proposed that the identity of the single most abundant bat species can also be indicative of disturbance. For example, the bat genus *Artibeus* is more abundant in undisturbed tropical rainforest because it is associated with high trees such as the amates (*Ficus* spp.) and jobos (*Spondias* spp.). In moderately disturbed conditions, the genus *Carollia* is dominant; in heavily disturbed habitat, the single most abundant genus is *Sturnira* (Amín and Medellín, *in press*).

Small, non-volant mammals such as rodents are functionally important groups in tropical forests and constitute between 15% and 25% of the mammal fauna in Neotropical forests (Voss and Emmons, 1996). Rodents influence plant communities by dispersing seeds, are important food sources for predators (Dirzo and Miranda, 1990), and are sensitive to habitat changes. Large Neotropical forest mammals such as tapir, jaguar, and monkeys may be especially sensitive to habitat change and human activity due to their rarity, susceptibility to hunting, and timidity (Reid, 1997).

METHODS

A list of species that could potentially occur in LTNP was compiled from range maps and distribution descriptions in Hall (1981), Emmons and Feer (1997), and Reid (1997); these were compared with the collection data we obtained.

Sampling of Rodents

We established 540-m transects near the focal point of each focal area and sampled rodents using 90 snap traps (Victor Rat Trap and Museum Special Mammal Trap) distributed in pairs at 45 stations located every 12 m along the transect. Traps were baited in the afternoon with a mixture of peanut butter, raisins, crushed oats, and bacon and were checked in the early morning (Jones et al., 1996). Traps were operated two to three nights per transect. Each captured rodent was identified and prepared in the field as a study skin and vouchers were placed in liquid formol. All specimens are stored in the Zoological Collections of the Museum of Natural History of the University of San Carlos.

Sampling of Bats

We used four mist nets measuring 2.6x12 m, placed in a "T" configuration to optimize capture effort (Kunz et al., 1996). Nets were placed in a different location each night, were opened at sunset (ca. 18:30 hrs), and were closed four hours later (ca. 22:30 hrs). This is the period during which bats are most active (Fleming et al., 1972; Willig, 1986). Each captured bat was identified to species using Medellín et al. (1997) and freed in the same site. In some cases, bats were prepared as study specimens and deposited in the Zoological Collections of the Museum of Natural History of the University of San Carlos.

The capture effort and the relative abundance were calculated with the method proposed by Medellín (1993), where captures per total net meters per hour are used to estimate relative abundance. To estimate bat species richness, we created species accumulation curves using EstimateS 5 software (Colwell, 1999), using the species-

richness estimator ACE (Abundance-based Coverage Estimator of species richness) that is based on those species with 10 or fewer individuals in the sample. We calculated the diversity of the localities using the Shannon-Wiener diversity index, and these values were compared among focal areas using a one way ANOVA (Zar, 1984). We also considered the bat community according to feeding habit. Each species was classified into different diet categories following Medellín (1994).

Other Mammals

The majority of large mammals are difficult to observe. We used tracks, fur, feces, carcasses, or skeletal remains and burrows (Aranda, 1981) to provide data on the presence and absence larger mammals. We conducted daily visual surveys along trails in order to sight mammals.

RESULTS AND DISCUSSION

Bats

We recorded 39% of the potential species (based on maps) of bats (23 species) for LTNP (Appendix 13). Two hundred and twenty-one individuals and 20 species were captured with a sampling effort of 684 meters of net for 57 hours (14 nights). The average effort per night was 3.7 hours with 44 meters of net (Table 8.2). We captured and identified three species without mist nets, which were not considered in the analyses.

The greatest capture rate was observed at the Río San Pedro focal area, which contrasts with the lowest values recorded in Flor de Luna and Río Chocop focal areas (Table 8.1). These low values probably represent "lunar phobia," i.e. a negative influence on bat activity due to the moon being full during sampling in both localities (Morrison, 1978). The bat species accumulation curve (Figure 8.1) shows that during the first eight nights, 19 species were collected. These

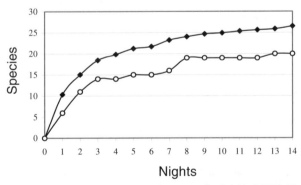

Figure 8.1. Species accumulation curve for bats in LTNP. The lower curve (open circles) indicates the observed accumulation, and the upper curve (solid diamonds) indicates the estimated accumulation from the abundance-based coverage estimator (ACE).

Table 8.1. Capture effort and success using mist nets in Laguna del Tigre National Park, Petén, Guatemala.

	San Pedro	Sacluc	Escondido	Flor de Luna	Chocop	Total	Average	Avg/ night
Net-m	144	144	96	132	144	660	132	44.00
Nights	3	3	2	3	3	14	2.8	0.93
Hours	12	12	8	12	12	56	11.2	3.73
# bats	100	38	38	19	25	220	44	14.67
# species	14	10	12	8	8	20	10.4	
Net-m/h	1728	1728	768	1584	1728	7536		
#bats/N-m/h	0.05787	0.02199	0.049479	0.01199	0.01446	0.029193		
#species/N-m/h	0.00810	0.00578	0.015625	0.00505	0.00463	0.002654		

Table 8.2. Relative abundance and dominance of bats in Laguna del Tigre National Park, Petén, Guatemala.

Rank	Species	Number of Individ.	Ind/N-m/h
1	Artibeus intermedius	48	0.006369
2	Carollia brevicauda	41	0.005441
3	Artibeus lituratus	30	0.003981
4	Sturnira lilium	20	0.002654
5	Dermanura phaeotis	17	0.002256
6	Pteronotus parnellii	16	0.002123
7	Artibeus jamaicensis	10	0.001327
8	Glossophaga commissarisi	8	0.001062
9	Dermanura watsoni	7	0.000929
10	Rhynchonycteris naso	5	0.000663
11	Tonatia evotis	4	0.000531
12	Desmodus rotundus	3	0.000398
13	Mimon bennettii	3	0.000398
14	Uroderma bilobatum	2	0.000265
15	Mormoops megalophylla	1	0.000133
16	Micronycteris schmidtorum	1	0.000133
17	Chrotopterus auritus	1	0.000133
18	Centurio senex	1	0.000133
19	Mimon crenulatum	1	0.000133
20	Tonatia brasiliense	1	0.000133

represented 95% of the total species captured in the study. In the next six nights, only one species was added. The ACE estimator of species richness was 26.5.

There was no significant difference in diversity of the bat communities of the four localities (F = 1.116, p>0.05, n = 12). Total diversity was H' = 1.471 for the four localities combined. Between the localities, Río Escondido had the highest value of diversity (Table 8.2). The highest evenness was found in Río Chocop (E = 0.94), perhaps due to the low number of species (S = 8) and to the low abundance (two to three individuals/species).

The 20 species of bats captured belonged to four families. The family Phyllostomidae was the most abundant, with 198 individuals of 17 species, and was present in all localities. Four species (three phyllostomids and one mormoopid) were common in all localities. The relative abundance (Table 8.3) shows that the dominant species of the community are *Artibeus intermedius*, *Carollia brevicauda*, and *A. lituratus*. These species comprised 54%

of the total bats (see also Appendix 13). A second, less abundant group consists of 11 species, some of which may be considered common (e.g. *Sturnira lilium*, *Dermanura phaeotis*), and others which are less often captured (e.g. *Uroderma bilobatum*). A third group of rare species consists of six species captured only once that together represent 2.7% of the bats captured (e.g. *Chrotopterus auritus*, *Mimon crenulatum*).

Among the more abundant species (*A. intermedius*, *C. brevicauda*, and *A. lituratus*), two are dominant in undisturbed habitat: *A. intermedius* and *A. lituratus*; the other, *C. brevicauda*, is dominant in moderately disturbed conditions. The abundance of each of these species is < 25% of the total for all bats captured, suggesting that the habitat is undisturbed according to the criteria of Amín and Medellín (*in press*).

Bats captured in this study represented a variety of trophic groups (Figure 8.2). Insectivores comprised 43% of species, followed by frugivores at 39%, carnivores at 8.7%, nectarivores at 4.3%, and sanguivores at 4.3%. In their studies of tropical forest bats, Wilson et al. (1997) and Amín (1996) reported that the frugivores were the largest trophic group (50%), followed by insectivores (23%). Our results may differ from theirs because our sampling points were close to bodies of water, where insectivores are generally more abundant (Kunz and Kurta, 1988).

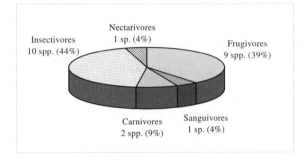

Figure 8.2. The percentage of bat species belonging to different trophic groups in LTNP.

Analysis of Bat Diversity by Sites (Table 8.3)

Río San Pedro Focal Area
In this locality we recorded the highest number of species (15 species) and captures (138 individual). This locality was subdivided into two points for sampling—the Río San Pedro and the Río Sacluc—because of the contrasting vegetation between them. Along the Río San Pedro, sampling was carried out in riparian forest (SPT1-3) and bosque alto (SPT6; refer to Table 1.2, Appendix 2); here we captured 100 individuals of 14 species in three nights. The most abundant species were *Artibeus intermedius*, *Carollia brevicauda*, *A. lituratus*, and *Sturnira lilium*. In Río Sacluc, the nets were placed in encinal (SPT4-5), and 38 individuals of 10 species were captured. The most abundant species here were *A. lituratus*, *C. brevicauda*, and *A. intermedius*. The capture success here was half that in the bosque alto habitat (Table 8.1). There was no statistically significant difference among the focal areas (San Pedro vs. Sacluc:

t 0.05(2),74.34=1.933, tobs=1.352, P >0.005), but the diversity values were slightly higher in Río San Pedro (Table 11). The only wrinkle-faced bat (*Centurio senex*) observed was captured at the Río Sacluc site. One black mastiff bat (*Molussus ater*) was found under a camping tent in the Las Guacamayas Biological Station (SP13), and one group of the proboscis bats (*Rhynchonycteris naso*; five to six individuals) was observed on a tree trunk directly above water. In the Laguna el Perú, we also observed a greater fishing bat (*Noctilio leporinus*).

Río Escondido
Here we sampled in bosque alto with some natural open areas (ET 1-2). The forest was burned ca. 20 years ago. We captured 39 individuals and 13 species. The most abundant species were *Artibeus intermedius* and *Sturnira lilium*. Richness values were similar to those at Río San Pedro, although abundance was lower. Capture success here was slightly higher than at Río Sacluc (Table 8.1). This locality exhibited the highest richness values (Table 8.2). We observed a group of greater white-lined bats (*Saccopteryx bilineata*) that was using a hollow log of matapalos (*Ficus* sp.) as a roost.

Laguna Flor de Luna
Bosque bajo consisting of tintales was sampled here (FT 1-3). We captured 19 individuals from eight species. During the first night of sampling, we obtained the highest number of individuals; the most abundant species were *Pteronotus parnellii* and *Sturnira lilium*. Capture success here was the lowest of all the focal areas (Table 8.1). Richness, abundance, and capture success decreased after the second night of sampling, probably due to the influence of a full moon (Morrison, 1978).

Río Chocop
Bosque alto was sampled here, and dominant plant species were *Mataiba* sp., *Chrysophila* sp., and some sapoteceas (*Manilkara sapota* and *Pouteria* sp.). We captured 25 individuals and eight species. The dominant species were *Artibeus intermedius* and *Pteronotus parnellii*. These results

Table 8.3. Richness and diversity of bats by focal area in Laguna del Tigre National Park, Petén, Guatemala.

H' = the Shannon-Wiener Index

Focal Area	# Ind.	# Spp.	Diversity (*H'*)	Evenness
Río San Pedro	100	14	1.6897	0.8270
Río Sacluc	38	10	1.3514	0.9050
Río Escondido	38	12	1.8475	0.9173
Laguna Flor de Luna	19	8	1.1769	0.8993
Río Chocop	25	8	1.2899	0.9404
Overall values	220	20	1.4711	0.8978

show a clear decrease in the capture rate of individuals as the sampling proceeded. Capture success was low, similar to Flor de Luna.

Rodents

This was the second most-often recorded group of mammals in this study (Appendix 13). We identified eight species of rodents in seven genera. Two species were registered by direct observation and not collected. The captured rodents constitute 47% of the potential rodent species for LTNP (based on previous records). The species captured were *Oryzomys couesi, Sigmodon hispidus, Ototylomys phyllotis, Peromyscus yucatanicus, Heteromys desmarestianus*, and *H. gaumeri*). Two of these are endemic to the Yucatán Peninsula (*Peromyscus yucatanicus and H. gaumeri*).

The Sigmodontinae (Muridae) consist of 21 genera and 61 species in southeastern Mexico and Central America (Reid, 1997). We found that most of the species captured during the sampling belong to this subfamily. The family Heteromyidae is represented by two genera in Central America, and we recorded one genus (*Heteromys*) and two of its species (all that could potentially occur) in LTNP. The other genus (*Liomys*) is distributed mainly on the Pacific Slope in dry deciduous forest from northern Mexico to Panama (Reid, 1997).

In a total of 1149 trap-nights, we obtained a total of 10 individuals and five species (Table 8.4). The highest species richness was found in the Río San Pedro/Río Sacluc focal area (3 sp.), in comparison with the other localities, where only one species was registered. The trapping success was constant for the majority of the localities, with the exception of Río Sacluc (encinal), where the lowest value was observed.

Our data and other sources suggest that different rodent species tend to associate with different vegetation types in the region. *Sigmodon hispidus* is associated with grasslands (Río Escondido); *Heteromys desmarestianus* (Río San Pedro), *H. gaumeri* (Río Chocop), *and Ototylomys phyllotis* (Río San Pedro) with bosque alto; and *Oryzomys couesi* (Río Sacluc) and *Peromyscus yucatanicus* (Flor de Luna) with bosque bajo and tintales (Medellín and Redford, 1992; Young and Jones, 1983).

We obtained a new record for Guatemala: the Yucatán deer mouse (*Peromyscus yucatanicus*; Muridae). This mouse is of moderate size, larger than *P. leucopus* when in sympatry (Young and Jones, 1983). This is an endemic of the Yucatán Peninsula, although its southern distribution limit has not yet been determined (Huckaby, 1980). The southernmost locality from which this species has been reported in the literature is 7.5 km west of Escárcega, Campeche, Mexico (Dowler and Engstrom, 1988). Four specimens of *P. y. badius* were collected from Calakmul, Campeche, and are deposited in the collection of Instituto de Biología, Universidad Autonoma de Mexico (IBUNAM 37362, 37364, 37365, 37366), with 115 km ESE as the last record. Our record constitutes a range extension of 116 km south from Escárcega and 134 km southeast from Calakmul.

Other Mammals

Eleven large mammal species were recorded in LTNP (Appendix 13). We confirmed the presence of several threatened and endangered species, including Baird's tapir (*Tapirus bairdii*), collared peccary (*Tayassu tajacu*), red brocket deer (*Mazama americana*), Central America spider monkey (*Ateles geoffroyi*), and Yucatán black howler monkey (*Alouatta pigra*).

The presence of the Central America spider monkey and Yucatán black howler monkey in all four focal areas suggests that there are relatively undisturbed habitats remaining in these areas. In Veracruz, Mexico, Silva-López et al. (1993) reported a reduction in the population density for both species in disturbed conditions and a greater relative abundance of spider monkeys with respect to howler monkeys in disturbed patches.

Río San Pedro was the only locality in which a tapir and red brocket deer were observed. Both species live in undisturbed habitats, although tapirs may tolerate moderately disturbed conditions (Fragoso, 1991). Nonetheless, tapirs are registered as threatened in Appendix I of CITES, and red brocket deer are registered in Appendix III of CITES (Reid, 1997). Hunting is the principal threat for tapir and has caused extinctions in some parts of Central America (March, 1994). To preserve this species in Guatemala, it will

Table 8.4. Capture effort and success for rodents in Laguna del Tigre National Park, Petén, Guatemala.

Focal Area	# Spp.	# Ind.	Capture effort (trap-nights)	Capture success (% of traps)
Río San Pedro	2	2	243	0.82
Río Sacluc	1	1	276	0.36
Río Escondido	1	2	180	1.10
Laguna Flor de Luna	1	4	270	1.15
Río Chocop	1	1	180	1.10
Overall values	5	10	1149	0.90

be necessary to conserve populations in the Maya Bio-sphere Reserve (Matola et al., 1997) and in LTNP in particular.

Although we did not record jaguar (*Panthera onca*) in LTNP, a recent rapid assessment identified areas containing viable populations of jaguars in Guatemala. The most important population in Guatemala was judged to reside inside the MBR, specifically in LTNP and Sierra del Lacandon National Park (McNab and Polisar, *in press*).

CONCLUSIONS AND RECOMMENDATIONS

Three conclusions emerge from this work and other studies conducted in the region:

- LTNP is an important refuge of high mammal diversity and harbors populations of several threatened and endangered species due to its great area, habitat heterogeneity, and relatively pristine state.

- Within LTNP, the Río San Pedro focal area is perhaps the most important conservation area because of its comparatively high bat abundance and richness, and the presence of threatened and endangered species.

- LTNP represents one of the few places in the region where longterm viable populations of large mammals such as tapir and jaguar can be supported.

- The principal factors that threaten the populations of wild mammals of LTNP are slash-and-burn agriculture, hunting, and the expansion of human settlements, although the impacts of these factors in different parts of the park are little understood. To improve our under-standing of these processes, a monitoring program evaluating the impact of humans on mammal populations in LTNP should be initiated. To organize such a monitoring program and manage mammal populations for the long term, it will be necessary to develop regional collaborations with Mexico and Belize with the objective of maintaining the existence of biological corridors in order to avoid the genetic isolation of populations. One focus of such a program should emphasize charismatic species such as the tapir and jaguar in order to guarantee their permanence in Guatemala.

LITERATURE CITED

Amín, M. 1996. *Ecología de comunidades de murciélagos en bosque tropical y hábitats modificados en la Selva Lacandona, Chiapas.* Tesis de Licenciatura. Universidad Nacional Autónoma de México.

Amín, M., and R. A. Medellín. *In press.* Bats as indicators of habitat disturbance. In *Single Species Approaches to Conservation: What Works, What Doesn't and Why* Island Press, Washington, DC, USA.

Aranda, J. M. 1981. Rastros de los mamíferos silvestres de México. Manual de campo. Instituto Nacional de Investigaciones sobre Recursos Bióticos, México.

August, P. V. 1983. The role of habitat complexity and heterogeneity structuring tropical mammal communi-ties. Ecology 64:1495-1507.

Colwell, R. K. 1999. *EstimateS 5. Statistical Estimation of Species Richness and Shared Species from Samples.* Web site: viceroy.eeb.uconn.edu/estimates.

Dirzo, R., and A. Miranda. 1990. Contemporary Neotropi-cal defaunation and forest structure, function, and diversity: A sequel to John Terborgh. Conservation Biology 4: 444-447.

Dowler, R. C., and M. D. Engstrom. 1988. Distributional records of mammals from the southwestern Yucatán Peninsula of Mexico. Annals of Carnegie Museum 57:159-166.

Eisenberg, J. F. 1989. *Mammals of the Neotropics: The Northern Neotropics: Panama, Colombia, Venezuela, Suriname, French Guiana. Vol 1.* University of Chicago Press, Chicago, IL, USA.

Emmons, L. H., and F. Feer. 1997. *Neotropical Rainforest Mammals: A Field Guide.* Second edition. The University of Chicago Press, Chicago, IL, USA.

Fenton, M. B., L. Acharya, D. Audet, M. B. C. Hickey, C. Merriman, M. K. Obrist, D. M. Syme, and B. Adkins. 1992. Phyllostomid bats (Chiroptera: Phyllostomidae) as indicators of habitat disruption in the Neotropics. Biotropica 24: 440-446.

Fleming, T. H., E. T. Hooper, and D. E. Wilson. 1972. Three Central American bat communities: structure, reproductive cycles, and movement patterns. Ecology 53: 555-569.

Fragoso, J. M. 1991. The effect of selective logging on Baird's tapir. Pp. 295-304 in *Latin America Mammalogy: History, Biodiversity, and Conservation* (M. A. Mares and D. J. Schmidly, eds.). University of Oklahoma Press, Norman, OK, USA.

Gardner, A. L. 1977. Feeding habits. Pp. 351-364 in *Biology of Bats of the New World: Family Phyllostomidae Part II* (R. J. Baker, J. K. Jones, and D. C. Carter, eds.). Special Publications, Museum, Texas Tech University, Lubbock, TX, USA.

Garrott, R. A., P. J. White, and C. A. Vanderbilt White. 1993. Overabundance: an issue for conservation biologists? Conservation Biology 7: 946-949.

Hall, E. R. 1981. *The Mammals of North America, Vol. 1.* John Wiley and Sons.

Huckaby, D. C. 1980. Species limits in the *Peromyscus mexicanus* group (Mammalian: Rodentia: Muridea). Contributions Sciences National History Museum, Los Angeles 326: 1-24.

Jones, C., W. J. McShea, M. J. Conroy, and T. H. Kunz. 1996. Capturing mammals. Pp. 115-155 in *Measuring and Monitoring Biological Diversity: Standard Methods for Mammals* (D. Wilson, J. R. Cole, J. D. Nichols, R. Rudran, and M. S. Foster, eds.). Smithsonian Institution Press, Washington, DC, USA.

Kunz, T. H., and A. Kurta. 1988. Capture methods and holding devices. Pp. 1-29 in *Ecological and Behavior Methods for Study of Bats* (T. H. Kunz, ed.). Smithsonian Institution Press, Washington, DC, USA.

Kunz, T. H., D. W. Thomas, G. C. Richards, C. R. Tidemann, E. D. Pierson, and P. A. Racey. 1996. Observational techniques for bats. Pp. 105-114 in *Measuring and Monitoring Biological Diversity: Standard Methods for Mammals* (D. Wilson, J. R. Cole, J. D. Nichols, R. Rudran, and M. S. Foster, eds.). Smithsonian Institution Press, Washington, DC, USA.

LaVal, R. K., and H. S. Fitch. 1977. Structure, movement, and reproduction in three Costa Rican bat communities. Occasional Papers, Museum of Natural History, University of Kansas 69:1-28.

Malcolm, J. R. 1997. Biomass and diversity of small mammals in Amazonian Forest fragments. Pp. 207-221 in *Tropical Forest Remnants: Ecology, Management, and Conservation of Fragmented Communities* (W. F. Laurace and R. O. Bierregaard, eds.). The University of Chicago Press, Chicago, IL, USA.

March, I. J. 1994. Situación actual del tapir en México. Centro de Investigaciones Ecológicas del Sureste (Serie Monográfica No. 1).

Matola, S., A. D. Cuarón, and H. Rubio-Torgler. 1997. Status and action plan of Baird's tapir (*Tapirus bairdi*). In *Tapir: Status and Conservation Action Plan* (D. M. Brooks, R. E. Bodmer, and S. Matola, eds.). The IUCN/SSC Tapir Specialist Group Report. Web site: http://www.tapirback.com/tapirgal/iucn-ssc/actions97/cover.htm.

McNab, R. B., and J. Polisar. *In press.* A participatory methodology for a rapid assessment of jaguar (*Panthera onca*) distributions in Guatemala. Wildlife Conservation Society.

Medellín, R. A. 1993. Estructura y diversidad de una comunidad de murciélagos en el trópico húmedo mexicano. Pp. 333-354 in *Avances en el Estudio de los Mamíferos Méxicanos* (R. A. Medellín and G. Ceballos, eds.). Publicaciones Especiales, Asociación Mexicana de Mastozoología, A. C. 1:1-464.

Medellín, R. A. 1994. Mammals diversity and conservation in the Selva Lacandona, Chiapas, México. Conservation Biology 8:788-799.

Medellín, R. A., H. Arita, and O. Sánchez. 1997. Identificación de los murciélagos de México. clave de campo. Asociación Mexicana de Mastozoología, A.C. Publicaciones Especiales No 2. México.

Medellín, R. A., M. Equihua, and M. A. Amín. *In press.* Bat diversity and abundance as indicators of disturbance in Neotropical rainforest. Conservation Biology.

Medellín, R. A., and K. H. Redford. 1992. The role of mammals in Neotropical forest-savanna boundaries. Pp. 519-548 in *Nature and Dynamics of Forest-Savanna Boundaries* (P.A. Furley, J. Procyor, and J.A. Ratter, eds.). Chapman & Hall.

Morrison, D. W. 1978. Lunar phobia in a tropical fruit bat, *Artibeus jamaicensis* (Chiroptera: Phyllostomidae). Animal Behaviour 26:852-855.

Reid, F. A. 1997. *A Field Guide to the Mammals of Central America and Southeast Mexico*. Oxford University Press, New York, NY, USA.

Silva-López, G., J. Benítez-Rodríguez, and J. Jimenéz-Huerta. 1993. Uso del hábitat por monos araña (*Ateles geoffroyi*) y aullador (*Alouatta palliata*) en áreas perturbadas. Pp. 421-435 in *Avances en el Estudio de los Mamíferos Méxicanos* (R. A. Medellín and G. Ceballos, eds.). Publicaciones Especiales, Asociación Mexicana de Mastozoología, A.C. 1:1-464.

Voss, R. S., and L. H. Emmons. 1996. Mammalian diversity in Neotropical lowland rainforest: a preliminary assessment. Bulletin American Museum of Natural History 230:1-115.

Williams, S. E., and H. Marsh. 1998. Changes in small mammal assemblage structure across a rain forest/open forest ecotone. Journal of Tropical Ecology 14:187-198.

Willig, M. R. 1986. Bat community in South America: a tenacious chimera. Revista Chilena de Historia Natural 59: 151-168.

Wilson, D. E. 1989. Bats. Pp. 365-382 in *Tropical Rain Forest Ecosystems: Biogeographical and Ecological Studies* (H. Lieth and M. J. A. Werger, eds.). Elsevier Science Publications, Amsterdam, the Netherlands.

Wilson, D. E., R. Baker, S. Solari, and J. J. Rodríguez. 1997. Bats: Biodiversity assessment in the Lower Urubamba Region. Pp. 293-301 in *Biodiversity Assessment and Monitoring of the Lower Urubamba Region, Peru: San Martin-3 and Cashiriari-2 Well Sites* (F. Dallmeier and A. Alonso, eds.). SI/MAB Series # 1. Smithsonian Institution/MAB Biodiversity Program, Washington, DC, USA.

Wilson, D. E., and D. M. Reeder. 1993. *Mammal species of the world: a taxonomic and geographic reference*. 2nd. ed. Smithsonian Institution Press, Washington, DC, USA.

Young, C. J., and J. K. Jones. 1983. *Peromyscus yucatanicus*. Mammalian Species 196: 1-3.

Zar, J. H. 1984. Biostatistical Analysis. 2nd. Ed. Prentice Hall. Englewood Cliffs, NJ, USA.

CHAPTER 9

THE ANTS (HYMENOPTERA: FORMICIDAE) OF LAGUNA DEL TIGRE NATIONAL PARK, PETEN, GUATEMALA

Brandon T. Bestelmeyer, Leeanne E. Alonso, and Roy R. Snelling

ABSTRACT

- We recorded 112 species and 39 genera of ants in Laguna del Tigre National Park (LTNP). These values are greater than those recorded in several other studies in the region.

- The genus *Pheidole* contained the largest number of species, but several arboreal genera were also highly diverse.

- Three ant species endemic to the Maya Forest region were recorded, including *Sericomyrmex aztecus*, *Xenomyrmex skwarrae*, and *Odontomachus yucatecus*. One extremely rare ant genus, *Thaumatomyrmex,* was discovered for the first time in Guatemala.

- Species richness was highest in relatively open or disturbed habitats, contrary to our expectations. This was likely due to the high microhabitat diversity of these sites. A disturbed habitat was relatively similar to a nearby pristine forest site, which is consistent with the suggestion that the recovery of Neotropical ant faunas from disturbance is rapid.

- Turnover rates of ant composition between sites is very high, precluding an assessment of species-habitat associations. The overall richness of the LTNP ant fauna is likely to be much higher than that recorded.

INTRODUCTION

Insects are believed to be the most speciose group of organisms on Earth (May, 1988). This group reaches its greatest diversity in tropical forests (Erwin and Scott, 1980), so the increasing rate of tropical deforestation may result in severe reductions in insect diversity worldwide. For this reason, surveys that document patterns of insect diversity among tropical forests are needed to prioritize the conservation of tropical regions and habitats within regions (Haila and Margules, 1996). Another consequence of the great diversity of insects is that it is often logistically difficult to consider all insects simultaneously in such surveys. Thus, ecologists often rely on information provided by one or more groups of insects to characterize patterns of insect diversity across landscapes and regions.

Ants (Hymenoptera: Formicidae) are one group that may be useful for characterizing the responses of terrestrial insects. Ants are extremely conspicuous, abundant, active, and diverse elements of Neotropical forests. Because ants are social insects, ants taken as a whole are very flexible in their responses to differences in their environments in time and space (Wilson, 1987). As a consequence of this flexibility, ants are very successful in even the most disturbed habitats. Ants may be found nesting and foraging in nearly all parts of terrestrial ecosystems, from deep in the soil column to the leaf litter-soil interface to twigs and leaves on the litter surface to bushes, epiphytes, and tree canopies.

The species composition and richness of ant communities, however, may reveal strong responses to natural habitat variation and habitat disturbances (Andersen, 1991; Bestelmeyer and Wiens, 1996; Vasconcelos, 1999) because different species may be closely associated with particular soil conditions (Johnson, 1992), microclimates (Perfecto and Vandermeer, 1996), and even the presence of particular plant species (Alonso, 1998), all of which may be altered by human economic activities. In tropical forests in particular, canopy cover is an important correlate of more proximate environmental features that determine the distribution and abundance of ant species (MacKay, 1991; Perfecto and Snelling, 1995). The responses of ants to such variation can be used to evaluate both the conservation value of particular areas as well as the effects of human land use on biodiversity.

No studies of ant communities have yet been reported for the forests of the Petén in Guatemala, and our primary objective in this rapid assessment was to provide a first glimpse of the richness and composition of the ant fauna in Laguna del Tigre National Park (LTNP) in a biogeographic context. We also sought to provide a preliminary assessment of the relationships between habitat variation and variation in ant community composition and richness. Specifically, we focused on the contrast between 1) closed-canopy habitats that characterize bosque alto, 2) relatively open-canopy habitats including encinales and tintales that are produced by natural processes, and 3) guamiles that are produced by secondary succession following deforestation. We hypothesized that open-canopy habitats would have lower species richness and altered community composition when compared to closed-canopy habitats in the context of this tropical forest, as was found by MacKay (1991) and Vasconcelos (1999). We did not, however, sample intensively in recently disturbed pasture habitats. Such an effort would be redundant as several studies have amply demonstrated the consequences of such drastic changes in vegetation on ant diversity (Greenslade and Greenslade, 1977; Quiroz-Robledo and Valenzuela-Gonzales, 1995; Vasconcelos, 1999). We did observe patterns of ant species dominance in grassland habitats, however, for comparison with other studies in the region. We use this body of information to highlight the regional importance of the park and to attempt to identify some of the processes that may support or threaten ant and other arthropod populations in the park. Our observations are intended to suggest fruitful avenues for more intensive studies in the park as well as to provide tentative information on patterns of ant species richness in the absence of such studies.

STUDY AREAS AND METHODS

Ants and environmental variables were sampled along transects in one or more habitat types in each focal area (refer to "Expedition Itinerary and Gazetteer"). No attempt was made to characterize the ant fauna of focal areas because each focal area contained a variety of distinct habitat types that could not be sampled adequately in the time provided. Ant species respond to habitat variation occurring at very fine scales, so our principal objective was to examine ant-habitat associations in distinct habitat types. Pitfall trapping was used to gather standardized samples for quantitative comparisons, following procedures in Bestelmeyer et al. (*in press*). One pitfall trap (75 mm diameter) was placed at 10 points along a 90-m transect at 10-m intervals (see Table 9.4). Traps were opened for about 48 hrs. Pitfall trapping

was performed at five terrestrial sample points: SPT1 (guamil habitat; see Table 1.2); SPT4 (encinal habitat); SPT6 (bosque alto, ramonal habitat); ET1 (recently burned ramonal); and FT1 (bosque bajo, tintal habitat).

In addition, Winkler litter extraction was performed at SPT1 and SPT6. Litter extraction was performed using the "mini-Winkler" design, following methods outlined in Bestelmeyer et al. (*in press*). All leaf litter and twigs within 1-m^2 quadrats at 10 stations per transect were sampled. Litter samples were taken at least 2 m from pitfall trap locations.

Tuna baits were used to sample scavenging/omnivorous ant species and to provide information on the foraging behavior of some species at ET1, ET2 (grassland habitat only), FT1, CT1 (bosque alto), and CT4 (bosque alto). A one-half-teaspoon-sized bait of canned tuna (ca. 3 g) was placed at 10 points (10-m spacing) along a 90-m transect. Baiting occurred in the late afternoon or at night during the periods of peak ant activity during the study. The number of individuals per ant species was recorded at each bait 45 minutes after bait placement or repeatedly over a two-hour period (10, 30, 60, 90, and 120 minutes after placement), depending upon the objectives of a particular baiting session.

Finally, direct sampling was used to record the presence of ant species in each focal area. Direct sampling was performed along 90-m transects in all sampled points, and opportunistic sampling occurred throughout the focal study area. Direct sampling involved searching the ground and litter for nests by sweeping away the leaf litter from a 1-m^2 plot (at least five plots/transect) and visually searching for ants; breaking open suitable fallen twigs, branches, and trunks; searching under stones; and searching on leaves, on and under tree bark, and in epiphytes up to ca. 2 m. These data were not used for comparisons because habitats and transects differed greatly in the composition of microhabitat elements and because different amounts of time were available to search in different sites.

Species and genera were preliminarily categorized according to functional groups based upon field observations and reports in the literature. The functional groupings follow Andersen (1995) and Bestelmeyer and Wiens (1996). Further, species were classified according the type of geographic distribution that they exhibited based on distribution information in Kempf (1972), Brandão (1991), Longino (online), and Ward (online). Six types of ranges were recognized here: Mayan Forest (forest in the Yucatán Peninsula of Mexico, northern Guatemala, and western Belize), northern Central American (southern Mexico–Honduras), Central American (southern Mexico–Colombia), broad Central American (southern United States–Central America), Neotropical (southern Mexico–northern Argentina), and broad Neotropical (southern United States–mid South America).

A densiometer was used to quantify forest canopy cover at each point along the transects at which pitfall sampling occurred.

Both richness and species composition were compared among sample points representing distinct macrohabitat types or variants. Measured species richness, as well as estimated true richness, was used to compare sample points given that habitats may have varied in the completeness of the species sampling. Richness was estimated using richness extrapolation techniques (Colwell and Coddington, 1994) and using EstimateS software (version 5.0.1; Colwell, 1997). Only incidence-based estimators were considered because ant abundance is often clumped due to their colonial nesting habit. First- and second-order jackknife estimators and the incidence-based coverage estimator (ICE) are reported here.

Patterns of species composition were evaluated in two ways. First, we compared each sample point in a pairwise fashion using similarity indices. Estimated Sorensen's indices were used that are based upon the expected number of species shared between samples from a coverage-based estimator (EstimateS 5.0.1, Colwell, 1997). Additionally, unmodified similarity values are provided in the form of Morisita-Horn indices. Second, we used hierarchical agglomerative clustering techniques to establish patterns of similarity in species composition among groups of sample points considered together. A centroid-based clustering procedure, namely Ward's minimum variance method, was used (PC-ORD 4.0; McCune and Mefford, 1999) following discussions in Legendre and Legendre (1998).

Finally, patterns of abundance of ants at baits were compared between two transects, one in grassland (ET2) and the other in bosque alto (FT1) in order to compare foraging and dominance patterns of ants between these distinct habitat types. Ant abundances at individual baits were transformed according to a six point abundance scale (0-5) following Andersen (1991).

RESULTS

Over all of the sampling points, 112 species and morphospecies from 39 genera of ants were recorded (Appendix 14). Of these, 71 species were gathered using systematic methods (pitfalls and litter extraction), and the remaining 41 species were gathered only via hand collecting or at tuna baits. Hand collecting alone registered 79 species. *Pheidole* was the most speciose genus (20 species and OTUs), followed by *Pseudomyrmex* (9), *Camponotus* and *Cephalotes* (7), *Pachycondyla* and *Solenopsis* (6), *Azteca* (5), and *Odontomachus* (4). Seven species were found in all four focal areas and can be considered to be common; these include *Camponotus atriceps*, *C. novogranadensis*, *C. planatus*, *Cyphomyrmex minutus*, *Dolichoderus bispinosus*, *Solenopsis geminata*, and *Wasmannia auropunctata*.

Thirty-three species are known or suspected to be exclusively terrestrial and 29 to be exclusively arboreal. The remaining species may occur in either microhabitat or their habitat affinities are unknown. One large epiphytic brome-liad, *Aechmea tillandsia*, was thoroughly searched for ants. *Camponotus planatus*, *Cephalotes minutus*, *C. scutulatus*, *Crematogaster brevispinosus* group, and *Pseudomyrmex oculatus* were recorded on this plant. Ants were collected along the entire 20-m length of a *Piscidia piscipula* tree that fell during a storm. *Camponotus atriceps*, *Ectatomma tuberculatum*, *Hypoponera* sp. B (nest in bark), *Pachycondyla stigma*, *Pheidole* sp. H (nest in epiphyte, *Epidendrum* sp.), *Pheidole punctatissima*, and *Tapinoma inrectum* were collected from this tree.

Species recorded in the sample that are restricted to northern Central America (so far as is known) include *Azteca foreli*, *Belonopelta deletrix*, *Cephalotes basalis*, *C. biguttatus*, *Hypoponera nitidula*, *Pseudomyrmex ferrugineus*, *P. pepperi*, *Rogeria cornuta*, and *Trachymyrmex saussurei*. *Odontomachus yucatecus*, *Sericomyrmex aztecus*, *Xenomyrmex skwarrae* may be classified as Mayan Forest endemics. *O. yucatecus* was collected in pitfall traps at the burned ramonal and guamil site, *S. aztecus* from pitfalls in the guamil, and *X. skwarrae* was collected by hand at the El Perú archaeological site. The rare ant *Thaumatomyrmex ferox* was collected in the ramonal habitat using Winkler litter-sifting. Other rare species include *Discothyrea cf. horni*, which was found in the guamil site using a Winkler device, and *Belonopelta cf. deletrix*, which was collected by hand in bosque alto near the Laguna Flor de Luna. An undescribed form of *Pseudomyrmex* (PSW-13) was captured in a pitfall trap at the encinal site. Only 930 individuals were collected with pitfall traps and Winklers, averaging 13 individuals per trap or quadrat. Ant activity in the litter and on the ground surface was very low. Baiting observations suggested that ground-dwelling ant species in some forested areas were nesting underground and were drawn from cracks in the soil-surface to tuna baits. Many other ant species nested or foraged in trees or in epiphytes (Appendix 14) and so were not recorded in ground-surface or litter samples.

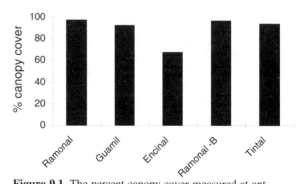

Figure 9.1. The percent canopy cover measured at ant pitfall stations in each sample point.

Canopy-cover measurements at pitfall transects indicate that the guamil and tintal habitats had only slightly lower canopy cover than the ramonal and burned ramonal habitats (Figure 9.1). Canopy cover values ranged from 92-97% in these habitats. The encinal had an appreciably lower canopy cover at ca. 67%.

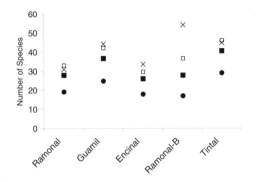

Figure 9.2. Ant species richness patterns among the sample points. Solid circle = measured richness, solid square = first-order jackknife estimator, open square = second-order jackknife estimator, x = incidence-based coverage estimator.

Measured richness was greatest in the tintal habitat, followed by the guamil habitat (Figure 9.2). Jacknife-richness estimators also suggest that richness was highest in these habitats. The ICE estimator indicated that the burned ramonal habitat was most rich, followed by the guamil and tintal habitats.

The expected similarity between habitats was low overall, ranging from 0.195-0.510 (Table 9.1). The encinal-tintal habitat comparison exhibited the greatest similarity value, whereas the ramonal and burned ramonal were the most distinct habitat pair. The ramonal-guamil pair and the encinal-burned ramonal pair also had high similarity values. The Morisita-Horn index, which includes a measure of abundance, produced a different pattern. Using this metric, the encinal and burned ramonal habitat were most similar, followed by the ramonal-guamil pairs and encinal-tintal pairs, whereas the other pairs showed generally low similarity. The cluster analysis revealed two distinct clusters, one containing the encinal, burned ramonal, and tintal habitats and guamil and ramonal in the other (Figure 9.3). The relatively high similarity between the encinal and burned ramonal observed in the Morisita-Horn indices is apparent in the cluster diagram.

Bait data from the bosque alto and grassland revealed large differences in the dominance of ant species (Figure 9.4). In the bosque alto transect, *Dolichoderus bispinosus* and *Pheidole* spp. increased in abundance at baits with time, while *Paratrechina* sp. B reached a peak at 30 min after bait placement and declined thereafter. In the grassland transect, *Solenopsis geminata* was extremely dominant and the abundance of *Pheidole* sp. A gradually declined with time.

The abundance of *Forelius* sp. and *Dorymyrmex* sp. increased through time as soil-surface temperatures increased during the baiting session.

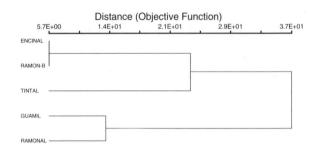
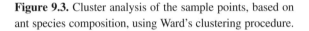

Figure 9.3. Cluster analysis of the sample points, based on ant species composition, using Ward's clustering procedure.

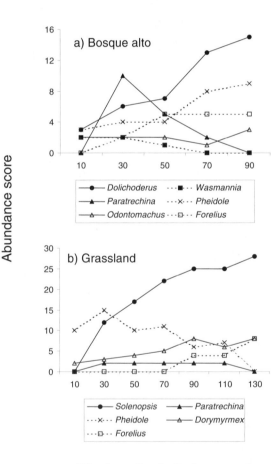

Figure 9.4. Patterns of ant dominance with time after bait placement on the ground surface at tuna-bait transects in a bosque alto and grassland habitats. Ant abundance scores are the sums of transformed abundances at all 10 baits (see text).

Table 9.1. Patterns of similarity of ant communities among habitat types based on pitfall trap data. Values in cells are Sorensen's indices, based upon the expected number of species shared between samples from a coverage-based estimator. Values in parentheses are Morisita-Horn indices based upon observed data.

	Ramonal	Guamil	Encinal	Ramonal-B
Ramonal	--			
Guamil	0.460 (0.215)	--		
Encinal	0.212 (0.020)	0.285 (0.178)	--	
Ramonal-B	0.195 (0.130)	0.280 (0.116)	0.409 (0.635)	--
Tintal	0.237 (0.054)	0.320 (0.072)	0.510 (0.201)	0.286 (0.114)

Table 9.2. A comparison of ant species richness and generic richness among Neotropical regions based upon published surveys.

These particular studies were selected to represent regions because they were broadly similar to the study presented here (i.e. an evaluation of landscape-level heterogeneity). Scale represents the maximum distance between samples. When values were not reported, a broad estimate of scale (followed by a "?") is provided based upon information in the text where this was possible. Entries are in descending order of latitude.

Species Richness	Generic Richness	Scale/# Habitats	Location
103	46*	w/in 10 km?/2	Veracruz, Mexico[1]
87	32	w/in 2 km?/3	Chiapas, Mexico[2]
112	39	ca. 80 km/5	LTNP, Guatemala[3]
109	31	w/in 35 km/4	Sarapaqui, Costa Rica[4]
127	49	?/elevation gradient	BCI, Panama[5]
156	49	?/3	Trombetas, Brazil[6]
184	45	?/3	Amazonas, Brazil[7]
104	30	ca. 10 km/4	Salta, Argentina[8]

*Generic richness values were adjusted if generic synonyms were recognized in species lists.
[1] Quiroz-Robledo and Valenzuela-Gonzales, 1995   [2] MacKay, 1991
[3] This study   [4] Roth et al., 1994
[5] Levings, 1983   [6] Vasconcelos, 1999
[7] Majer and Delabie, 1994   [8] Bestelmeyer and Wiens, 1996

DISCUSSION

Ant activity during the sampling period was low, so it is likely that the species richness values are greatly underestimated. Many ant species may have been completely inactive above ground due to the generally harsh and unproductive conditions that characterize the dry season (see also Janzen and Schoener, 1968; Levings, 1983; Bestelmeyer and Wiens, 1996). During tropical dry seasons, some ant species may descend from the litter stratum into the soil column where they cannot be sampled using either pitfall traps or leaf-litter extraction (Levings and Windsor, 1982; Harada and Bandeira, 1994). Thus, direct sampling supplemented by tuna baiting (which draws ants through cracks in the soil surface) appears to have been the most efficient combination of sampling procedures under the conditions present during this study, especially when preparation and post-processing time are taken into account.

In spite of these limitations, the number of ants recorded in LTNP compares favorably with the results of other studies (Table 9.2). Although the number of habitat types sampled and the scales over which samples were taken differ among these studies, some comparisons may be made with caution. The species and generic richness values in this study are similar to or greater than those recorded in studies performed in the tropical forests of Mexico and Costa Rica, but are less than those from studies in Panamanian and the Brazilian Amazon forests. Overall, this pattern suggests a bell-shaped richness curve through the Neotropical forests, with a peak in the southern Central America through the Brazilian Amazon. Further studies employing standardized collecting techniques and sampling designs will be needed to determine the validity of this pattern. If a unimodal (or other) relationship is supported, it might serve as a basis from which to anticipate and evaluate ant biodiversity in tropical localities.

Few species appear to be endemic to the Mayan forests, although comparable data from other studies are lacking. Several genera were commonly found in other ant surveys in Central American forests but were not found here, including *Adelomyrmex*, *Leptogenys*, and *Oligomyrmex*, and to a lesser extent *Megalomyrmex* and *Proceratium*. On the other hand, several genera were recorded in this study that have not often been recorded in other regional studies, including *Belonopelta, Discothyrea, Thaumatomyrmex,* and *Xenomyrmex*. The record of *Thaumatomyrmex ferox* is especially noteworthy. This ant was referred to as the "miracle ant" by Wilson and Hölldobler (1995) for its rarity and beauty. Fewer than 100 specimens are known to exist in museums, and this represents the first record that we know of from Guatemala.

Contrary to our expectations, ant richness was not greatest in the closed-canopy habitats, but was greatest in habitats that were slightly more open and, in the case of the guamil, more disturbed (Figure 9.2). The richness of the relatively open encinal habitat was similar to that in the closed-canopy ramonal habitat. Furthermore, ant community composition of the open or disturbed habitats did not differ systematically from that of more closed-canopy or "pristine" habitats (i.e. ramonal). This result may be accounted for by the fact that previous studies of the effects of forest disturbance, including those of MacKay (1991) and Vasconcelos (1999), considered habitats that were more recently or more severely disturbed than those considered here. MacKay (1991), for example, observed a 50% reduction in species richness in slashed-and-burned Chiapan forest and attributed this loss to the loss of arboreal species. In contrast, the disturbed sites in this study had abundant canopy cover (Figure 9.1). The recovery of mature-forest ant communities may take as little as 13 to 24 years after disturbance (Roth et al., 1994; Vasconcelos, 1999), and the guamil habitat studied here was estimated to be disturbed >20 years ago (Francisco Bosoc, pers. comm.). Further, our study also considered ant communities occurring in naturally open forest types in the region. Populations of many ant species now present in the forests of the Petén may be insensitive to moderate differences in vegetation structure because these ants have been exposed to historical increases in regional aridity as well as widespread anthropogenic disturbances (Méndez, 1999). For similar reasons, the greater ant richness observed in the guamil and tintal habitats may be due to the fact that the ants inhabiting these sites are better able to cope with dry-season conditions. The observed richness of the ramonal habitat might be greater than disturbed sites if it were sampled in a wetter time of year or if it were compared to more recently disturbed sites. Unfortunately, the statistical power of this study is too low to identify particular species that may decline when bosque alto is converted to guamil. It is likely that litter-dwelling cryptic species and some litter-inhabiting specialist predators and arboreal species (Appendix 14) would be most affected (see Bestelmeyer and Wiens, 1996).

The relatively high similarity between the ramonal and guamil habitats may be governed by their spatial proximity rather than habitat structure or disturbance history. Such an effect was noted by Feener and Schupp (1998) and Bestelmeyer (2000). The similarity of the burned ramonal and encinal habitats cannot be attributed to spatial proximity, and it is unclear what may have caused this pattern.

The relative dominance of *Solenopsis geminata* in the grassland habitat (Figure 9.4) has been observed in other studies conducted in Central America (Perfecto, 1991; Perfecto and Vandermeer, 1996). This ant is favored by the warm, well-insolated conditions in deforested habitats. The presence of *Forelius* and *Dorymyrmex*, which are dominant arid-zone genera, in the grassland and other disturbed habitats was surprising. We were not able to locate records of these genera in similar habitats in studies conducted further south in Central America. The presence of these genera in the Petén forests may be related to the relative aridity of many of the natural microhabitats in the region and to their proximity to arid areas to the north. *Dorymyrmex, Forelius,* and *Solenopsis* are characteristic of Nearctic deserts, and their combined dominance in the grassland underscores the immense shift in environmental conditions here when compared to the forest (see also Quiroz-Robledo and Valenzuela-Gonzáles, 1995).

CONCLUSIONS AND RECOMMENDATIONS

It is fruitful to view patterns of ant diversity with respect to the kinds of microhabitats available for different ant species to occupy . Generally, ant richness appears to be at a maximum when several kinds of microhabitats—suitable leaf litter and twigs, patches of bare ground, fallen limbs and logs, epiphytes, bushes, and trees of various species, for example—are abundant in an area. The reduction or elimination of these ant habitats in an area results in declines in ant richness and increased dominance by highly competitive ant species. Furthermore, frequent natural or anthropogenic disturbances will eliminate ant species from areas even when suitable habitats are available.

Evidence presented here and in studies cited above suggests that the recovery of suitable microhabitats for tropical forest ants following disturbance may be rapid. The recovery of both the forest and its microhabitats is contingent upon the degree of agricultural use and soil erosion. Soil erosion may be significant in Petén forests due to the thin soils on slopes and the crests of hills (Beach, 1998). In addition, the recolonization of ants to microhabitats necessarily depends upon the presence and proximity of source populations. The relationship between source population distribution and colonization rates have yet to be considered in detail, but it is likely that there are threshold amounts of deforestation beyond which colonization by ant species will be limited (Roth et al., 1994), metapopulation dynamics are disrupted, and region-wide extinctions may occur (Hanski and Gilpin, 1996). Rapid assessment surveys and short-duration studies of the effects of habitat alteration on ant communities cannot hope to detect such thresholds or regional effects. The best we can do in this case is to be cognizant of the potential for these effects. Despite the relatively good colonization abilities of ants and some other insect species, deforestation threatens the extinction of their populations. At this point, however, ant diversity within the park is high, and every effort should be made to maintain it.

The management of forests within LTNP for ant conservation should emphasize the preservation of a variety of habitats. One way to achieve this is to conserve the breadth of topographic gradients, which are generally recognized as important determinants of ant diversity in tropical landscapes (e.g. Olson, 1994) and of biodiversity in LTNP in particular (Mendez et al., 1998). Epiphytes and other canopy micro-habitats available in tall trees may contain many unique ant species (Longino and Nadkarni, 1990), thus bosques altos should be targeted for additional studies and conservation.

Ant community composition is very patchy and turnover rates are very high (Table 9.1), so more intensive studies will be needed to disentangle the effects of spatial pattern and environmental factors on ant diversity. Ants respond in important and informative ways to landscape variation and would provide a useful focal taxon in more extensive surveys and monitoring. Ants respond to processes such as topographic variation, flooding patterns, and succession and can thus indicate environmental patchiness and dynamics that are important for land managers. Understanding the relationships between animal distributions, drainage, and fire will be needed to identify priority conservation areas within the park's terrestrial habitats.

SPECIFIC CONSERVATION RECOMMENDATIONS

- Highly complex bosques altos in areas such as El Perú and Laguna Guayacán that appear to have suffered relatively little previous damage should be protected vigorously and examined in greater detail—especially the canopy ant communities.

- The conservation of a variety of habitat types—where habitat types are defined according to vegetation, topography, and soil characteristics—is likely to be a good strategy for conserving ants and other terrestrial arthropods. In LTNP, remote sensing technologies can aid this process.

- The ant fauna of bosques altos and other sites should be surveyed at a wetter time of year, such as at the beginning of the wet season. Malaise traps should be used to capture volant sexual forms during this period, in addition to traditional techniques. This will likely lead to a more complete record of the park's ant (and insect) fauna.

- Additional survey and monitoring efforts should examine the role of drainage and topography, soil variation, and fire history on ant and other animal communities in the park.

- Replicated surveys in natural habitat types such as encinales and selva baja are needed to provide more detailed assessments of the complementarity of ant faunas between habitat types in the park.

LITERATURE CITED

Alonso, L. E. 1998. Spatial and temporal variation in the ant occupants of a facultative ant-plant. Biotropica 30: 210-213.

Andersen, A. N. 1991. Sampling communities of ground-foraging ants: pitfall catches compared with quadrat counts in an Australian tropical savanna. Australian Journal of Ecology 16: 273-279.

Andersen, A. N. 1995. A classification of Australian ant communities, based on functional groups which parallel plant life-forms in relation to stress and disturbance. Journal of Biogeography 22: 15-29.

Beach, T. 1998. Soil catenas, tropical deforestation, and ancient and contemporary soil erosion in the Petén, Guatemala. Physical Geography 19: 378-405.

Belshaw, R., and B. Bolton. 1993. The effect of forest disturbance on the leaf litter ant fauna of Ghana. Biodiversity and Conservation 2: 656-666.

Bestelmeyer, B. T. 2000. *A multiscale perspective on ant diversity in semiarid landscapes.* Dissertation. Colorado State University, Fort Collins, CO, USA.

Bestelmeyer, B. T., and J. A. Wiens. 1996. The effects of land use on the structure of ground-foraging ant communities in the Argentine Chaco. Ecological Applications 6:1225-1240.

Bestelmeyer, B.T., D. Agosti, L. Alonso, C. R. Brandão, W. L. Brown Jr., J. H. C. Delabie, and R. Silvestre. *In press.* Field techniques: an overview, description, and evaluation of sampling techniques. In *Measuring and Monitoring Biodiversity: Standard Techniques for Ground-Dwelling Ants* (D. Agosti, J. Majer, T. Schultz, and L. Alonso, eds.). Smithsonian Institution Press, Washington, DC, USA.

Brandão, C. R. F. 1991. Adendos ao catálogo abreviado das formigas da região neotropical Hymenoptera: Formicidae. Revista Brasileira de Entomología 35:319-412.

Colwell, R. K. 1997. *EstimateS 5. Statistical Estimation of Species Richness and Shared Species from Samples.* Web site: http://viceroy.eeb.uconn.edu/estimates.

Colwell, R. K., and J. A. Coddington. 1994. Estimating terrestrial biodiversity through extrapolation. Philosophical Transactions of the Royal Society of London 345:101-118.

Erwin, T. L., and J. C. Scott. 1980. Seasonal and size patterns, trophic structure, and richness of Coleoptera in the tropical arboreal ecosystem: the fauna of the tree *Luehua seemanni* Triana and Planch in the Canal Zone of Panama. Coleopterists Bulletin 34: 305-322.

Feener, D. H., and E. W. Schupp. 1998. Effects of treefall gaps on the patchiness and species richness of neotropical ant assemblages. Oecologia 116:191-201.

Greenslade, P. J. M., and P. Greenslade. 1977. Some effects of vegetation cover and disturbance on a tropical ant fauna. Insectes Sociaux 24: 163-182.

Haila, Y., and C. R. Margules. 1996. Survey research in conservation biology. Ecography 19:323-331.

Hanski, I., and M. Gilpin, eds. 1996. *Metapopulation Dynamics: Ecology, Genetics, and Evolution.* Academic Press, New York, NY, USA.

Harada, A. Y., and A. G. Bandeira. 1994. Estratificacão e densidade de invertebrados em solo arenoso sob floresta primária e plantios arbóreos na Amazonia central durante estacão seca. Acta Amazonica 24: 103-118.

Janzen, D. H., and T. W. Schoener. 1968. Differences in insect abundance and diversity between wetter and drier sites during a tropical dry season. Ecology 49:96-110.

Johnson, R. A. 1992. Soil texture as an influence on the distribution of the desert seed-harvester ants *Pogonomyrmex rugosus* and *Messor pergandei.* Oecologia 89:118-124.

Kempf, W. W. 1972. Catálogo abreviado das formigas da região neotropical Hymenoptera: Formicidae. Studia Entomológica 15:3-344.

Legendre, P., and L. Legendre. 1998. *Numerical Ecology,* 2nd English ed. Elsevier, Amsterdam, the Netherlands.

Levings, S. C. 1983. Seasonal, annual, and among-site variation in the ground ant community of a deciduous tropical forest: some causes of patchy species distributions. Ecological Monographs 53: 435-455.

Levings, S. C., and D. M. Windsor. 1982. Seasonal and annual variation in litter arthropod populations. Pp. 355-387 in *The Ecology of a Tropical Forest: Seasonal Rhythms and Long-term Changes* (E. G. Leigh Jr., A. S. Rand, and D. M. Windsor, eds). Smithsonian Insitution Press, Washington, DC, USA.

Longino, J. T. Online. www.evergreen.edu/user/serv_res/research/arthropod/AntsofCostaRica.html).

Longino, J. T., and N. Nadkarni. 1990. A comparison of ground and canopy leaf litter ants (Hymenoptera: Formicidae) in a neotropical montane forest. Psyche 97:81-93.

May, R. M. 1988. How many species are there on Earth? Science 241:1141-1148.

MacKay, W. P. 1991. Impact of slashing and burning of a tropical rainforest on the native ant fauna (Hymenoptera:Formicidae). Sociobiology 3: 257-268.

McCune, B., and M. J. Mefford. 1999. *PC-ORD. Multivariate Analysis of Ecological Data, version 4.0.* MjM Software Design, Gleneden Beach, OR, USA.

Méndez, C. 1999. How old is the Petén tropical forest? Pp. 31-34 in *Thirteen Ways of Looking at a Tropical Forest: Guatemala's Maya Biosphere Reserve* (J. D. Nations, ed.). Conservation International, Washington, DC, USA.

Olson, D. M. 1994. The distribution of leaf litter invertebrates along a Neotropical altitudinal gradient. Journal of Tropical Ecology 10: 129-150.

Perfecto, I. 1991. Dynamics of *Solenopsis geminata* in a tropical fallow field after ploughing. Oikos 62: 139-144.

Perfecto, I., and R. R. Snelling. 1995. Biodiversity and the transformation of a tropical agroecosystem: ants in coffee plantations. Ecological Applications 5: 1084-1097.

Perfecto, I., and J. Vandermeer 1996. Microclimatic changes and the indirect loss of ant diversity in a tropical agroecosystem. Oecologia 108: 577-582.

Quiroz-Robledo, L., and J. Valenzuela-Gonzalez. 1995. A comparison of ground ant communities in a tropical rainforest and adjacent grassland in Los Tuxtlas, Veracruz, Mexico. Southwestern Entomologist 20: 203-213.

Roth, D. S., I. Perfecto, and B. Rathcke. 1994. The effects of management systems on ground-foraging ant diversity in Costa Rica. Ecological Applications 4: 423-436.

Vasconcelos, H. L. 1999. Effects of forest disturbance on the structure of ground-foraging ant communities in central Amazonia. Biodiversity and Conservation 8: 409-420.

Ward, P.S. Online. entomology.ucdavis.edu/faculty/ward/pseudo.html.

Wilson, E. O. 1987. Causes of ecological success: the case of the ants. The sixth Tansley lecture. Journal of Animal Ecology 56: 1-9.

Wilson, E. O., and B. Hölldobler. 1995. *Journey to the Ants*. Belknap Press, Cambridge, MA, USA.

EXPEDITION ITINERARY AND GAZETTEER

Dates	Events
8 April	Arrival to *Las Guacamayas Biological Station** (17°14.75' N, 90°32.80' W)
9-14 April	Studies at Focal Area 1; *Río San Pedro/Sacluc sites*
15 April	Transfer to El Naranjo, sampling of *Rapids site* (17° 16.99' N, 90°42.80' W)
16-19 April	Sampling at Focal Area 2; *Río Escondido* (17^0 18.78' N; 90^0 52.15' W)
20-22 April	Sampling of Focal Area 3; *Flor de Luna* (17°35.97' N, 90°53.84' W)
23-25 April	Sampling of Focal Area 4; *Río Chocop* (17°36.13' N, 90°24.53' W)
26 April	Return to Las Guacamayas Biological Station
27-29 April	Preliminary report and specimen preparation
29 April	Presentation of preliminary results to CONAP
30 April	Return to Flores for departure of participants

* Refer to Map 3 for site locations.

Principal RAP Sampling Sites in Laguna del Tigre National Park, Petén, Guatemala

The following four principal sites, or focal areas, were surveyed during the RAP expedition to LTNP. Field work was undertaken from one camp at each site. Not all taxonomic groups were surveyed at each site. See Appendices 1 and 2 for specific sampling location coordinates and macrohabitat types.

Focal Area 1: Las Guacamayas Biological Station/Río San Pedro and Río Sacluc
(17°14.75' N, 90°32.80' W)
The Las Guacamayas Biological Station is located at the confluence of the ríos San Pedro and Sacluc. We traveled by pickup trucks to the road's end at the Río Sacluc and were then transported by boat seven km to the station. The station is outfitted with dormitories, a kitchen and eating area, and bathrooms with showers.

Aquatic habitats in the area include deep rivers bordered by sawgrass marsh or riparian forest and springs feeding the San Pedro. Terrestrial habitats in the area include unflooded, tall forests on hills or *bosques altos* (one very close to the station), regenerating forests or *guamiles*, varzea-type forest near the village of Paso Caballos to the east, and a drier oak forest (*encinal*) along the Río Sacluc. The area immediately around the station is comparatively undisturbed due in part to the presence of the station, which has served to discourage illegal settlement. Several settlements occur within the park's boundaries along the San Pedro to the west. We observed no settlements on the Sacluc between the boat landing and the station.

The Rapids
Although not part of our original sampling plan, we passed a series of rapids on our way to El Naranjo. In part of the rapids we found that the bottom consisted of a reef made of bivalve shells (mollusks) mixed with living individuals. This area was quite unique and constitutes an important discovery.

Focal Area 2: Río Escondido
(17⁰ 18.78' N; 90⁰ 52.15' W)

We camped along the west side of the Río Escondido just south of the Laguna del Tigre Biotope. The name of the river, "hidden river," is appropriate; following the river north from its confluence with the San Pedro, it appears to dead-end in a lagoon surrounded by a great marsh. In fact, the river continues northward but as a shallow, narrow canal passing through a great wetland. Eventually, the river deepens and widens and becomes navigable. We took advantage of an abandoned settlement to provide shelter and set camp. Aquatic habitats here are diverse, including deep stretches of river, marshland canals, oxbow lagoons, and small inlets. Terrestrial habitats found here include previously burned bosque alto, guamiles, grasslands resulting from slash-and-burn practices, and a tall riparian forest at the northern limit of our sampling here where the river becomes narrow. Eight families live along the Río Escondido to the north of our camp, and some settlements probably occur within the boundaries of the biotope.

Focal Area 3: Laguna Flor de Luna
(17°35.97' N, 90°53.84' W)

Here we camped at the end of a road alongside the Laguna Flor de Luna. This road passes a military checkpoint and the Xan oil fields. Aquatic habitats here included only lagoons. Samples of fish tissue and benthic sediment examined in toxicological analyses were taken in lagoons at different distances from the Xan fields. Terrestrial habitats were sampled around Flor de Luna and included bosque alto and inundated forest (bosque bajo) which was dominated by the tinto tree (*Haematoxylum campechianum*) in areas closest to the lagoon (tintales). We observed no settlements near Flor de Luna, but several exist near lagoons to the southeast.

Focal Area 4: Río Chocop
(17°36.13' N, 90°24.53' W)

This camp was located at a CONAP monitoring post where the Río Chocop passes the road. Another military checkpoint is located on the route to the station; this is intended to prevent the illegal colonization of the area. Aquatic habitats found in the area included two small, shallow rivers—the Río Chocop and the nearby Río Candelaria—as well as several lagoons. Terrestrial habitats sampled included bosque bajo/tintales and bosques altos. Areas along the road around Santa Amelia are occupied by several families, have been severely disturbed, and are regenerating or are currently used for crops or pastures.

GLOSSARY

Aguada: Small limestone sinks that forms ponds that may not contain water during the dry season.

Akalché: Larger limestone sinks that form lagoons and often contain water throughout the year.

Arroyo: Spanish for "stream."

Bosque alto: Upland forests with tall trees, a dense canopy, little undergrowth, and abundant leaf litter.

Bosque bajo: A short (15-20 m) statured forest with an open canopy relative to bosque alto, and an abundant, dense undergrowth of palms, shrubs, and herbs.

Caño: A body of stagnant, turbid water through which marshes are connected to a river.

Community: A group of species, usually delimited by taxonomic affiliation, that occupy a locality.

Complementarity: A measure of the degree to which species composition (i.e. the identity of species) differs between two areas. A high complementarity indicates that few species are shared and that species richness of the two sites together is much higher than in either one alone.

Encinales: Oak-dominated forests.

Endemic: Native to, and restricted to, a particular geographic region.

Fragility: A qualitative measure of the degree to which the communities or ecosystems are able to resist or recover from a perturbation.

Guamil: Local vernacular term for an area of young (usually 1-20 years) regenerating vegetation where forest vegetation has been disturbed.

Karstic: Pertaining to irregular limestone strata that are permeated by underground streams and sinks.

Laguna: Spanish for "lagoon."

Miocene: A geologic epoch during the Tertiary period, from ca. 26 to 5 million years before present.

Morphospecies: Informal species groups, used when organisms cannot be assigned to a formally described species.

Río: Spanish for "river."

Varzea: Lowland inundated forests.

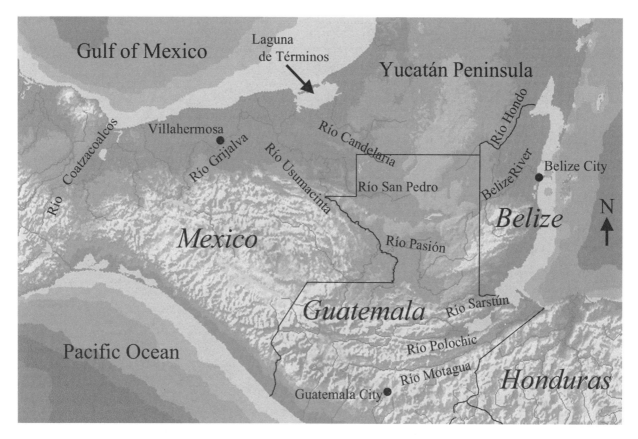

Map 1. The primary rivers of Guatemala, Belize, and southern Mexico, with RAP survey sites located on and near the Río San Pedro.
Mapa 1. Los ríos principales de Guatemala, Belice y el sur de México, con los lugares de muestreo del RAP localizados en o cerca del Río San Pedro.

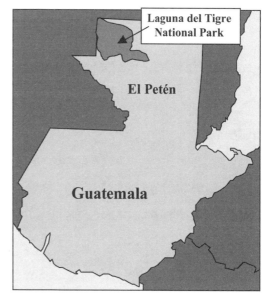

Map 2. Location of the Petén region and Laguna del Tigre National Park within Guatemala.
Mapa 2. Localización de la región del Petén y el Parque Nacional Laguna del Tigre en Guatemala.

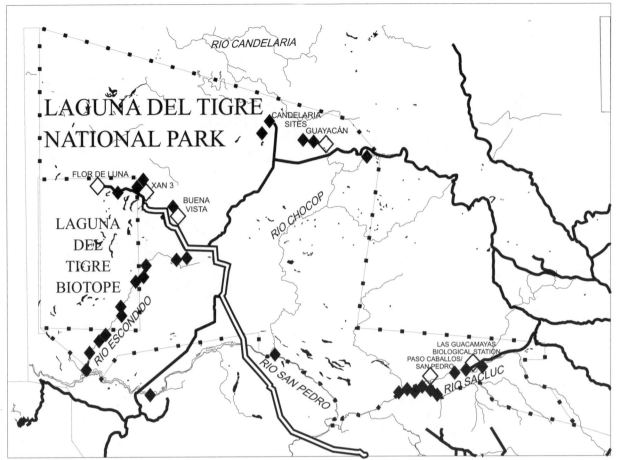

Map 3. Laguna del Tigre National Park and Laguna del Tigre Biotope boundaries, with sites sampled during the 1999 RAP expedition.

Mapa 3. Parque Nacional Laguna del Tigre y las fronteras Biotopos de la Laguna del Tigre, con las localidades de muestreo durante la expedicón RAP en el 1999.

▫▫ Laguna del Tigre National Park and Biotope limits/*Parque Nacional Laguna del Tigre y los del Biotopo*

N Roads/*Carreteras*

/\/ Rivers/*Ríos*

/N/ Pipeline of Basic Resources/*Oleoducto de Basic Resources*

◇ Sites where toxicology samples were taken/*Lugares donde se tomaron muestras toxicología*

◆ Aquatic and terrestrial sampling points/*Puntos de muestreo aquáticos y terrestres*

N

Fuente: ProPetén
Mayo 1996
Escala 1:666,666

Aerial photos produced by/*Fotos aereas producidas por*: Conservation International
In cooperation with/*En cooperación con*: USAID and Intel Corporation
Researchers/*Investigadores*: Dana Slaymaker, University of Massachusetts; John Musinsky, CI

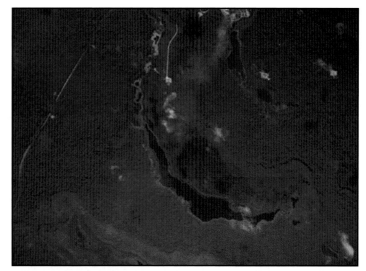

Oil operations near lagoons in Laguna del Tigre National Park.
Operaciones petroleras cerca de lagunas en el Parque Nacional Laguna del Tigre.

Encroachment and oil road near the Río San Pedro where the rapids, including the molluscan reef, is located.
Invasiones y camino petrolero cerca del Río San Pedro donde se localizan los rápidos, incluyendo el arrecife de moluscos.

C. Barrientos

A unique molluscan reef system was discovered at the rapids area of the Río San Pedro near the town of El Naranjo.
Un sistema único de arrecife de moluscos fue descubierto en el área de los rápidos en el Río San Pedro cerca de El Naranjo.

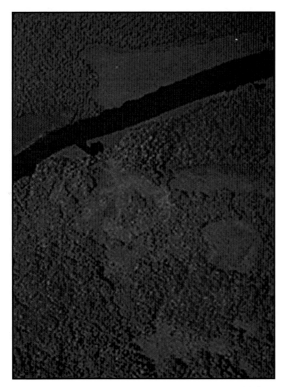
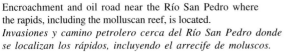

A. Queral-Regil

Mangroves along the Río San Pedro are the furthest inland population in Guatemala.
Manglares a lo largo del Río San Pedro, únicos individuos que se han localizado tierra adentro en Guatemala.

The Río San Pedro, flanked by inundated areas and sibal.
El Río San Pedro, rodeado por áreas inundadas y sibal.

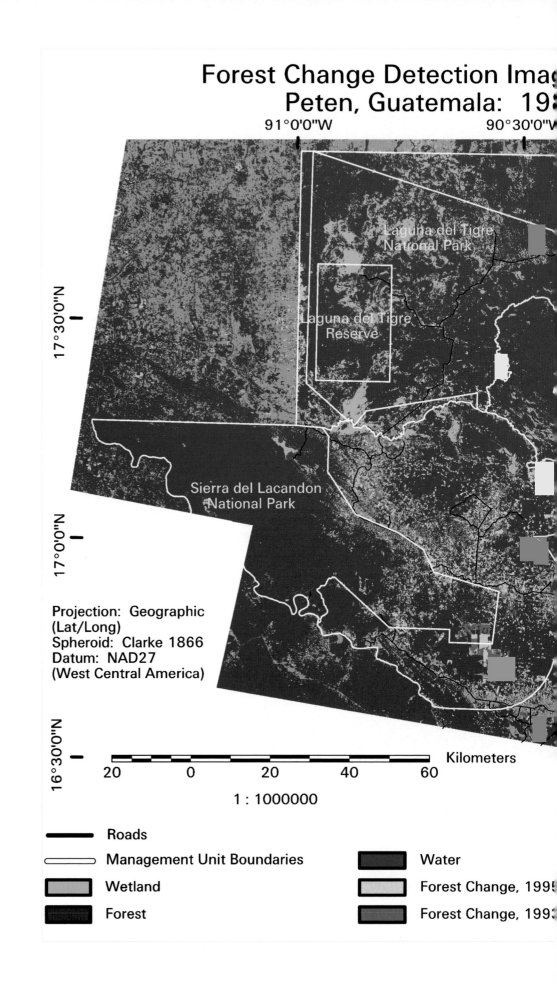

Forest Change Detection Ima[g]
Peten, Guatemala: 19[:]

91°0'0"W   90°30'0"[W]

17°30'0"N

Laguna del Tigre
National Park

Laguna del Tigre
Reserve

17°0'0"N

Sierra del Lacandon
National Park

Projection: Geographic
(Lat/Long)
Spheroid: Clarke 1866
Datum: NAD27
(West Central America)

16°30'0"N

20   0   20   40   60   Kilometers

1 : 1000000

— Roads

⊏⊐ Management Unit Boundaries   ▮ Water

▮ Wetland   ▮ Forest Change, 1995

▮ Forest   ▮ Forest Change, 199[3]

The Maya Biosphere Reserve
990, 1993, 1995, 1997

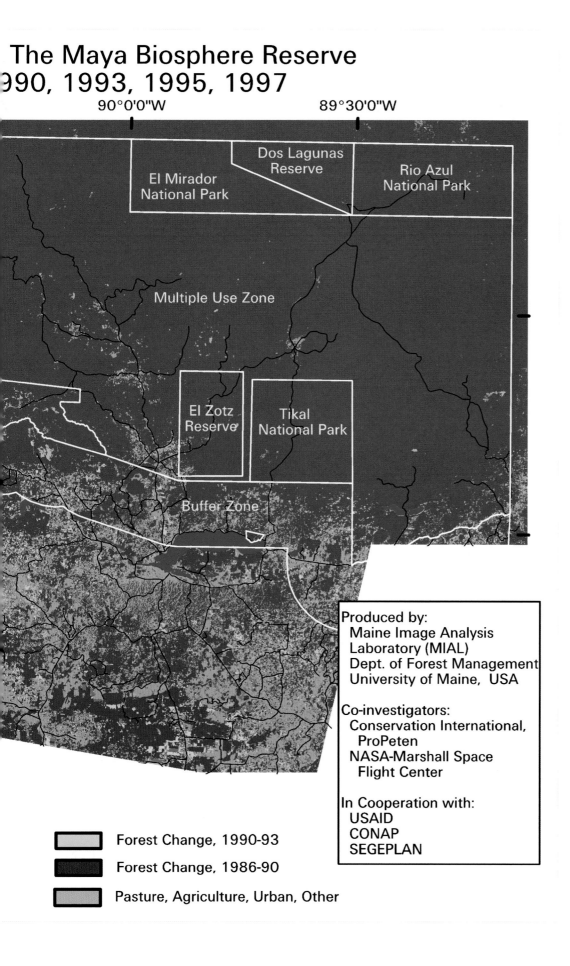

90°0'0"W

89°30'0"W

El Mirador National Park

Dos Lagunas Reserve

Rio Azul National Park

Multiple Use Zone

El Zotz Reserve

Tikal National Park

Buffer Zone

Produced by:
  Maine Image Analysis
  Laboratory (MIAL)
  Dept. of Forest Management
  University of Maine, USA

Co-investigators:
  Conservation International,
    ProPeten
  NASA-Marshall Space
    Flight Center

In Cooperation with:
  USAID
  CONAP
  SEGEPLAN

Forest Change, 1990-93

Forest Change, 1986-90

Pasture, Agriculture, Urban, Other

Oil operations within Laguna del Tigre National Park.
Operaciones de petróleo dentro del Parque Nacional Laguna del Tigre.

An example of tintal (bosque bajo), which is a relatively short-statured forest with dense undergrowth.
Un ejemplo de tintal (bosque bajo), un bosque con árboles de baja estatura y con maleza densa.

One of the twenty bat species (here, *Carollia brevicauda*) found during the RAP expedition to LTNP.
Una de las veinte especies de murciélagos (aquí, Carollia brevicauda) encontrada en la expedición RAP en el PNLT.

The burning of dry forest and sawgrass marshes (sibales) poses a great threat to the biodiversity of LTNP.
La quemazón de bosque seco y sibales es una amenaza muy seria a la biodiversidad del PNLT.

RAP scientists study aquatic plants growing along a natural spring in LTNP.
Científicos de RAP estudian plantas acuáticas creciendo a lo largo de un manantial natural del PNLT.

Petenia splendida (Blanco) exhibits color polymorphism in the park and is subject to heavy fishing pressure.
Petenia splendida (Blanco) *muestra polimorfismo cromático en el parque y es sujeto a la pesca excesiva.*

RAP scientists study the Morelet's crocodile (*Crocodylus moreletii*), which is understudied and may be in danger of extinction.
Científicos de RAP estudian el cocodrilo de Morelet (Crocodylus moreletii), el cuál no ha sido estudiado mucho y está en peligro de extinción.

TABLA DE CONTENIDO

PARTICIPANTES

Leeanne E. Alonso, Ph.D. (mirmecología [hormigas])
Programa de evaluación rápida
Conservation International
2501 M Street, NW, Suite 200
Washington, DC 20037 USA
email: l.alonso@conservation.org

Ana Cristina Bailey (entomología)
Laboratorio de Ecología Sistemática
Universidad del Valle de Guatemala
18 Avenida 11-95 Zona 15, VH. III
Apdo. Postal 82
01901 Ciudad Guatemala, Guatemala
email: abailey@uvg.edu.gt

Christian Barrientos (ictiología, protocolo RAP)
CI-ProPetén
Calle Central
Ciudad Flores
El Petén, Guatemala
email: cbarrientos@conservation.org.gt

Brandon T. Bestelmeyer, Ph.D. (mirmecología,
jefe de equipo internacional)
Departamento de Biología
MSC 3AF, Box 30001
New Mexico State University
Las Cruces, NM 80003 USA
email: bbestelm@jornada.nmsu.edu

Marcos Callisto, Ph.D. (limnología)
Universidade Federal de Minas Gerais ICB,
Depto. Biología General
CP. 486, CEP. 30.161-970
Belo Horizonte, MG, Brasil
email: callisto@mono.icb.ufmg.br.gt

Francisco Castañeda Moya (herpetología, jefe del equipo
nacional, protocolo RAP)
CI-ProPetén
Calle Central
Ciudad Flores
El Petén, Guatemala
email: fmoya@conservation.org.gt

Miriam Lorena Castillo Villeda (ornitología)
CI-ProPetén
Calle Central
Ciudad Flores
El Petén, Guatemala
email: mcastillo@conservation.org.gt

Barry Chernoff, Ph.D. (ictiología)
Departamento de Zoología
Field Museum
Roosevelt Road at Lakeshore Drive
Chicago, IL 60605 USA
email: chernoff@fmnh.org

Javier García Esquivel (protocolo RAP)
CONAP
Vía 5 4-50, Zona 4
Edificio Maya
4º Nivel,
Ciudad Guatemala, Guatemala

Karin Herrera (limnología)
Universidad de San Carlos de Guatemala
2º Nivel, Edificio T-12
Ciudad Universitaria
Zona 12
Ciudad Guatemala, Guatemala
email: kherrera@usa.net

Herman A. Kihn, M.Sc. (ictiología)
Centro de Estudios Conservacionistas (CECON)
Unidad de Investigación
Vida Silvestre
Avenida Reforma 0-63, Zona 10
Ciudad Guatemala, Guatemala

Oscar Lara, M.Sc. (herpetología, protocolo RAP)
CONAP
4a. calle 6-17 z. 1
Ciudad Guatemala, Guatemala
email: seconap@guate.net
Escuela de Biología
Universidad de San Carlos de Guatemala
Edificio T-10
Ciudad Universitaria, zona 12
Ciudad Guatemala, Guatemala
email: oslara@usac.edu.gt

Blanca León, Ph.D. (botánica)
Museo de Historia Natural
Universidad Nacional Mayor de San Marcos
Av. Arenales 1256, Apartado 14-0434
Lima-14 Perú
email: leon@umbc.edu

Mario Mancilla (protocolo RAP)
Asociación para la Conservación de la Naturaleza
CANANKAX
Barrio el Redentor
Colonia Itzá, Frente al Salón Social San Benito
Petén, Guatemala
e-mail: canankaxpet@guate.net

Julio Morales Can (botánica)
Universidad de San Carlos de Guatemala
Edificio T-13, Ciudad Universitaria
Zona 12
Ciudad Guatemala, Guatemala
email: quinchobarrilete@yahoo.com

Jorge Ordóñez (entomología)
Proyecto Fauna
CI-Guatemala
7a-avenida 3-33
Zona 9, Edificio Torre Empresarial
4o nivel, oficina 406
Ciudad de Guatemala, Guatemala
email: jorge.e.ordonez@usa.net
        jorge.e.ordonez@citel.com.gt

Ismael Ponciano (protocolo RAP)
Centro de Estudios Conservacionistas (CECON)
Unidad de Investigación
Vida Silvestre
Avenida Reforma 0-78, Zona 10
Ciudad Guatemala, Guatemala

Sergio G. Pérez (mastozoología)
Colecciones Zoológicas
Museo de Historia Natural
Universidad de San Carlos de Guatemala
Calle Mariscal Cruz 1-56
Zona 10
Ciudad Guatemala, Guatemala
email: museo@usac.edu.gt

Alejandro Queral-Regil, M.Sc. (herpetología)
11700 Old Columbia Pike, Suite 1902
Silver Spring, MD 20904 USA
fax: 202-547-6009
email: alejandro.queral@sierraclub.org

Carlos Rodríguez Olivet (protocolo RAP)
CI-Guatemala
7a-avenida 3-33
Zona 9, Edificio Torre Empresarial
4o nivel, oficina 406
Ciudad de Guatemala, Guatemala
email: crodriguez@citel.com.gt

Edgar Selvin Pérez (ornitología)
Escuela de Biología
Universidad de San Carlos de Guatemala
Edificio T-10
Ciudad Universitaria, zona 12
Ciudad Guatemala, Guatemala
email: selvin_perez@latinmail.com

Christopher W. Theodorakis, Ph.D. (toxicología)
Texas Tech University
The Institute of Environmental and Human Health
P.O. Box 41163
Lubbock, TX 79409-1163 USA
fax: 806-885-4577
email: chris.theodorakis@tiehh.ttu.edu

Philip W. Willink, Ph.D. (ictiología)
Fish Division
Field Museum
Roosevelt Road at Lakeshore Drive
Chicago, IL 60605 USA
email: pwillink@fmnh.org

Heliot Zarza (mastozoología)
Instituto de Ecología, UNAM
Circuito Ext. s/n junto al Jardín Botánico, CU
Ap. Postal 70-275, 04510
México D.F., México
fax: 56 22 98 95
email: hzarza@nosferatu.ecologia.unam.mx

PERFIL DE LA ORGANIZACION

CONSERVATION INTERNATIONAL

Conservation International (CI) es una organización internacional sin fines de lucro, localizada en Washington, DC. CI actúa con la creencia de que el patrimonio nacional natural de la Tierra debe mantenerse si las generaciones futuras desean enriquecerse espiritual, cultural y económicamente. Nuestra misión es conservar los procesos biológicos y ecológicos que dan apoyo a la vida de la tierra para demostrar que las sociedades humanas pueden vivir en armonía con la naturaleza.

Conservation International
2501 M Street, NW, Suite 200
Washington, DC 20037 USA
202-429-5660 (Teléfono)
202-887-0193 (Fax)
http://www.conservation.org

CONSERVATION INTERNATIONAL - PROPETEN

En respuesta a las crecientes amenazas a la biodiversidad de Guatemala en el Departamento de Petén, en el norte del país, CI estableció una oficina en Guatemala, localmente conocida como ProPetén, en 1991. CI-ProPetén trabaja con comunidades locales para conservar la diversidad biológica incrementando la conciencia ambiental y desarrollando alternativas económicas para las comunidades locales. ProPetén se esfuerza en demostrar que las comunidades locales pueden vivir dentro de los bosques que aún existen en Petén sin destruirlos. Los esfuerzos de CI-ProPetén se enfocan en las 1.6 millones de hectáreas (4 millones de acres) de la Reserva de la Biósfera Maya (RBM). La reserva contiene la mayor parte de los bosques de la Selva Maya, la cual se extiende a Belice y México.

Establecida en 1990, la RBM incluye un gran número de ecosistemas. El más importante es el Parque Nacional Laguna del Tigre (PNLT), el humedal de agua dulce más grande de Centro América. El PNLT ha sido una de las prioridades de ProPetén en los últimos tres años. Al trabajar muy de cerca con colaboradores locales, nacionales e internacionales, ProPetén ha ayudado a crear una infraestructura de manejo viable para el parque y ha proporcionado lineamientos de políticas a las autoridades nacionales que mitiguen los impactos de las actividades humanas en la región. Tanto el personal científico de ProPetén como su Estación Biológica Las Guacamayas, ubicada en la frontera sur del PNLT, fueron utilizadas durante la Expedición RAP.

CI-ProPetén
Calle Central
Ciudad Flores
El Petén, Guatemala
502-926-1370 (Teléfono)
502-926-0495 (Fax)

CECON: CENTRO DE ESTUDIOS CONSERVACIONISTAS

El Centro de Estudios Conservacionistas (CECON) es una institución de la Universidad de San Carlos de Guatemala, ubicada en la Ciudad de Guatemala. CECON tiene como propósitos principales los siguientes: 1) Llevar a cabo estudios biológicos básicos en Guatemala y utilizar dichos estudios como base para los esfuerzos nacionales de conservación, los que incluyen definir políticas que el Gobierno Nacional puede utilizar para un mejor manejo de las áreas protegidas de Guatemala; 2) manejar racional y técnicamente un sistema de biotopos y otras áreas protegidas para la conservación de la vida silvestre y 3) dirigir un esfuerzo nacional para promover los temas biológicos con énfasis en la conservación.

CECON maneja cuatro biotopos protegidos en la Reserva de la Biósfera Maya (RBM): Laguna el Tigre-Río Escondido, Cerro Cahuí, Naachtún-Dos Lagunas, y Zotz-San Miguel La Palotada. El Biotopo Laguna del Tigre se encuentra ubicado en el parque más grande, Parque Nacional Laguna del Tigre y es considerado uno de los hábitats más críticos de Guatemala. En un esfuerzo por divulgar y utilizar más información como una base para futuros esfuerzos de protección contra la actividad humana en la región, CECON desempeñó un papel crítico en la Expedición RAP del Parque Nacional Laguna del Tigre.

Centro de Estudios Conservacionistas
Avenida Reforma 0-63, Zona 10
Ciudad Guatemala, Guatemala
502-334-7662 (Teléfono)
502- 334-7664 (Fax)

CONAP: CONSEJO NACIONAL DE ÁREAS PROTEGIDAS

Fundado en 1989 como una oficina de la Presidencia de Guatemala, el Consejo Nacional de Áreas Protegidas (CONAP) está a cargo de administrar todas las áreas protegidas de Guatemala. Estas áreas comprenden por lo menos ochenta y un unidades que contienen menos de dos millones de hectáreas o cerca del dieciocho por ciento del territorio de Guatemala. La unidad más grande dentro de este sistema de áreas protegidas es la Reserva de la Biósfera Maya (RBM), ubicada en la parte norte de Petén. Bajo el proyecto de la Biósfera Maya y otras grandes iniciativas en Petén, el gobierno de Guatemala —a través del CONAP—ha hecho de la región una prioridad para las inversiones nacionales e internacionales. En 1996, el CONAP declaró al Parque

Nacional Laguna del Tigre (PNLT) y al Parque Nacional Sierra del Lacandón (PNSL) sus prioridades más grandes. Desde entonces, el CONAP ha trabajado con agencias locales e internacionales—CI-ProPetén entre ellas—para poner en orden los sistemas de manejo y para que reconcilien la necesidad económica humana con la conservación efectiva.

Consejo Nacional de Areas Protegidas (CONAP)
4a calle 6-17 Zona 1
Ciudad Guatemala, Guatemala
502-220-1821 (teléfono)
502-220-1827 (fax)
Agradecemos a Juventino Gálvez, Oscar Lara y Javier García Esquivel miembros del Consejo Nacional de Areas

CONAMA: COMISION NACIONAL DEL MEDIO AMBIENTE

La Comisión Nacional del Medio Ambiente (CONAMA) es una institución gubernamental bajo el poder ejecutivo del gobierno de Guatemala. La agencia asesoria, coordina, y facilita las acciones necesarias para proteger y mejorar el medio ambiente. Para alcanzar estas metas, la institución trabaja con diferentes ministerios del gobierno, con la Secretaría General de Planeación Económica, diferentes agencias autónomas y semi-autónomas, municipalidades, y el sector privado del gobierno.

Comisión Nacional del Medio Ambiente (CONAMA)
7a. Avenida 7-09, Zona 13
Ciudad Guatemala, Guatemala 01013
502-440-7916 (teléfono/fax)
502-440-7917

CÄNAN K'AAX: ASOCIACION GUATEMALTECA PARA LA CONSERVACION NATURAL

La Asociación Guatemalteca para la Conservación Natural (CÄNAN K'AAX) es una organización sin fines de lucro cuyos miembros son, en su mayor parte, de la región del Petén. CÄNAN K'AAX fue formada para responder a los desafíos de conservación y administración de áreas protegidas en el Petén y la Selva Maya usando una perspectiva local. La asociación deriva su nombre de la palabra Maya Itzá que significa "Guardiánes de la Selva."

La filosofía de la asociación está basada en el principio de una co-existencia sana entre los seres humanos y el medio ambiente. Para realizar esta filosofía, sus acciones están basadas en el desarrollo y fortalecimiento de la capacidad y actitudes de la sociedad civil y el gobierno para así administrar los recursos naturales de una manera sustentable, y en particular con referencia a áreas protegidas. Actualmente, la asociación se dedica a la co-administración del Parque Nacional Laguna del Tigre conjuntamente con CONAP.

Barrio el Redentor
Colonia Itzá, Frente al Salón San Benito
Petén, Guatemala
502-926-3732 (teléfono)
502-926-3733

ESTACION BIOLOGICA LAS GUACAMAYAS

La Estación Biológica Las Guacamayas cuyo nombre se origino por la especie Ara macao (Guacamayo rojo) que habita el Parque Nacional Laguna del Tigre y que se encuentra en peligro de extinción, fue establecida originalmente en 1996, convirtiendose en el eje de operaciones de Propetén/CI en el área este del PNLT.

Se concibió para funcionar como un centro de investigación y capacitación dentro del PNLT, teniendo como objetivo general el apoyo al manejo del PNLT a través de la generación de información científica sobre los recursos naturales, promoviendo de esto forma de uso sostenible de los mismos.

Su implementación es uno estrategico encaminado a preservar la mayor diversidad posible de ecosistemas, hábitats y especies en esta parte de la zona núcleo.

Su misión es servir como base para la conservación en el Este del Parque Nacional Laguna del Tigre, a través de actividades de investigación de campo a estudiantes guatemaltecos. Actualmente se encuentra en negociación la cooperación con universidades de Estados Unidos.

Para más información envíe sus preguntas a:

CI-ProPetén
Calle Central
Ciudad Flores
El Péten, Guatemala
502-926-1370 (teléfono)
502-926-0495 (fax)

RECONOCIMIENTOS

Protegidas por su participación, dedicación y cooperación para hacer de esta expedición una realidad y un éxito. Además, agradecemos a Herman Kihn y al Centro de Estudios Conservacionistas de la Universidad de San Carlos de Guatemala por coordinar el uso del Biotopo de la Laguna del Tigre y por participar en la expedición. Gracias además, a Mario Mancilla y a CÄNAN K'AAX por su participación.

Al personal de CI-ProPetén, especialmente a Miriam Castillo, Rosita Contreras, Jorge Ordóñez y Juan Manuel Burgos, quienes organizaron la logística de la expedición. A Imelda López Ascencio y a Gloria Reyes Mendoza quienes nos alimentaron de una forma extremadamente maravillosa, y siempre con muchas tortillas deliciosas! A Manuel Alvarado quien nos facilitó lanchas y lancheros. Agradecemos también a Francisco Bosoc, Baudilio Choc, Raquel Quixchán, Víctor Cohuog, David Figueroa, Carlos Girón, Federico Godoy, Ramón Manzanero, César Morales, Ramiro Lucero, Román Santiago y José Soza por sus constantes esfuerzos en el campo, su buen humor y amistad, y por compartir su conocimiento sobre este hermoso lugar en el que viven. Sin estos amigos, nuestro trabajo no hubiera sido posible.

Agradecemos a las siguientes personas por su ayuda en la identificación de especies entre otros: Mary Anne Rogers y Kevin Swagel (Capítulo 4); Rodrigo Medellín, Fernando Cervantes y Yolanda Hortelano (Capítulo 8) y William P. MacKay, John T. Longino y Phil S. Ward (Capítulo 9).

Finalmente, agradecemos a Jim Nations de CI-Washington, a Carlos Rodríguez y Carlos Soza de CI-ProPetén por su guía y apoyo en este esfuerzo.

Esta investigación fue financiada por una donación de la Agencia Internacional para el Desarrollo (USAID) y el Global Environment Facility, otorgada a CI-ProPetén y por una donación de Rufford Foundation para el Programa de Evaluación Rápida (RAP) de Conservation International. Además agradecemos a Jay Fahn, Leslie Lee, David y a Janet Shores por apoyar los estudios de AquaRAP.

UNA EVALUACION BIOLOGICA DEL PARQUE NACIONAL LAGUNA DEL TIGRE, PETEN, GUATEMALA

1) Fechas de los estudios
Expedición RAP: Del 8 al 30 de abril de 1999

2) Descripción de la ubicación
El Parque Nacional Laguna del Tigre (PNLT) se ubica dentro de la Selva Maya, un extenso sistema de bosque seco que se extiende de la parte sur de México a toda la parte norte de Guatemala y Belice. El PNLT ocupa más de 338,566 hectáreas de la Reserva de la Biosfera Maya (RBM), la cual es el sistema más grande de áreas protegidas y manejadas de la Selva Maya. Dentro de la RBM, el parque es el área núcleo protegida más grande. La parte oeste del parque se caracterizán por zonas inundadas o sujetas a inundación con numerosas lagunas. Dichas áreas de humedales fueron incluidas como " humedales de importancia internacional" bajo la convención RAMSAR en 1990 y 1999. La parte este posee bosque alto entremezclado con bosque bajo. Existe también un área de bosque transicional formado por bosque alto, bosque bajo y sabana entre mezclado sin patrón definido. El río más grande dentro del parque, el Río San Pedro, define su límite al sur. Este río es un tributario del Río Usumacinta, el cual drena la Cuenca del Usumacinta al Golfo de México. Otros ríos que se encuentran en el parque son tributarios del Río San Pedro, incluyendo al Río Sacluc en la parte este y el Río Escondido al oeste. El Río Candelaria es un afluente independiente de la Laguna de Términos.

3) Motivación para los Estudios RAP
A pesar de la importancia estratégica del PNLT como un área núcleo de conservación dentro de la RBM, el parque se encuentra extremadamente amenazado debido a la presencia de las áreas núcleo para la explotación del petróleo. La colonización humana del parque junto con los caminos construidos para las operaciones del petróleo también son una presión para los ecosistemas del parque. Una colonización adicional está llevándose a cabo a lo largo de ríos, particularmente en el Río San Pedro y el Río Escondido. Entre algunas de las consecuencias de esta colonización se pueden incluir la deforestación por la agricultura de corte y quema; la quema no intencional de áreas como ciénagas en áreas bajas, la caza, pesca y contaminación no controladas. Los efectos de dichas actividades en la diversidad biológica tienden a volverse más severas a menos que se tome una acción inmediata para preservar la integridad de las áreas núcleo designadas de la RBM. Nuestros estudios resaltan la importancia regional del PNLT, indagan la naturaleza específica de las amenazas a su diversidad biológica y sugieren las opciones de manejo para las áreas dentro del parque.

4) Resultados principales
Se observó una gran variedad de hábitats tanto acuáticos como terrestres, así como grupos taxonómicos en el curso del muestreo de RAP, incluyendo ciénagas, lagunas, ríos, arroyos, bosques en regeneración y bosques antiguos. Cada hábitat fue evaluado con respecto a fitoplancton, insectos acuáticos, peces, anfibios, reptiles, hormigas, aves y mamíferos. La heterogeneidad ambiental fue reconocida de forma diferente por cada grupo taxonómico. Para algunos taxones, las ciénagas eran importantes; para otros lo eran las lagunas, ríos o los bosques complejos.

En total, se observaron 647 especies. Registramos extensiones de rango para una especie de mamífero, tres de aves y dos de hormigas. Este estudio RAP proporcionó las primeras listas de especies de hormigas y fitoplancton para el PNLT. De forma extraordinaria, el parque alberga a dieciséis especies de invertebrados que son de importancia internacional y que son enumerados por la Unión Mundial para la conservacion de la Naturaleza (UICN) y/o la Convención sobre el Tráfico Internacional de Especies en Peligro de Flora y Fauna Silvestre (CITES). Además, fue descubierto en el Río San Pedro un hábitat extremadamente raro de arrecife de agua dulce compuesto de moluscos bivalvos vivos y muertos. Este arrecife albergaba una única combinación de plantas y animales.

En general, la diversidad dentro de los taxones y la representación de especies endémicas, raras o en peligro fue increíble, haciendo invaluable el valor del parque para la conservación en Centro América y el potencial del parque para mantener la integridad ecológica a largo plazo. Adicionalmente, una evaluación ecotoxicológica de dos especies de peces revelaron el impacto de contaminantes mutagénicos en algunos de los sistemas acuáticos del parque. Este resultado confirma la necesidad de más estudios toxicológicos intensivos.

Cantidades de especies registradas:

Fitoplancton:	71 especies
Insectos acuáticos asociados con *Salvinia auriculata*:	44 especies
Plantas:	130 especies
Peces:	41 especies
Aves:	173 especies
Reptiles y anfibios:	36 especies
Mamíferos:	40 especies
Hormigas:	112 especies

Nuevos registros para la región o Guatemala:

Aves:	3 especies
Reptiles:	1 especies
Mamíferos:	1 especies
Hormigas:	2 especies

5) Recomendaciones de Conservacion

El PNLT, en general, necesita protección inmediata de la deforestación y de los incendios para preservar sus altos niveles de biodiversidad. Existen varias áreas de particular importancia dentro del parque y que se encuentran amenazadas: sabanas anegadas en y alrededor del Biotopo Laguna del Tigre, ciénagas de llanos bajos a lo largo de los ríos San Pedro y Escondido y las lagunas junto al camino Xan-Flor de Luna. Las áreas cerca de la Estación Biologica Las Guacamayas, incluyendo la región cerca de El Perú, que disfrutan de protección relativamente buena; deben mantenerse. Las áreas que actualmente no se encuentran dentro de los límites del parque pero que son únicas y deberían protegerse incluyen al Río Sacluc, los bosques anegados (inundados) cerca de Paso Caballos, ciénagas de llanos bajos al sur del Río San Pedro, manglares junto al Río San Pedro y un sistema de arrecife de agua dulce al este de El Naranjo.

RESUMEN EJECUTIVO

INTRODUCCION

Dentro del Parque Nacional Laguna del Tigre (PNLT) y el Biotopo Laguna del Tigre, se encuentran parte de las áreas núcleo de conservación dentro de la Reserva de la Biósfera Maya (RBM). Estas áreas contienen entre sí las últimas grandes huellas de la Selva Maya de Centro América. Esta selva y los diversos hábitats que ésta contiene se encuentran en peligro de desaparecer en este siglo dadas las tasas actuales de deforestación. El valor del PNLT surge de su estado relativamente primitivo y su ubicación central dentro de la RBM. Con el advenimiento de la construcción de las carreteras y otro desarrollo en el área, las selvas, humedales y ríos del parque—y la biodiversidad que estos hábitats albergan—se encuentran bajo una creciente presión de la colonización humana y el desarrollo de las instalaciones de extracción de petróleo dentro del parque. Los objetivos principales de la Expedición del Programa de Evaluación Rápida (RAP) de Conservation International (CI) fueron aumentar la base de datos de la biodiversidad del PNLT, resaltar la importancia regional del parque, evaluar la naturaleza de las amenazas de los animales y plantas dentro del parque, y proporcionar información a los administradores de recursos naturales y a los encargados de la toma de decisiones. Otros objetivos incluyeron atraer la atención internacional a un recurso científico importante, Estación Biológica Las Guacamayas (EBG), una estación de investigación ubicada cerca del límite sur del parque, y capacitar a estudiantes locales en la identificación de especies y en técnicas de monitoreo ecológico. El trabajo ecológico se enfocó en la documentación de los valores de la biodiversidad de varios taxa acuáticos y terrestres en el parque, investigando las relaciones entre los taxa y diferentes hábitats y áreas del parque para asegurarse del valor de conservación de estas áreas y de evaluar los efectos potenciales de la creciente deforestación y desarrollo petrolero en la biodiversidad del PNLT. A lo largo de este proyecto, CI ha trabajado muy de cerca con científicos de CI-ProPetén (Programa CI de Guatemala), el Consejo Nacional de Áreas Protegidas (CONAP) y el Centro de Estudios Conservacionistas (CECON) de la Universidad de San Carlos de Guatemala. Este Proyecto fue posible debido a generosas donaciones de USAID para CI-ProPetén y de la Fundación Rufford para el RAP.

Esta expedición RAP fue única en su énfasis de proporcionar simultáneamente datos tanto para los sistemas acuáticos como terrestres. Para caracterizar los patrones de la diversidad de especies, muestreamos los siguientes grupos taxonómicos: plantas acuáticas, peces, fitoplancton e insectos acuáticos, reptiles acuáticos, aves acuáticas, ribereñas hormigas, murciélagos y roedores. Cada uno de estos grupos taxonómicos representan componentes importantes de biodiversidad en la región y reflejan una amplia variación ecológica que afectan la biodiversidad en general. Se encontraron contaminantes de dos especies de peces utilizando novedosas técnicas moleculares. Este componente nos permitió evaluar las amenazas debido a la contaminación en la biodiversidad del parque y en las poblaciones humanas relacionadas con el parque.

El equipo RAP examinó a los taxa dentro de las cuatro áreas focales del parque que albergan varios tipos de hábitats acuáticos y terrestres. Estas áreas incluyeron sitios cerca del la EBG junto a los Ríos San Pedro y Sacluc; dentro del Biotopo Laguna del Tigre junto a Río Escondido; cerca de la Laguna Flor de Luna y la petrolera ubicada en el pozo Xan, donde se concentraron los estudios ecotoxicológicos y finalmente en las cercanías de los Ríos Chocop y Candelaria (consulte el Mapa 3).

RESUMENES DE CAPITULOS

Limnología y calidad del agua

Los limnólogos juntamente con los botánicos proporcionaron una clasificación de primer plano en diversos macrohábitats acuáticos del PNLT. Se clasificaron a ocho tipos diferentes de hábitats representando fuentes de agua, ríos profundos, ciénagas y lagunas. Las características fisicoquímicas de estos hábitats son la base de la diversidad acuática del parque. El pH de las aguas del parque osciló de ácido (6.78) a alcalino (8.30) y sus aguas de turbias y oscuras a acuamarinas y de eutrópicas a oligotrópicas. Esta diversidad se basa en la geología cárstica común de la región. Uno de los descubrimientos más inesperados fue la presencia de un arrecife de moluscos de agua fresca en el Río San Pedro cerca del pueblo El Naranjo. Este arrecife, el cual puede servir para filtrar contaminantes del río, alberga una única mezcla de invertebrados y por esto es de un gran valor biológico. La calidad del agua del parque generalmente fue buena. Las bacterias coliformes fue detectado en un 37% de las muestras, pero la bacteria fecal coliforme se detectó en únicamente el 13% de las muestras y fueron asociados con el uso de de los cuerpos de agua de parte de los colonizadores.

La composición del fitoplancton también puede utilizarse para evaluar la calidad del agua y generalmente refleja la variación en las características fisicoquímicas de los hábitats acuáticos. Se registraron cincuenta y nueve categorías y 71 especies y morfoespecies. Las diatomeas (Bacillarioficea), seguido de clorofitas (Cloroficeae) y cianofitas (Cianoficeae) dominaron las muestras. La mayoría de taxones eran heterogéneos en sus distribuciones. Varias lagunas y puntos de muestreo en los Ríos Escondido y San Pedro albergaban especies de fitoplancton que son indicadores de contaminación (por ejemplo: *Microcystis aeruginosa* y *Synedra ulna*), pero altas densidades de estas especies se observaron sólo en dos puntos en el Río San Pedro, un sitio en el Río Escondido, y en la Laguna la Pista. Los análisis multivariables revelaron que la única fuente más importante de variación que determina la composición de fitoplancton era el pH, y el segundo lo era la conductividad. Ninguno era variable; sin embargo, ambos explican mucho sobre la variación.

En un estudio intensivo de asociación de plantas e insectos, se encontraron 44 morfoespecies y 26 familias de insectos que habitaban en una sola especie de planta acuática, *Salvinia auriculata*. Este resultado resalta la gran diversidad de insectos en los sistemas litorales acuáticos del parque.

Plantas acuáticas

Los botánicos del RAP registraron 67 familias, 120 clases y 130 especies de macrófitas acuáticas. En general, la composición de esta flora es similar a áreas vecinas en la Península de Yucatán. Dos tipos generales de vegetación, ciénagas (sibal, principalmente plantas herbáceas) y bosques ribereños (caracterizados por plantas madereras), forman un mosaico dentro del PNLT. Las ciénagas, dominadas por la hierba espesa parecida al llano, *Cladium jamaicense*, se encontró que tenían la diversidad más alta de plantas acuáticas y terrestres y parecen encontrarse en un estado constante de sucesión.

Ciperaceae y Fabaceae fueron las familias más diversas en el PNLT y *Utricularia*, una planta acuática que se alimenta de pequeños invertebrados, fue el género con más diversidad de especies. El árbol conocido como pucté (*Bucida buceras*) domina el bosque ribereño. Las asociaciones de pucté con otras especies de plantas definen cinco tipos de hábitats ribereños en el parque. Se encontraron unos cuantos individuos de manglar costero (*Rhizophora mangle*) junto al Río San Pedro, que pueden ser residuos de Pleistoceno, cuando esta área se encontraba en su mayoría bajo el océano. Esta población es la más continental en la región de Yucatán. Debido a que estos árboles se encuentran fuera del parque, puede ser que los manglares no estén protegidos de la destrucción de su hábitat. Entre los hábitats terrestres es raro encontrar el hábitat de encinal dominado por roble (*Quercus oleiodes*). El hábitat de arrecife de moluscos mencionado anteriormente alberga la rara planta acuática *Vallisneria americana*.

Peces

Se encontró que la diversidad de peces ha cambiado poco desde un estudio que se realizó hace más de sesenta años. Se registraron cuarenta y un especies, y tres de estas son registros nuevos para la Cuenca del Río San Pedro. Dos especies no habían sido descritas. En general, muchos de los taxones encontrados en el parque son endémicos a la cuenca del Usumacinta y la protección del parque ayudaría a proteger el 25% de esta fauna. El área más rica en especies fue la del Río San Pedro, pero la mayoría de las especies estaba homogéneamente distribuidas a lo largo del parque. Las lagunas, por ejemplo, albergaban 12 especies, sin tener en cuenta el disturbio en la vegetación a su alrededor. Varios hábitats son particularmente importantes para las poblaciones de peces en el PNLT. Las ciénagas de llanos bajos compuestos de *Cladium jamaicense* (sibal) son sitios de reproducción para cíclidos y pez gato , los cuales son abundantes en los Ríos San Pedro y Escondido. Un tipo de varzea o bosque anegado, hábitat cerca de Paso Caballos, es un criadero importante para *Brycon* y que generalmente es un hábitat poco conocido en Centro América. Finalmente, pequeños riachuelos son hábitats únicos pero que se ven amenazados tanto por la contaminación como por la sedimentación debido a la deforestación. La sobre pesca y la pérdida del hábitat por incendios son serias amenazas para esta fauna económica y estéticamente importante.

Toxicología

Aparte de evaluar la diversidad de peces, los miembros del equipo examinaron los efectos de compuestos tóxicos en dos especies de peces de amplia distribución: *Cichlasoma synspilum* y *Thorychthys meeki*. Se examinaron peces en forma individual en una laguna adyacente al pozo petrolero Xan 3 y a distancias variables (de 5 km a 67 km) desde la fuente de contaminantes bajo la hipótesis de que la instalación de petróleo es el origen de los compuestos tóxicos. Para caracterizar la exposición a los compuestos tóxicos, se utilizaron dos medidas celulares: Fragmentación de filamentos de ADN y división de cromosomas. Estas medidas indican la exposición a compuestos tales como hidrocarbonos aromáticos policíclicos (PAH), un contaminante asociado con la extracción de petróleo. Estas técnicas son únicas en el sentido de que éstas pueden medir el efecto de la exposición a compuestos tóxicos de forma directa, indicando así el grado al cual se ven afectadas por la contaminación las poblaciones de diferentes organismos.

Algunos peces en el sitio Xan 3 mostraron signos de deformación de aletas, lo que indica un stress en estos individuos, posiblemente debido a la exposición de contaminantes. La división de cromosomas en *C. synspilum* fue mayor en Xan 3 que en el sitio más distante (en el Río San Pedro cerca de EBG) pero no fue mayor en Xan 3 que en el sitio más cercano. Sin embargo, no se notó ninguna variación aparente en la fragmentación de ADN en *C. synspilum*. En contraste, *T. meeki* exhibió altos valores de rompimimiento de ADN en Xan 3, Laguna Buena Vista y en la Laguna Guayacán y bajos valores en el Río San Pedro y en la Laguna Flor de Luna, la cual fue el tercer sitio más cercano a Xan 3 y más cerca que Guayacán. Sin embargo, la fragmentación de cromosomas no varió en esta especie.

Las evaluaciones de la concentración de PAH en los sedimentos en donde se recolectaron los peces revelaron que las concentraciones en Xan 3 eran similares a aquéllas en los sitios de referencia. Las concentraciones de PAH eran mayores en la Laguna Buena Vista que en cualquier otro sitio. Parece ser más una fuente de contaminación que está actuando en las poblaciones de peces, pero tomará más trabajo determinar la fuente o identidad de los contaminantes. Es posible que otros tipos de contaminantes, los cuales no medimos, estén afectando las poblaciones de peces, y no está clara la forma en que estos efectos puedan relacionarse con la extracción de petróleo u otras influencias humanas. No obstante, no puede descartarse la extracción de petróleo como explicación de los efectos tóxicos. Por ejemplo, los patrones de redistribución de los contaminantes y sus efectos, puede relacionarse con las conexiones hidrológicas temporalmente variables con respecto a la temporización de la liberación de contaminantes, así como el comportamiento de las especies animales examinadas. Nuestros resultados apoyan la noción de que la contaminación sí está sucediendo en el PNLT y que es necesario conducir estudios más intensivos.

Aves

La fauna de aves del PNLT es excepcionalmente rica y y consiste de varias especies de importancia internacional. Como consecuencia, los humedales de la parte oeste del PNLT se incluyeron en la lista de Ramsar de los humedales mundialmente importantes en 1990 y 1999. Observamos a 173 especies, incluyendo 11 nuevos registros para el parque. Se han enumerado un total de 256 especies para el parque. Por primera vez en 10 años, se observó un jabirú (*Jabiru mycteria*), la cual es el ave más grande en América y es muy rara. Adicionalmente, fueron documentadas las extensiones de importante distribución para el mosquerito de cola de tenedor (*Tyrannus savanna monachus*), la codorniz de pecho gris (*Laterallus exilis*) y el mosquerito vermillón (*Pyrocephalus rubinus*). La parte oeste del parque, con sus sabanas anegadas y lagunas poco profundas (playones), contiene hábitats únicos para aves de los humedales. Adicionalmente, son sitios importantes el Río Escondido y la Laguna La Lámpara. Los especialistas de Selva incluyen, el gran guaco (*Crax rubra*), reyezuelos (Trogloditidae) y aves trepadoras de árboles (Dendrocolaptidae) como abundantes en el bosque alto de las colinas y en los bosques ribereños cercanos (galería). En general, los patrones de la topografía y la estructura de la vegetación se encuentran muy relacionados con los patrones de la variación en la composición de las aves dentro del PNLT. La conservación de los tipos de hábitat dentro del PNLT, particularmente los humedales y bosques altos, es clave para conservar su gran diversidad de aves.

Reptiles acuáticos

Los reptiles fueron investigados en aguas cercanas a lo largo del parque, con un enfoque hacia las poblaciones del cocodrilo pantanero (*Crocodylus moreletii*). Este animal se clasifica como un animal "deficiente en datos"—es decir, que no se ha estudiado lo suficiente—por la Unión Mundial para la Naturaleza (IUCN) y puede estar en peligro de extinción debido al exceso de caza y la pérdida de hábitat. En 87.14 km de la orilla de playa estudiada se observaron 130 cocodrilos. Las lagunas tenían densidades más altas que los ríos. De las lagunas, el área de Flor de Luna y La Pista tenían las densidades más altas. Las densidades combinadas en las lagunas junto al camino Xan-Flor de Luna son las más altas hasta ahora registradas en Guatemala. La Laguna El Perú y el Río Sacluc tenían una representación relativamente alta de adultos, los cuales son esenciales para el mantenimiento de las poblaciones de reproducción en el parque. De la misma manera que varias especies de peces, los cocodrilos utilizan los hábitats de ciénagas de llanos bajos (sibal) para el establecimiento de nidos, cobertura y alimento. Por lo tanto, este hábitat es particularmente importante. En general, las poblaciones de cocodrilos en el parque son saludables, y los programas de monitoreo deberían iniciarse de inmediato.

Adicionalmente, se observaron 14 especies de anfibios y 22 especies de reptiles. El anfibio más abundante fue la rana de franjas oscuras (*Leptodactylus melanonotus*) y el reptil más abundante fue el cutete rayado (*Basiliscus vittatus*), una especie endémica de Centro América. La bella, en peligro, tortuga de río americana o tortuga blanca (*Dermatemys mawii*) fue observada durante la expedición. El equipo de herpetología también capturó a la culebra vientre punteado listada (*Coniophanes quinquevittatus*), una especie que nunca ha sido documentada antes en esta área. Registros adicionales de la nota incluyen la rana de ojos rojos (*Agalychnis callidryas*), la Rana de Baudin (*Smilisca baudinii*), jicotea (*Trachemys scripta*) y la tortuga guao (*Staurotypus triporcatus*).

Mamíferos

A pesar del corto período de tiempo y de las difíciles circunstancias para el estudio de los mamíferos, se registraron 40 especies, lo que es un 30% del número potencial de especies en el PNLT. Se observaron veintitrés especies de murciélagos, dos especies de primates, tres especies de carnívoros, una especie de perisodáctilo, tres especies de artiodáctilos y ocho especies de roedores. Entre los murciélagos, fueron comunes tres especies, incluyendo a *Artibeus intermedius* (un gran murciélago que se alimenta de frutas), *Carollia brevicauda* y *A. lituratus*. Dos roedores endémicos de Yucatán: el ratón venado (*Peromyscus yucatanicus*)—un nuevo registro para Guatemala—y el pequeño ratón espinoso (*Heteromys gaumeri*). No hubo diferencias aparentes en la diversidad de mamíferos entre las cuatro localidades observadas en el parque. Eran abundantes las poblaciones de monos aulladores mexicanos (*Alouatta pigra*) y los monos araña de Geoffroy (*Ateles geoffroyi*) a lo largo de estas localidades, lo que sugiere que muchas áreas del parque se encuentran relativamente inalteradas desde el punto de vista de los mamíferos. También se observaron en el parque otras especies raras y en peligro de extinción incluyendo el tapir de Baird (*Tapirus bairdii*), pecarí de collar (*Tayassu tajacu*) y el venado rojo (*Mazama americana*). Estos se observaron cerca de la EBG junto al Río San Pedro, lo que sugiere que esta área puede ser especialmente importante para la conservación de mamíferos en el PNLT.

Hormigas

Así como para los mamíferos, el tiempo disponible y la época del año hizo difícil obtener una muestra completa de las hormigas. No obstante, se registraron 112 especies y 39 géneros. Estos valores son mayores que los que se registraron en muchos otros estudios en la región. Como en los sitios neotrópicos, el género *Pheidole* se encontraba en mayor número de especies, aunque, también fueron altamente diversos varias géneros arbóreos. Se registraron tres especies endémicas de la región de la Selva Maya, incluyendo *Sericomyrmex aztecus*, *Xenomyrmex skwarrae* y *Odontomachus yucatecus*. Un género extremadamente raro de hormiga, *Thaumatomyrmex,* fue descubierto por primera vez en Guatemala. La riqueza de especies fue la mayor en hábitats relativamente abiertos o alterados, contrario a nuestras expectativas. Esto se debió a la alta diversidad de microhábitats de estos sitios. Un hábitat alterado era relativamente similar a un sitio cercano de bosque pristino, el cual coincide con la suposición de que la recuperación de la faunas de hormigas neotrópicas es rápida. En general, la composición de las hormigas fue muy irregular y fueron muy altas las tasas de cambio de especies entre sitios, impidiendo una evaluación de las asociaciones de los hábitats y especies, sugiriendo que la riqueza en general era mucho mayor que la que se registró. Se necesitan estudios más intensivos para asegurarse cuáles hábitats tienen la mayor representación de especies raras o únicas, pero nuestros resultados preliminares sugieren que los bosques altos albergan un gran número de especies de hormigas de follaje y de hojarasca, por lo que posee un gran valor para la conservación.

UNA COMPARACION DE LOS RESULTADOS DEL RAP CON ESTUDIOS ANTERIORES

Un estudio anterior llevado a cabo por CI-ProPetén (Méndez et al., 1998) tuvo algunos de los mismos objetivos que nuestro RAP pero se enfocó en varios sitios diferentes y taxones distintos. Los taxones observados por Méndez et al. (1998) pero que no se observaron en este RAP incluyeron comunidades de mariposas y de plantas terrestres. También se estudiaron las aves y los anfibios. Méndez et al. consideró la forma en que las características topográficas varían en diferentes zonas del PNLT correspondientes a patrones de diversidad biológica en una inclinación oeste-este de relieve topográfico creciente a través del parque. Su objetivo era proporcionar prioridad a los hábitats del parque. Ellos reconocieron tres componentes geográficos del parque en esta amplia escala: 1) el área oeste del parque, lo que consiste de una matriz de follaje abierto con parches de selva de bosque alto alto, de follaje cerrado; 2) un área central que es similar al área oeste pero con sabana anegada al norte, y 3) un área este con una mayor cobertura de bosque alto y parches de selva abierta, ciénagas y bosque de roble (encinales). Méndez et al. encontró que la mayoría de especies de plantas eran especies de roble, y que este grupo contenía un gran número de especies encontradas en la parte este pero en ninguna otra parte del parque (es decir, especies "exclusivas"). La mayoría de los bosques altos en los cuales residen estos árboles se encontraban en colinas de 100 a 170 metros. Méndez et al. registró 219 especies de aves, 60 de las cuales eran de alguna manera restringidos en hábitat. Ellos encontraron que el área central tenía el mayor número de especies exclusivas de aves. La riqueza más grande de anfibios se encontró en el área central, y esto se atribuyó a la naturaleza de transición dentro del parque. De las 97 especies de mariposas que se registraron, la riqueza de las especies fue mayor en las áreas central y este, y el área este albergaba el mayor número de especies exclusivas.

Finalmente, los análisis de grupos basados en la composición de las especies revelaron que la parte este del parque agrupó hacia afuera de las áreas oeste y central del parque para todos los taxones examinados, indicando las características únicas del área este. Méndez et al. concluyó que la inclinación topográfica estaba relacionada con los patrones de la biodiversidad del PNLT. Aunque el área este, la parte topográficamente más variada del parque era la más rica en general en especies exclusivas, se le proporciona la protección más grande en el parque. El área oeste, por el otro lado, es la menos rica en especies exclusivas pero se encuentra menos protegida. Méndez et al. consideró que el área oeste es extremadamente frágil debido a la extrema vulnerabilidad de los ricos sistemas acuáticos que se encuentran allí y la amenaza impuesta por el desarrollo del petróleo a esas especies de los sistemas.

En muchos casos, los resultados del RAP coinciden con los de Méndez et al. El bosque alto y varios hábitats acuáticos presentes en el área focal de Río San Pedro que se encuentran en la parte este del PNLT fueron reconocidos por su importancia para los grupos tales como peces, mamíferos y hormigas. Los hábitats de humedales y lagunas encontrados en las partes central y oeste del parque (es decir, las áreas focales de Río Escondido y Flor de Luna), sin embargo, también se reconocieron como esenciales para la conservación de aves acuáticas y el cocodrilo pantanero. Adicionalmente, algunas áreas que no se encuentran actualmente localizadas dentro de los límites del parque deberían considerarse en el manejo de la región. En general, el valor de los hábitats en lo particular depende de los taxones considerados. Visto de esta manera, y considerando la condición del PNLT como un área núcleo dentro de la RBM, hay justificación para la protección de todos los hábitats del parque.

RECOMENDACIONES DEL ESTUDIO Y DE CONSERVACION

Nuestras observaciones nos hacen creer que el PNLT es un área núcleo crítica dentro de la RBM. Es una reserva importante de hábitats de bosque seco y de humedales que albergan las comunidades bióticas y especies individuales de gran valor conservacionista regional y global. Nuestros estudios en el PNLT reflejan las necesidades de información específicas que sólo podrán ser obtenidas por estudios más intensivos en el parque. No obstante, podemos ofrecer varias recomendaciones específicas para el manejo del parque. La base para las siguientes recomendaciones puede encontrarse en el texto del resumen ejecutivo y en los siguientes capítulos. Las recomendaciones no se enumeran en orden de prioridad. Primero proporcionamos recomendaciones para una futura investigación científica en el parque, haciendo notar que la EBG está en una ubicación ideal desde la cual se pueden realizar estas investigaciones luego vienen las recomendaciones de conservación.

Prioridades de investigación científica:

*   Investigación adicional determinando las relaciones hidrológicas a lo largo del parque y sus dinámicas estacionales que serán un primer paso necesario para generar un entendimiento de los sistemas acuáticos del parque, incluyendo el flujo de los contaminantes y de las dinámicas de dispersión y población de las especies.

*   Considerando la intensa dinámica estacional de las áreas de bosque seco tropical tal como el PNLT, deben llevarse a cabo estudios tanto de sistemas terrestres y acuáticos, en diferentes épocas del año, particularmente durante la estación húmeda cuando los sistemas acuáticos están mejor conectados y cuando se están reproduciendo muchos organismos terrestres.

*   Se necesita proporcionar estudios más detallados de las relaciones entre la topografía, suelos (particularmente la profundidad del suelo), y las comunidades de plantas y animales para proporcionar observaciones a escala más fina de patrones de biodiversidad para propósitos de orden por prioridad. Al considerar el alto grado de incursión de los colonizadores en el parque, dichos estudios ayudarán a la asignación óptima de protección para hábitats en las áreas invadidas.

Recomendaciones específicas de conservación para el PNLT:

- Reducir la tasa de deforestación y de quema de los hábitats tales como ciénagas de llanos bajos (sibales), con la meta primordial de parar la deforestación y de liberar completamente la quema dentro del PNLT. La integridad ecológica del PNLT, como un área núcleo de la RBM, debe mantenerse si la RBM va a persistir y prosperar.

- Inmediatamente conservar áreas determinadas del bosque alto primario, no alterado. Deben mantenerse los niveles actuales de protección asignados a las áreas cerca de la EBLG, junto al Río San Pedro y las áreas boscosas más allá de esta parte del área protegida, tal como El Perú o cerca de la Laguna Guayacán, las cuales deben ser monitoreadas cuidadosamente. La alteración humana de los bosques del área del Río San Pedro debería minimizarse para favorecer a las grandes poblaciones de mamíferos.

- Proteger varios hábitats únicos dentro del parque, y para propósitos de manejo, incluir dentro del parque otros hábitats terrestres y acuáticos que ahora se encuentran fuera del parque. Estos incluyen los bosques anegados tipo varzea cerca de Paso Caballos y junto al Río Sacluc, los hábitats de la ciénaga de llanos bajos (sibal) junto al margen sur del Río San Pedro y junto al Río Escondido, el sistema de arrecifes de moluscos en el área de los "rápidos" en el Río San Pedro cerca de El Naranjo, el manglar restante junto al Río San Pedro, y la vegetación de encinos junto al Río Sacluc. Es de particular importancia la ampliación de los canales al hábitat de la ciénaga de llanos bajos junto al Río Escondido, asociado con el pasaje de lanchas que debea ser controlado.

- Iniciar un programa de monitoreo de calidad de agua dentro del parque. Son de particular interés las áreas que utilizan las poblaciones humanas, incluyendo las fuentes de los tributarios junto al Río San Pedro y a las lagunas en el camino de Xan-Flor de Luna. De forma específica, deberían examinarse los niveles y efectos de la contaminación y la actividad agrícola.

- Investigar los determinantes de stress y de genotoxicidad en los peces—revelada en esta evaluación—en Xan 3 y en otras lagunas. Aún quedan varias preguntas sin responder: ¿Qué contaminantes están involucrados?, ¿Cuáles son las fuentes de estos contaminantes?, ¿Cuál es el rol de la hidrología en determinar la redistribución de los contaminantes?, ¿De qué forma el comportamiento de las especies

- Monitorear las poblaciones de especies dentro del parque para evaluar las tendencias en su salud. De particular preocupación lo son aquellas poblaciones que son pequeñas y/o explotadas por los humanos. Específicamente, deberían monitorearse las dinámicas de los peces que se pueden consumir tales como *Petenia splendida* y *Brycon guatemalensis* (especialmente en el fuertemente explotado Río San Pedro). Un programa de conservación y de monitoreo para cocodrilos (*Crocodylus moreletii*) debería enfocarse en sitios en donde actualmente son abundantes, incluyendo el Río Sacluc, la Laguna el Perú y la Laguna la Pista. Finalmente, las poblaciones de aves deberían monitorearse, especialmente en los hábitats acuáticos en las partes central y oeste del parque debido a que esta área es de importancia internacional así como las amenazas impuestas a estos hábitats debido a la extracción del petróleo y a la colonización humana.

La condición, las amenazas y las acciones conservacionistas que recomendamos se resumen por tipo de hábitat en la Tabla i. Esperamos que estas recomendaciones proporcionen alguna guía en el manejo del PNLT y motivamos a los grupos interesados para que contacten a los autores de este informe si desean clarificación adicional.

LITERATURA CITADA

Méndez, C., C. Barrientos, F. Castañeda y R. Rodas. 1998. *Programa de Monitoreo, Unidad de Manejo Laguna del Tigre: Los Estudios Base para su Establecimiento.* CI-ProPetén, Conservation International, Washington, DC, USA.

Tabla i. Resumen de la importancia de conservación, amenazas, condición general del hábitat y recomendaciones de manejo para los principales tipos de hábitat del Parque Nacional Laguna del Tigre y el Biotopo

Tipos de hábitat de importancia	Ubicación (es)	Importancia para la conservación	Amenazas	Condición del hábitat	Recomendaciones
Ríos, tributarios, y caños	Río San Pedro/ Sacluc y Río Escondido	Gran riqueza de plantas acuáticas, peces	Contaminación, especialmente de tributarios, exceso de pesca	Variable, más deficiente cerca de los asentamientos	Monitorear, regular las prácticas de desperdicio humano y presión de la pesca
Rápidos/ Arrecifes de agua dulce	Río San Pedro	Único sistema de arrecife de agua dulce de moluscos bivalvos y diversidad asociada	Contaminación, exceso de pesca de moluscos	Aún bueno	Incluir en el PNLT, el monitoreo y la regulación de pesca de moluscos y la contaminación del río
Lagunas	Flor de Luna, Río Chocop	Altas densidades de cocodrilos pantanero, aves acuáticas únicas	Contaminación, contaminación por petróleo	Variable, una laguna cerca de un pozo petrolero que puede estar contaminado	Prohibir el uso humano de ciertas lagunas, monitorear los niveles de contaminación
Lagunas de recodo	Río Escondido	Gran riqueza de plantas acuáticas	Erosión debido al paso de las lanchas en los canales	Variable	Regular el uso de lanchas y posiblemente el tamaño o la velocidad de la lancha en el parque
Ciénagas de llanos cortos (sibal)	Todos	Gran riqueza de aves acuáticas, criaderos para especies de peces	Quema	Variable, muchos daños por los incendios	Incluir más tierras de ciénagas en el parque, evitar la esparción de los incendios en las ciénagas, regular la quema en el parque
Bosque tipo varzea	Río San Pedro/Sacluc	Criaderos para especies de peces	Colonizaciones	Desconocido	Incluirlo en el PNLT, se necesitan más estudios
Bosque alto	Todos	Gran riqueza de hormigas, mamíferos	Deforestación, quema	Variable, muchos daños por los incendios	Quema controlada y proteger parches relativamente primitivos
Encinal	Río Sacluc	Plantas únicas, al igual que animales únicos	Deforestación, fragmentación	Desconocido	Incluir áreas junto al Río Sacluc en el PNLT

CAPITULO 1

INTRODUCCION A LA EXPEDICION DEL RAP AL PARQUE NACIONAL LAGUNA DEL TIGRE, PETEN, GUATEMALA

Brandon T. Bestelmeyer

PARQUE NACIONAL LAGUNA DEL TIGRE, PETEN, GUATEMALA

El Parque Nacional Laguna del Tigre (PNLT) es una zona núcleo de la Reserva de la Biósfera Maya (RBM), y las amenazas que enfrenta reflejan la necesidad tanto de investigación como de esfuerzos de conservación en el parque. El concepto de reservas de biósfera mantiene que la conservación y el desarrollo pueden equilibrarse por un diseño y manejo cuidadoso de las redes de reservas. La función de las áreas núcleo, al contrario de las zonas extractivas o de amortiguamiento en las cuales puede ocurrir la actividad económica humana, es preservar a los ecosistemas representativos mínimamente afectados que sirven como una meta para esfuerzos regionales de monitoreo y que protegen poblaciones silvestres. De acuerdo con Batisse (1986), "La protección contra cualquier acción que pueda poner en peligro el rol de conservación asignado al área núcleo debe... asegurarse completamente". El objetivo de este RAP es evaluar el valor del PNLT como un área núcleo de la RBM y evaluar las amenazas para determinar su rol como área núcleo.

El parque es parte de la Selva Maya, la cual contiene los ecosistemas de bosque tropical restantes de Centro América. El bosque se expande de la parte sur de México (especialmente los estados de Chiapas, Campeche y Quintana Roo) hasta la parte norte de Guatemala y Belice formando una unidad natural biogeográfica y cultural (Nations et al., 1998). La RBM, de 16,000 km², se encuentra en el departamento de El Petén en la parte norte de Guatemala contiene el sistema más grande de áreas manejadas y protegidas en la Selva Maya. El PNLT, con un área de 338,566 (incluye Biotopo) hectáreas es el área núcleo más grande e importante dentro de esta reserva.

Sin importar su condición como áreas protegida, el parque se encuentra muy amenazado por la colonización y el desarrollo y se deforesta a una tasa de 0.57% al año (datos de 1995 a 1997; Sader, 1999). Como consecuencia el 50% de la selva se ha perdido y una de las estimaciones de las proyecciones es que el 98% de la selva original en El Petén desaparecerá para el año 2010 si las tasas actuales de deforestación no se reducen (Canteo, 1996).

CARACTERISTICAS ECOLOGICAS DEL PNLT

El parque se encuentra ubicado dentro de una zona caracterizada como bosque seco tropical (Murphy y Lugo, 1995; Campbell, 1998). Debido a que los bosques húmedos tropicales han recibido mucha atención de parte de los ecologistas y conservacionistas, los ecosistemas de bosque seco tropical—bosques secos neotrópicos en particular— son menos estudiados, utilizados de forma intensiva por humanos y extremadamente en peligro por el desarrollo a nivel mundial (Mooney et al., 1995; Gentry, 1995). Por esta razón, la Selva Maya tiene importancia global.

La topografía del área es baja y plana, con las elevaciones máximas sin exceder los 300 m. El área está compuesta de piedra caliza de la edad Miocena, realzando el paisaje cárstico junto a los cursos de los grandes ríos, colinas bien drenadas con suelos superficiales rodeados por áreas bajas con suelos profundos (conocido como bajos) y unos pocos riachuelos pero numerosos agujeros de agua que surgen de las depresiones de la piedra caliza (Wallace, 1997). Los hundimientos menores de piedra caliza son conocidos como aguadas y los más grandes como akalchés o lagunas (Campbell, 1998). Se cree que los humedales de la parte oeste aún permanecen en Centro América (Wallace, 1997). Estos humedales también son los más grandes de Guatemala y son reconocidos bajo la convención de Ramsar por su importancia internacional.

Como en todos los ecosistemas de bosque seco tropical, el contraste entre las estaciones secas y húmedas es la característica ecológica temporal de mayor importancia (Murphy y Lugo, 1995). La precipitación anual promedio de la región es de 1600 mm por año, pero pueden ocurrir precipitaciones ocasionales en la estación seca de enero a abril, cuando las temperaturas pueden exceder 40° C. Este período representa un reto importante para muchos de los organismos del trópico húmedo en el parque. Adicionalmente, la estacionalidad en la precipitación interactúa con la topografía para producir variaciones extremas en condiciones ambientales. Por ejemplo, los suelos bajos pueden estar inundados durante la estación húmeda pero extremadamente secos durante la estación seca (Whitacre, 1998). El agua persiste en grandes lagunas a lo largo del año (Campbell, 1998) y por lo tanto son características del paisaje y ecología del parque.

Posiblemente el eje más importante de la variación ambiental terrestre natural es entre los bosques bajos inundados estacionalmente y los llanos altos bien drenados ubicados en barrancos de piedra caliza. Whitacre (1998) reporta que tanto la composición arbórea como la composición de aves varía dramáticamente entre estas condiciones topográficas, de tal manera que áreas con topografía similar separadas por varias millas son mucho más parecidas que áreas de topografía diferente tan sólo a cientos de millas de distancia. Los árboles de bosques altos tienden a ser más altos con abundante hojarasca y poca maleza. Los árboles de bosques bajos tienden a ser más bajos (15 m a 20 m) y poseen mayor densidad de arbustos, palmas y hierba. Esta variación tiende a ser una fuerte influencia en los patrones de la biodiversidad en la región (Méndez et al., 1998).

Los sistemas acuáticos dentro del parque son menos entendidos que los sistemas terrestres. El curso de agua más importante en el parque es el Río San Pedro, el cual es un tributario mayor del Río Usumacinta. El Río Sacluc es un tributario del Río San Pedro y drena áreas al sur este del parque. El Río Escondido es otro tributario del Río San Pedro que drena áreas dentro y alrededor del Biotopo de la Laguna del Tigre en la parte oeste del PNLT. El Río Chocop drena el centro del parque al Río San Pedro y el Río Candelaria es un afluente independiente de la Laguna de Términos. Se desconocen las relaciones hidrológicas entre estos ríos y las muchas lagunas—especialmente uniones subterráneas, las cuales pueden ser un componente importante en paisajes cársticos. Adicionalmente, existen algunas uniones estacionalmente variables similares entre los cuerpos de agua con conexiones en la estación húmeda que no son aparentes en la estación seca, durante la cual se llevó a cabo nuestro muestreo. Dichas uniones deberían tener consecuencias importantes tanto para dinámicas de dispersión de poblaciones animales y vegetales así como para la dispersión de contaminantes. Estas características

deberían ser una prioridad importante para una investigación futura en el parque.

Los cambios históricos en la estructura climática y de la vegetación han dejado también una impresión importante en la ecología de los bosques de El Petén. La región ha estado sujeta a mucha inestabilidad climática a lo largo del Holoceno. Por ejemplo, Leyden (1984) relaciona que la vegetación en la región estaba dominada por formaciones de pino y cedro a la altura del último período glacial y que un período y de mayor aridez ocurrió hace casi 2200 a 1140 años atrás. Los cambios más recientes a la vegetación son asociados con el surgimiento de la cultura Maya. Durante la parte final del período Clásico (600 a 850 A.D.), los Mayas alcanzaron densidades máximas de hasta 200 personas por km^2 (Culbert y Rice, 1990), y esto sugiere que la conversión extensiva agrícola del paisaje fue necesaria para sustentarse. Como consecuencia, los bosques tropicales que observamos hoy día probablemente no tienen más de 1000 años de antigüedad (Méndez, 1999). El entendimiento de los patrones ecológicos actuales dentro de la Selva Maya requiere que conozcamos la influencia de estos procesos históricos.

HISTORIA DEL PNLT

El PNLT es administrado por el Consejo Nacional de Áreas Protegidas (CONAP), la agencia del gobierno de Guatemala a cargo de los recursos naturales. CONAP es el ente responsable de proteger y conservar el patrimonio natural de Guatemala, tareas que logra a través de promover el uso sustentable de los recursos naturales y a través de conservar las poblaciones de vida silvestre y los ecosistemas a lo largo del país. El Biotopo de la Laguna del Tigre, un área núcleo localizada dentro del parque, es administrada por la Universidad de San Carlos de Guatemala a través del centro local de conservación e investigación , Centro de Estudios Conservacionistas (CECON).

Ponciano (1998) relata la siguiente historia del biotopo y del parque. El CONAP fue fundado en 1989 y una de sus primeras acciones fue incorporar los ya establecidos Parque Nacional Tikal, el Biotopo Laguna del Tigre, el Biotopo El Zotz y el Biotopo Dos Lagunas a la RBM. Adicionalmente, fueron creados los parques nacionales Laguna del Tigre y Sierra del Lacandón. La creación de la RBM fue catalizada por dos eventos: 1) la propuesta de la red de reservas de la "Ruta Maya" para la conservación y el turismo en un artículo de National Geographic, y 2) la amenaza impuesta por la migración y el desarrollo debido a un proyecto de desarrollo de una carretera que conecta a Flores con Cadenas en el sur. El gobierno alemán estaba financiando este proyecto, pero basado en una evaluación de impacto ambiental, decidió suspender los fondos hasta que se creó la reserva. CONAP pudo crear la reserva debido a que tenía un fuerte apoyo político y debido a que en esta época existía un gobierno democrático. La reserva tenía como objetivo albergar estrictamente áreas núcleo protegidas (parques nacionales y biotopos), zonas de usos múltiples (por ejemplo, extractivas) entre estas zonas núcleos y una zona de amortiguamiento en la cual pueden aparecer tierras privadas (Gretzinger, 1998). El parque fue registrado como un sitio de humedales importantes bajo la convención de Ramsar en 1990 y 1999.

RETOS DE CONSERVACION PARA EL PNLT

El crecimiento y la migración a Petén ha complicado el manejo sustentable y la conservación de la biodiversidad. El crecimiento de la población se estima en una tasa anual del 7% al 10% , principalmente debido a la migración de agricultores afectados por la pobreza en el sur del país (Nations et al., 1998), así como la repatriación de los ciudadanos guatemaltecos refugiados en México quienes huyeron del país durante la guerra civil que duró 30 años, la cual terminó en 1996 (Ponciano, 1998). Como consecuencia de estas presiones, la colonización sucedió dentro de las áreas núcleo y ha probado ser difícil de manejar a lo largo de la reserva. Los recientes inmigrantes han traído consigo las prácticas agrícolas (técnicas de corte y quema a gran escala) de las tierras altas, que no son sostenibles en el contexto de los bosques de tierras bajas. La tala no regulada e ilegal de especies maderables valiosas y otros productos amenazan con la disminución de la capacidad del bosque de dar apoyo a las poblaciones humanas a través de prácticas no destructivas.

En situaciones similares en otros lugares, se ha necesitado que los derechos de la tenencia de la tierra y las concesiones de recursos naturales sean otorgados a nivel de la comunidad, que existia fortalecimiento de las políticas de manejo y que hallan mercados ampliados para productos forestales con el objeto de frenar la pérdida del bosque. Sin embargo, ha sido muy difícil implementar las actividades regulatorias en El Petén. La extracción manejada de árboles comercialmente valiosos tales como la caoba (*Swietenia macrophylla*) y el cedro español (*Cedrela odorata*), así como los productos forestales no maderables tales como el chicle (utilizado en goma de mascar), xate (un helecho utilizado en arreglos florales) y potpurrí, derivado de productos forestales, pueden proporcionar ingresos a los peteneros para el establecimiento de un bosque semi intacto. Los estudios de Whigham et al. (1998) de aves en Quintana Roo sugieren que la tala de madera selectiva no afecta a las comunidades de forma adversa (consulte también a Whitacre, 1998). Esto puede ser en parte, debido a las adaptaciones de las aves a disturbios periódicos naturales en la región, tal como aquéllos causados por huracanes. Adicionalmente, el manejo de las prácticas de limpieza y plantación en los

paisajes agrícolas (Warkentin et al., 1995) junto con la preservación de áreas intactas de bosque antiguo fuera de los ciclos de la agricultura (Whitacre, 1998) pueden ayudar a preservar la biodiversidad de Petén. Estas alternativas de manejo deben explorarse además para desarrollar la estrategia prioritaria para conservar la biodiversidad de Petén.

Un reto adicional de desarrollo surge del hecho de que 40% de la reserva se traslapa con áreas definidas con potencial de desarrollo petrolero (Ponciano, 1998). Basic Petroleum Inc. tenía derechos para desarrollar operaciones petroleras dentro del parque antes de la creación de la biósfera en 1990. Rosenfeld (1999) reporta que a Basic se le han otorgado los derechos de exploración para la búsqueda de petróleo en aproximadamente 55% del área dentro del PNLT. Se desconocen los efectos directos de las operaciones petroleras de Basic en el parque, pero queda claro que la creación de los caminos a áreas que no han sido colonizadas con anterioridad han exacerbado la tala de bosques dentro del parque (Rosenfeld, 1999). Los acuerdos entre los intereses de conservación y el Ministerio de Energía y Minas acerca de la forma de cómo proceder con el desarrollo todavía no han sido determinados.

En respuesta a las amenazas impuestas por el desarrollo petrolero y la colonización humana dentro del parque, se inició un programa más grande de investigación y monitoreo en el PNLT de parte de CI-ProPetén en colaboración con el CONAP y el CECON (Méndez et al., 1998). Este estudio documentó por primera vez la gran diversidad de plantas, mariposas diurnas, reptiles, anfibios, y aves dentro del parque. El estudio de la vegetación llevado a cabo en este estudio es la base el terreno para el esfuerzo del estudio terrestre presentado aquí, y los datos de Méndez et al. (1998) como conjunto, complementan la información de flora y fauna acuática y los datos terrestres de hormigas y mamíferos. Se espera que nuestros esfuerzos, en combinación con este estudio anterior, ayuden a llamar la atención al valor de la biodiversidad del PNLT, proporcionen dirección e ímpetu para estudios más intensivos en la región, y ayuden a los administradores de las tierras en la conservación y el manejo de la biodiversidad del PNLT.

DISEÑO GENERAL DEL ESTUDIO

Llevamos a cabo este estudio al final de la estación seca para facilitar la movilidad a través del parque y para muestrear mejor ciertos grupos tales como peces y aves acuáticas. El muestreo se concentró alrededor de cuatro áreas focales (ver el Mapa 3). El primero se enfocó en la Estación Biológica Las Guacamayas, la cual se localiza en la confluencia entre el Río San Pedro y el Río Sacluc. El muestreo se llevó a cabo cerca de las fuentes de agua junto al Río San Pedro y en las partes cercanas del Río Sacluc. La segunda área focal estaba localizada junto al Río Escondido al oeste, el cual fluye en el

Río San Pedro desde el norte y pasa a través del Biotopo de la Laguna del Tigre. La tercera área focal se centró en un sistema de lagunas cerca de la esquina noreste del Biotopo de la Laguna del Tigre. Estas lagunas se encuentran cerca de los campos petroleros de Xan que actualmente operan dentro del parque. Los estudios toxicológicos se enfocaron aquí en lagunas localizadas a varias distancias de los pozos de petróleo de Xan. La cuarta área se centraba en el Río Chocop, la cual incluye otras lagunas y el Río Candelaria. En conjunto, estas áreas focales representan un espacio de hábitats que se encuentran dentro del parque, incluyendo áreas que están amenazadas por la contaminación y el desarrollo agrícola.

Dentro de cada área focal, se colectó en varios puntos de muestreo (o puntos de georeferencia) que representan diferentes macrohábitats que se encuentran en el área. Las muestras fueron recolectadas por cada grupo dentro de unos pocos cientos de metros de un punto de muestreo. Los puntos de muestreo generalmente ocurren a solo kilómetros del punto que define el centro de cada una de las áreas focales; este es el punto focal (consulte los asteriscos en el Apéndice 1 y el Apéndice 2). Este diseño permite análisis jerárquicos de relaciones entre los taxones en dos escalas espaciales.

Aquí se utiliza una clasificación de macrohábitats acuáticos y terrestres como el fundamento de las comparaciones entre taxones y entre regiones dentro del parque. Estos tipos de macrohábitats se describen a continuación.

Macrohábitats acuáticos

Existen pocos estudios de hábitats dentro del parque. En este estudio muestreamos hábitats acuáticos y organismos acuáticos entre tres a doce puntos de muestreo en cada área focal. El Apéndice 1 presenta una clasificación preliminar de macrohábitats acuáticos por los miembros del equipo de RAP. Las discusiones de las características de estos macrohábitats pueden encontrarse en los Capítulos 2 y 3, y se resumen en la Tabla 1.1.

Macrohábitats terrestres

Los tipos de macrohábitat terrestre se basan en las clasificaciones de vegetación reconocidas por Méndez et al. (1998). Ecologos terrestres examinaron entre uno y tres puntos de muestreo dentro de cada área focal (Apéndice 2). En la Tabla 1.2 pueden encontrarse descripciones detalladas de los diferentes macrohábitats terrestres.

Notese que fué posible muestrear los grupos acuáticos en muchos más puntos que los grupos terrestres. Esto se debe a las diferencias en el esfuerzo que se necesita para muestrear de forma adecuada los diferentes tipos de organismos en cada punto.

Tabla 1.1. Descripciones de macrohábitats acuáticos en el Parque Nacional Laguna del Tigre, clasificados por primera vez en este estudio.

Tipo de macrohábitat	Características
Sistemas fluviales (ríos)	**Secciones acuáticas**
Bahías de ríos (bahías)	Porciones de agua de movimiento lento junto a las orillas de los ríos.
Ríos permanentes (permanentes)	Partes profundas de los ríos que nunca se secan, usualmente con lecho compuesto de piedra caliza.
Ríos rápidos (rápidos)	Partes de los ríos de movimientos rápidos, frecuentemente con substratos rocosos, y en el caso de San Pedro, arrecifes conformados de bivalvos.
Ríos estacionales (someros)	Ríos pequeños o estacionales o estrechos de ríos que son estacionalmente secos. Estos pueden tener una variedad de substratos y partes más profundas que pueden contener agua una vez al año pero pueden estar muy estancados en la estación seca.
Sistemas no fluviales	
Tributarios (tributarios, riachuelos)	Los pequeños riachuelos que alimentan a los ríos frecuentemente se alimentan de las fuentes. Aguas de movimientos rápidos altamente oxigenadas.
Caños	Ambientes lóticos angostos conectados a los ríos desde las ciénagas. Las aguas son turbias casi estancadas.
Lagunas en recodos (lagunas meándricas)	Lagunas formadas por recodos definidos de ríos que se desprenden de los cursos de los ríos y existen como lagunas aisladas a lo largo de éste.
Lagunas	Las lagunas formadas por depresiones en piedra caliza, también conocidas como alkachés o aguadas. Las aguas pueden ser claras o turbias.

ELABORACION DE INFORMES DE LOS RESULTADOS

En los siguientes capítulos que sigue, se proporcionan los resultados para cada taxón o sub estudio de RAP como manuscritos independientes. Cada estudio y la presentación resultante necesariamente emplean una organización levemente diferente, dependiendo de la dificultad en divulgar y relacionar información acerca de los taxones focales. Sin embargo, varias características analíticas se comparten en todos los capítulos basados en los taxones para evaluar la diversidad de un grupo. Algunas o todas estas medidas pueden utilizarse en un capítulo en lo particular, dependiendo de la utilidad dada a los datos disponibles. Primero, la riqueza de las especies se mide como el número de especies. Segundo, la diversidad de las especies (en el sentido estricto) se mide como la distribución de abundancia entre las especies. Una distribución más pareja entre las especies y un número mayor de especies generalmente lleva a valores de diversidad más altos. Tercero, de forma similar, la composición de las especies en un sitio, se mide utilizando los índices de similitud o utilizando las técnicas de ordenamiento o agrupamiento. Dos sitios que son diferentes en la abundancia de especies, tienen una mayor diversidad combinada que los dos hábitats que son relativamente más parecidos. Esto se refleja en los diagramas de agrupamiento y ordenamiento por las distancias que se encuentran entre las muestras en el diagrama.

Y finalmente, utilizamos la presencia y abundancia de especies global o regionalmente raras o endémicas, tales como las que son protegidas por acuerdos internacionales incluyendo los apéndices de la Convención Internacional de Tráfico de Especies de Flora y Fauna Silvestre en Peligro (CITES)y la Lista roja de la Unión Mundial para la Naturaleza (UICN). Esto requiere de información en la distribución a gran escala los taxa y puede ser que no se encuentre disponible para ciertos grupos que no se han estudiado lo suficiente tales como el plancton.

Tabla 1.2. Descripciones de macrohábitats terrestres examinados durante la expedición RAP al Parque Nacional Laguna del Tigre (basado en Méndez et al., 1998).

Tipo de macrohábitat	Características
Bosque alto	Caracterizado por una maleza relativamente cerrada un llano escaso dominado por palmas y arbustos *Trepadores*. Asociaciones que incluyen al *ramonal*, dominados por el árbol *Brosimum alicastrum*. El bosque alto ocupa un área de 19,354 hectáreas (5.72% del parque).
Bosque bajo	El Bosque bajo con un follaje más denso debido a inundaciones periódicas. Dominado por árboles tales como pucté y tinto, mezclado con roblehicpo (*Coccoloba* sp.) y cojché (*Nectandra membranacea*). También son dominantes las palmas incluyendo a botán (*Sabal morisiana*) y escoba (*Crysophila argentea*).
Encinal	Residuos de un bosque dominado por roble (*Quercus oleoides*) existen cerca de la confluencia de los ríos San Pedro y Sacluc. Los árboles asociados incluyen el pucté (*Bucida buceras*) y el tinto (*Haematoxylum campechianum*) en áreas anegadas. Este tipo ocupa un área de 2,367 hectáreas (0.70% del parque).
Prado	Áreas que recientemente han sido limpiadas por corte y quema y han sido colonizadas con llanos.
Guamil	Bosques en regeneración (secundarios) con llanos densos, algunas veces con falta de árboles tupidos en su totalidad. Incluye guarumo (*Cecropia petata*), acacia (*Acacia cornifera*), bayal (*Desmoncus* sp.), y palmas con elementos característicos.
Bosques fluviales	Parece como un mosaico de parches de bosque alto y bosque bajo en áreas que generalmente se inundan en la estación húmeda. La asociación más abundante es el pucté y la *Pachira* sp. Consulte también la sección de macrófitas acuáticos.
Tintal	Una clase de bosque bajo que está casi completamente compuesto por tinto (*Haematoxylum campechianum*), frecuentemente se encuentra cerca de los márgenes de las lagunas.

LITERATURA CITADA

Batisse, M. 1986. Developing and focusing the biosphere reserve concept. Nature and Resources 22: 2-11.

Campbell, J. A. 1998. *Amphibians and Reptiles of Northern Guatemala, the Yucatán, and Belize*. University of Oklahoma Press, Norman, OK, USA.

Canteo, C. 1996. Destrucción de Biósfera Maya avanza año con año. Siglo Veintiuno, November 21. Guatemala City, Guatemala.

Culbert, T. P. y D. S. Rice, eds. 1990. *Precolumbian Population History in the Maya Lowlands*. University of New Mexico Press, Albuquerque, NM, USA.

Gentry, A. H. 1995. Diversity and floristic composition of Neotropical dry forests. Pp. 146-190 in *Seasonally Dry Tropical Forests* (S. H. Bullock, H. A. Mooney, and E. Medina, eds.). Cambridge University Press, New York, NY, USA.

Gretzinger, S. P. 1998. Community-forest concessions: an economic alternative for the Maya Biosphere Reserve in the Petén, Guatemala. Pp. 111-124 in *Timber, Tourists, and Temples: Conservation and Development in the Maya Forest of Belize, Guatemala, and Mexico* (R. B. Primack, D. Bray, H. A. Galletti, and I. Ponciano, eds.). Island Press, Washington, DC, USA.

Leyden, B. W. 1984. Guatemalan forest synthesis after Pleistocene aridity. Proceedings of the National Academy of Sciences 81:4856-4859.

Méndez, C. 1999. How old is the Petén tropical forest? Pp. 31-34 in *Thirteen Ways of Looking at a Tropical Forest: Guatemala's Maya Biosphere Reserve* (J. D. Nations, ed.). Conservation International, Washington, DC, USA.

Méndez, C., C. Barrientos, F. Castañeda, and R. Rodas. 1998. *Programa de Monitoreo, Unidad de Manejo Laguna del Tigre: Los Estudios Base para su Establecimiento*. CI-ProPetén, Conservation International, Washington, DC, USA.

Mooney, H. A., S. H. Bullock, and E. Medina. 1995. Introduction. Pp. 1-8 in *Seasonally Dry Tropical Forests* (S. H. Bullock, H. A. Mooney, and E. Medina, eds). Cambridge University Press, New York, NY, USA.

Murphy, P. G. y A. E. Lugo. 1995. Dry forests of Central America and the Caribbean. Pp. 9-34 in *Seasonally Dry Tropical Forests* (S. H. Bullock, H. A. Mooney, and E. Medina, eds.). Cambridge University Press, New York, NY, USA.

Nations, J. D., R. B. Primack y D. Bray. 1998. Introduction: the Maya forest. Pp. xiii-xx in *Timber, Tourists, and Temples: Conservation and Development in the Maya Forest of Belize, Guatemala, and Mexico* (R. B. Primack, D. Bray, H. A. Galletti, and I. Ponciano, eds.). Island Press, Washington, DC, USA.

Ponciano, I. 1998. Forestry policy and protected areas in the Petén, Guatemala. Pp. 99-110 in *Timber, Tourists, and Temples: Conservation and Development in the Maya Forest of Belize, Guatemala, and Mexico* (R. B. Primack, D. Bray, H. A. Galletti, and I. Ponciano, eds.). Island Press, Washington, DC, USA.

Rosenfeld, A. 1999. Oil exploration in the forest. Pp. 68-76 in *Thirteen Ways of Looking at a Tropical Forest: Guatemala's Maya Biosphere Reserve* (J. D. Nations, ed.). Conservation International, Washington, DC, USA.

Sader, S. 1999. Deforestation trends in northern Guatemala: a view from space. Pp. 26-30 in *Thirteen Ways of Looking at a Tropical Forest: Guatemala's Maya Biosphere Reserve* (J. D. Nations, ed.). Conservation International, Washington, DC, USA.

Wallace, D. R. 1997. Central American landscapes. Pp. 72-96 in *Central America: A Cultural and Natural History* (A. G. Coates, ed.). Yale University Press, New Haven, CT, USA.

Warkentin, I. G., R. Greenberg y J. Salgado Ortiz. 1995. Songbird use of gallery woodlands in recently cleared and older settled landscapes of the Selva Lacandona, Chiapas, Mexico. Conservation Biology 9: 1095-1106.

Whigham, D. F., J. F. Lynch y M. B. Dickinson. 1998. Dynamics and ecology of natural and managed forests in Quintana Roo, Mexico. Pp. 267-282 in *Timber, Tourists, and Temples: Conservation and Development in the Maya Forest of Belize, Guatemala, and Mexico* (R. B. Primack, D. Bray, H. A. Galletti, and I. Ponciano, eds.). Island Press, Washington, DC, USA.

Whitacre, D. F. 1998. The Peregrine Fund's Maya Project: ecological research, habitat conservation, and development of human resources in the Maya forest. Pp. 241-266 in *Timber, Tourists, and Temples: Conservation and Development in the Maya Forest of Belize, Guatemala, and Mexico* (R. B. Primack, D. Bray, H. A. Galletti, and I. Ponciano, eds.). Island Press, Washington, DC, USA.

CAPITULO 2

LOS HABITATS ACUATICOS DEL PARQUE NACIONAL LAGUNA DEL TIGRE, PETEN, GUATEMALA: CALIDAD DEL AGUA, POBLACIONES DE FITOPLANCTON E INSECTOS ASOCIADOS CON LA PLANTA *SALVINIA AURICULATA*

Karin Herrera, Ana Cristina Bailey, Marcos Callisto y Jorge Ordóñez

RESUMEN

- Los cuerpos de agua en el Parque Nacional Laguna del Tigre (PNLT) fueron clasificados de acuerdo con ocho tipos de macrohábitat (refiérase a la Tabla 1.1). Las propiedades físico-químicas, calidad de agua, flora de fitoplancton e insectos asociados con la planta acuática *Salvinia auriculata* se reportan por cada cuerpo de agua examinado.

- El pH de las aguas del parque oscilaba de ácido (6.78) a alcalino (8.30), los cuerpos de agua de turbios y oscuros a acuamarinos y de eutrópicos a oligotrópicos. La calidad del agua del parque generalmente fue buena. La bacteria coliforme se detectó en el 37% de las muestras, pero la bacteria coliforme fecal se detectó en solamente el 13% de las muestras.

- Se descubrió un arrecife de agua dulce con moluscos en el Río San Pedro, cerca del pueblo de El Naranjo.

- Se registraron cincuenta y nueve géneros y 71 especies y morfoespecies de fitoplancton, diatomeas (Bacillariophyceae), seguido por clorofitas (Chlorophyceae) y cianofitos (Cyanophyceae), dominaron las muestras. Análisis multivariados revelaron que la única fuente más importante de variación que determina la composición de fitoplancton fue el pH y la segunda fue la conductividad.

- Se encontraron cuarenta y cuatro morfoespecies y 26 familias de insectos en *Salvinia auriculata*. Este resultado resalta la gran diversidad de insectos en los sistemas acuáticos litorales del parque.

- El arrecife de moluscos debería conservarse y se debería iniciar un programa de monitoreo de calidad del agua.

INTRODUCCION

Los resultados del estudio limnológico del Parque Nacional Laguna del Tigre (PNLT) se presentan, incluyendo lo siguiente: 1) descripciones del macrohábitat acuático encontrados en el transcurso de este estudio (referirse a la Tabla 1.1 y al Apéndice 1); 2) una evaluación de la calidad del agua usando variables bioticas tales como la presencia de bacteria colifome y *Escherichia coli*, así como medidas abióticas; y 3) una evaluación de los patrones de diversidad del fitoplancton dentro del parque. Además de estos componentes limnológicos, examinamos la composición de insectos asociados con plantas flotantes, y específicamente con *Salvinia auriculata*.

Este estudio limnológico tiene como propósito servir como la base para estudios más intensos de monitoreo de los sitemas acuáticos en el parque. El monitoreo de organismos acuáticos como el plancton e insectos, tiene ciertas ventajas sobre análisis químicos para determinar la salud de ecosistemas acuáticos (UNESCO, 1971). Insectos bénticos, por ejemplo, mantienen una composición relativamente estable que es determinada temporalmente en respuesta a condiciones más o menos promedio en los ambientes acuáticos en los que se encuentran, mientras que análisis químicos solo reflejan la calidad del agua en el momento en que se realizó el muestreo, lo cual puede ser muy variable, dependiendo de la hora del día, o la temporada del año (McCafferty, 1998). El uso de plancton e insectos acuáticos para monitorear ecosistemas puede representar una buena opción para registrar cambios ambientales en el PNLT, como se ha hecho en otros sistemas de agua dulce en Guatemala (Basterrechea, 1986, 1988, 1991; Herrera, 1999). Además, estos organismos son importantes por sí mismos. El plancton es un componente fundamental de la cadena alimenticia acuática, e insectos acuáticos son el grupo principal de organismos que convierten material vegetal a tejido animal en ecosistemas de agua dulce. Así, estos organismos forman la base para la existencia de muchos organismos incluyendo peces, anfibios y aves (Merritt y Cummins, 1984).

Aunque varios estudios han considerado los invertebrados acuáticos en Guatemala, incluyendo estudios del Río Polochic por el departamento de biología de la Universidad del Valle, y por Basterrechea y Torres (1992), en PNLT, este es el primer estudio en Guatemala que considera las relaciones entre insectos y las plantas flotantes *Salvinia auriculata*. *S. auriculata* es una planta acuática relativamente común y fácil de muestrear en sistemas acuáticos en el PNLT (ver León y Morales Can, este volumen) y por lo tanto sirve como un sustrato ideal para colectar insectos acuáticos y considerar sus relaciones con el ambiente. El objetivo de este componente del estudio limnológico es examinar la

factibilidad de más estudios intensivos de insectos sobre plantas acuáticas como una herramienta de monitoreo para PNLT.

METODOS

Calidad del agua

Se tomaron muestras de agua y sedimento en 33 puntos de muestreo en cinco áreas focales, usualmente cerca de los márgenes de los cuerpos de agua (ver profundidades de muestras en la Tabla 2.1). Se obtuvieron los valores de temperatura, pH, conductividad eléctrica y de oxígeno disuelto utilizando instrumentos portátiles. La profundidad en cada punto se midió utilizando un medidor de profundi-

Tabla 2.1. Resultados limnológicos en cada punto de muestreo (ver texto).

Punto de muestreo	Profundidad de muestreo (m)	Profundidad máxima (m)	Visibilidad (m)	Temp. (°C)	pH	Oxígeno (mg/L - % de saturación)	Conductividad MS/cm
SP1	2.10	2.20	1.95	27.5	7.15	5.80 - 84.5	0.81
SP2	3.30	3.90	2.55	25.9	7.03	6.10 - 78.0	1.02
SP3	1.75	5.40	1.75	24.8	6.89	4.72 - 58.0	1.22
SP4	2.12	6.60	2.12	27.5	7.33	7.36 - 95.0	1.55
SP5	1.75	7.10	1.75	28.3	7.33	7.67 - 110	1.32
SP6	1.65	6.10	1.65	29.5	7.20	7.77 - 103	1.98
SP7	0.80	4.10	0.80	28.1	7.40	*	1.51
SP8	2.60	2.60	1.45	29.5	7.41	*	1.48
SP9	2.33	9.00	1.45	29.5	7.43	*	1.42
SP10	0.95	7.85	0.95	29.6	7.36	8.09 - 103.7	1.66
SP11	3.00	5.10	1.30	29.9	7.47	*	1.44
SP12	0.50	3.10	2.50	29.3	7.67	*	1.37
E1	1.45	1.45	1.45	29.3	7.76	*	0.63
E2	0.85	0.85	0.85	29.8	7.27	*	1.57
E3	1.50	1.50	1.30	29.7	7.20	*	1.54
E5	0.40	0.40	0.40	29.6	7.02	*	1.62
E6	0.70	0.70	0.70	29.1	6.99	*	1.70
E7	2.80	2.80	1.30	29.4	7.46	*	1.90
E8	3.75	3.75	2.00	29.2	7.50	*	1.81
E9	0.20	0.20	0.20	27.3	7.57	*	0.73
E10	0.40	0.40	0.40	29.5	7.73	5.88 - 70.0	1.49
E11	0.50	0.50	0.50	25.4	8.30	4.24 - 53.5	1.95
F1	0.80	0.80	0.70	29.0	8.23	7.42 - 96.9	0.33
F2	1.05	1.05	1.0	29.3	7.96	6.36 - 46.5	1.98
F5 A	1.05	1.05	1.05	26.5	7.84	0.02 - 0.5	0.92
F5 B	0.9	0.9	0.9	26.8	7.84	0.95 - 9.5	0.91
F5 C	0.45	0.45	0.45	26.5	7.84	1.45 - 19.3	0.90
F5 D	0.30	0.30	0.30	26.3	7.84	1.40 - 17.4	0.89
C1	0.40	0.40	0.40	26.3	7.06	0.85 - 8.30	1.27
C2	1.85	1.85	1.85	30.0	7.16	8.51 - 106.5	1.06
C3	1.50	1.50	1.50	28.7	7.20	6.78 - 91.0	0.63
Ca1	1.45	1.45	1.45	25.3	6.84	2.80 - 40.7	1.41
Ca2	0.35	0.35	0.35	25.9	6.78	2.66 - 33.5	0.67

(* datos no disponibles).

dad hecho a mano. Se midió la transparencia de la columna del agua utilizando un disco de Secchi. La alcalinidad se determinó utilizando el método Gran, modificado por Carmouze.

Adicionalmente, se determinó la presencia o ausencia de microorganismos que indican la contaminación de agua por excreta humana, incluyendo los coliformes fecales y *Escherichia coli*. Fue posible realizar solamente una prueba cualitativa (Ready Cult Coliforms 100) y se reporta en este documento. La adición de un sustrato cromogénico que se une con la bacteria coliforme y un sustrato fluorescente MUG que es altamente específico para *E. coli* permite la detección simultánea de coliformes totales y fecales, y *E. coli*. La presencia de las coliformes totales se indicaba si la solución cambiaba a azul verdoso. La presencia de coliformes fecales se detectó por medio de una fluorescencia azul y por medio de una reacción positiva al Indol.

Fitoplancton

Las muestras de plancton se tomaron en la superficie del agua. Las muestras se obtuvieron por medio de la filtración de 100 L o 50 L de agua por medio de una red de 200 um; las muestras se preservaron en lugol. Se utilizaron las cámaras Sedgwick-Rafter (magnificación de 100x y 450x) para contar e identificar las especies de fitoplancton y morfoespecies. Los patrones de las especies y riqueza genérica se compararon entre los puntos de muestreo y los patrones de la composición de especies y la abundancia entre los puntos de muestreo se exploraron utilizando técnicas multivariadas. Se utilizó escala multidimensional no métrica (NMMS; McCune y Mefford, 1999) para caracterizar los patrones y cambio en la composición y abundancia del fitoplancton en el PNLT. De las múltiples técnicas de multivariables disponibles, esta técnica es especialmente adecuada para caracterizar las diferencias en la composición de especies entre las muestras.

Insectos acuáticos asociados con *Salvinia auriculata*

Para muestrear los insectos acuáticos que habitan en *Salvinia auriculata*, utilizamos un cuadrante de 25x25-cm. En cada punto de muestreo, donde se observaba *S. auriculata*, colocamos el cuadrante sobre una parte de la planta al azar y recolectamos todas las partes de la planta que caían dentro del cuadrante. Colocamos la *S. auriculata* recolectada en una bolsa plástica con algo de agua y, en el laboratorio, lavamos cuidadosamente el material de la planta con una red de 250-um. Descartamos todas las partes grandes de la planta y trabajamos con el material pequeño, incluyendo partes de las raíces y los insectos atrapados en la red. Esta muestra se fijó en una solución de 80% de formol. Los insectos en las muestras se separaron e identificaron por género y, si era posible, por morfoespecie.

RESULTADOS Y DISCUSION

A continuación, presentamos los resultados generales del muestreo coliforme, patrones de la diversidad de fitoplancton y una descripción de los insectos acuáticos recolectados de *Salvinia auriculata*. Así mismo, una descripción de las características de macrohábitat, calidad de agua, diversidad de fitoplancton e insectos acuáticos para cada área focal.

Coliformes: Consideraciones generales

Los resultados de las pruebas cualitativas para la presencia o ausencia de bacterias coliformes en los puntos de muestreo se presentan en la Tabla 2.2. Será necesario continuar esta investigación con mayor detalle para lograr las conclusiones con relación a la administración de los recursos de agua en el PNLT. Sin embargo, estos resultados proporcionan un punto de vista del impacto de las actividades humanas en los

Tabla 2.2. Resultados de una determinación cualitativa de presencia coliforme en los puntos de muestreo.

Punto de muestreo	Total de coliformes	Coliformes fecales
SP1	–	–
SP2	–	–
SP3	–	–
SP4	–	–
SP5	–	–
SP6	–	–
SP7	–	–
SP8	–	–
SP9	–	–
SP10	–	–
SP11	–	–
SP12	–	–
E1	+	–
E2	+	–
E3	–	–
E5	+	–
E6	–	–
E7	–	–
E8	+	–
E9	–	–
E10	+	+
E11	+	+
F1	+	+
F2	+	+
F5	–	–
C1	+	–
C2	–	–
C3	–	–
CA1	+	–
CA2	+	–

cuerpos de agua en el parque. Aproximadamente el 37% de los puntos de muestreo revelaron la presencia de coliformes totales y el 13.3% de los puntos revelaron coliformes fecales. Los patrones de calidad de agua dentro del parque se discuten posteriormente.

Fitoplancton: Consideraciones generales

Considerando la naturaleza rápida del régimen de muestreo, podemos proporcionar únicamente indicaciones muy generales en relación a la distribución del fitoplancton en PNLT. La estrategia de muestreo determina en gran parte el subgrupo de especies registradas en una ubicación en particular, como la de este estudio, la mayoría de especies de fitoplancton registradas fueron características de aguas superficiales. Sin embargo, en algunos casos, observamos fitoplancton que está íntimamente asociado con el sustrato (béntico), como *Pinnularia*.

En este estudio se identificaron seis clases, 59 géneros y un total de 71 especies y morfoespecies de fitoplancton (Apéndice 3). No se identificaron dos algas (una clorofita y una diatomea). Las diatomeas (Bacillariophyceae), seguido por clorofitas (Clorophyceae) y cianofitos (Cyanophyceae), predominaron en las muestras. Este patrón de abundancia siguió nuestras expectativas en base a condiciones reportadas previamente en estudios realizados en El Petén. Se observaron tres clases adicionales de fitoplancton: Euglenophyceae, Dynophyceae y Chrysophyceae. La mayoría de fitoplancton registrado aquí es cosmopolita, mientras que otros son observados con más frecuencia en ambientes específicos.

La presencia de determinadas especies de fitoplancton indicó condiciones ambientales determinadas en algunos puntos de muestreo. Por ejemplo, cianofitas a menudo indican contaminación humana. En la mayoría de los puntos de muestreo ubicados a lo largo del Río San Pedro; Río Escondido (excepto en los puntos E8 y E9); y lagunas Santa Amelia, Buena Vista, Flor de Luna y La Pista, se presentó *Microcystis aeruginosa* lo que indica la presencia de contaminantes asociados con agua. Sin embargo, es importante notar que las densidades de esta cianofita sobrepasó los 1000 en organismos/L solamente en SP2, SP7, E10 y C3 (refiérase al Apéndice 1). En el resto de los puntos de muestreo, las densidades fueron menores a los 400 organismos/L. *M. aeruginosa* se ha registrado en otros cuerpos de agua dentro de otros departamentos de Guatemala, como en el Lago de Izabal (Basterrechea, 1986) y es la especie de fitoplancton más común en el Lago de Amatitlán, el cual es eutrófico (AMSA, 1998). En estos casos, las densidades de *M. aeruginosa* son mucho mayores que las encontradas dentro del PNLT y es importante mencionar que estos cuerpos de agua existen bajo diferentes condiciones y que se muestrearon en forma diferente de aquellas en el PNLT.

Generalmente existen pocos estudios disponibles para otras comparaciones. *Synedra ulna* es otra especie que indica contaminación y podría proporcionar un buen indicador en futuros estudios dentro del PNLT. También será útil combinar el muestreo del fitoplancton con información sobre las concentraciones de hidrocarbono para proporcionar más vínculos concretos entre los indicadores de contaminación y fitoplancton.

Se registraron otros géneros de fitoplancton en el PNLT, los cuales se han reportado en los lagos de Amatitlán e Izabal, incluyendo *Anabaena, Ankistradesmus, Ceratium, Cyclotella, Euglena, Fragilaria, Synedra, Lyngbya, Oscillatoria, Melosira, Merismopedia, Pediastrum, Staurastrum, Scenedesmus, Tabellaria* y *Volvox. Coelastrum reticulatum* y *Golenkinia radiata* son especies que se reportaron tanto en el Lago de Izabal como en este estudio. La mayoría de los ocho géneros encontrados en el Lago Petén Itzá—*Cosmarium, Staurastrum, Cocconeis, Dinobryon, Navicula, Agmenellum, Microcystis aeruginosa* y *Lyngbya*— se encontraron en el PNLT (Basterrechea, 1988).

Encontramos una mayor riqueza genérica que la encontrada por Basterrechea y Torres (1992) en el Río Escondido-Biotopo de la Laguna del Tigre (22 géneros), aunque la temporada y métodos de muestreo fueron diferentes. Basterrechea y Torres muestrearon áreas solamente en lagunas dentro del biotopo, así como en el Lago Petén Itzá; ellos reportaron 26 géneros en general en este muestreo: Petén Itzá (7 géneros), La Carpa (3 géneros), Pozo Xan (7 géneros), El Toro (7 géneros) y El Remate (12 géneros). Sandoval (1997) reportó 18 géneros de fitoplancton en el Lago Petén Itzá. La mayoría de estos géneros se identificaron en el PNLT. En algunos puntos, Sandoval reportó el dominio de bacilariofitas, las cuales fueron comunes en este estudio.

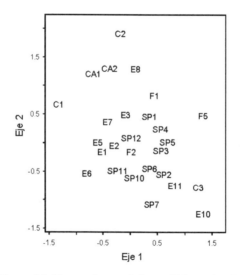

Figura 2.1. Una escala no métrica multidimensional ordenando puntos de muestra en base a las densidades de fitoplancton (ver texto).

Los dos primeros ejes del orden NMMS (Figura 2.1) explicaron el 58.8% de la variación en la matriz original con distancia reducida y los ejes fueron significativamente diferentes de aquéllos generados por datos al azar (permutaciones de Monte Carlo: P=0.035). El orden indica que la composición de fitoplancton de los Ríos San Pedro y Escondido fueron ampliamente similares. Todos, excepto un punto de muestreo (C3) en el área focal de Río Chocop/Río Candelaria fueron diferentes a los otros. Esta diferencia en la composición se relaciona con la riqueza generalmente baja en todas las muestras de Chocop/Candelaria excepto C3, que era relativamente rica en especies (Apéndice 3). A pesar de la similitud ecológica entre las lagunas Guayacán (C2) y La Pista (C3) (ver a continuación), así como su proximidad espacial, estas lagunas albergaban distintas floras de fitoplancton. Los índices del eje NMMS se encontraban en decadencia en las variables ambientales de la Tabla 2.1 para explorar qué variables ambientales fueron las mejor relacionadas con patrones de composición de plancton. El eje 1 indica correlatividad positiva solamente con pH (R^2=0.14; F=4.13; df=1, 25; P=0.053) y el eje 2 indica correlatividad positiva con la conductividad (R^2=0.15; F=4.46; df=1, 25; P=0.045). Ninguna de las variables ambientales pudo explicar en mejor forma la variación en la composición de especies.

En la mayoría de los casos, las densidades de fitoplancton obtenidas no fueron particularmente altos (Apéndice 4). Será importante continuar monitoreando estos y otros sitios dentro del PNLT para evaluar los cambios en las cargas nutrientes y otros patrones fisico-químicos y sus efectos sobre la productividad del fitoplancton.

Insectos acuáticos asociados con *Salvinia auriculata*

S. auriculata fue común solamente en el área focal del Río San Pedro para el muestreo de insectos. Los insectos se muestrearon del Arroyo Yalá (SP1), Paso Caballos (SP2), La Pista (SP3), El Sibalito (SP4), El Caracol (SP6), Río San Juan (SP7) y La Caleta (SP8) (ver Apéndice 5). En estas muestras, identificamos 5 órdenes, 26 familias y 44 morfoespecies. El orden con el mayor número de familias representadas fue Coleóptera (10) y Ephemeróptera fue representada por el menor número (2; Figura 2.2).

Es importante mencionar que debido a que las plantas son móviles, ocasionalmente existen conexiones entre plantas acuáticas y las orillas; por lo tanto, muchas de las especies que habitan en las plantas acuáticas flotantes también pueden ser semiacuáticas o terrestres. Dichos procesos pueden explicar la presencia de algunos Coleópteros como los Scolítidos. A pesar de la duración limitada y la extensión del muestreo, la riqueza de las especies de insectos en *S. auriculata* fue considerable e indica que se garanticen más estudios descriptivos intensos y monitoreo de este sistema.

Área focal del Río San Pedro
Clasificación de macrohábitats
En las 12 estaciones de muestreo a lo largo del Río San Pedro, se encontraron cinco tipos de macrohábitats diferentes (refiérase al Apéndice 1).

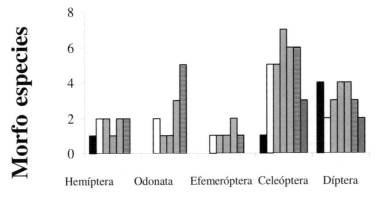

Figura 2.2. Patrones de riqueza de órdenes de insectos que habitan la *Salvinia auriculata* en muestras del área focal del Río San Pedro del PNLT.

Características del hábitat, calidad del agua y diversidad del fitoplancton

El Río San Pedro se caracteriza por una velocidad relativamente baja. En el Río Sacluc (SP11), los niveles del agua fueron un poco bajos. Las muestras de sedimento revelaron erosión considerable del márgen del río, probablemente debido al constante paso de las lanchas.

Algunas características abióticas generales de las muestras de los ríos San Pedro y Sacluc se presentan en la Tabla 2.3 e incluyen la siguiente información:

- La profundidad máxima del canal central se encontró frente a la Estación Biológica Las Guacamayas (7.10 m; Tabla 2.1) y la mínima se encontró frente al Arroyo Yalá (2.20 m).
- Las muestras se tomaron en la zona litoral de los ríos en profundidades que oscilan de 0.95 m a 3.30 m.
- Los valores de transparencia indicaron que la zona eufótica alcanzó el fondo del río en un 50% de los puntos de muestreo.
- La temperatura del agua fue alta, varía entre 24.8° C y 29.6° C.
- Las concentraciones de oxígeno disueltas fueron comparativamente altas en casi todos los puntos de muestreo.
- Las aguas del Río San Pedro poseen una conductividad eléctrica comparativamente alta. Esta característica se debe probablemente al paisaje cárstico por el cual fluye el río. Estas aguas pueden clasificarse como "difíciles" debido a las altas concentraciones de carbonato de calcio, lo que a su vez, puede relacionarse con los valores de alta conductividad observados.
- Sorpresivamente, el pH del agua es neutral a levemente alcalino. Tomando en cuenta las características geomorfológicas de esta región carstica, esperábamos observar un pH fuertemente alcalino (>8.5). Las cantidades relativamente grandes de material orgánico en el agua y las altas temperaturas en la misma pueden ocasionar una actividad acelerada de los descomponedores (bacterias, hongos y fermentos). La producción de ácidos orgánicos en estas aguas alcalinas pueden producir un pH casi neutral (pH 8). Los manantiales con corrientes rápidas en el arroyo La Pista (SP3) tenían agua fría, un pH más ácido y mayor conductividad eléctrica comparadas con las de San Pedro.
- La prueba cualitativa para observar el total de coliformes, coliformes fecales y *E. coli* fue negativa en todas las muestras tomadas.

Tabla 2.3. Descripción de puntos de muestreo a lo largo del Río San Pedro

Tipo de macrohábitat	Puntos de muestreo
Río permanente	SP 1, 4, 5, 9, 11
Tributarios	SP 2, 3, 6, 7
Caños	SP 8
Ensenadas del río	SP 10
Rápidos del río	SP 12
Tipos de fondos	
Sedimentos con partículas grandes de materia orgánica	SP1, 3
Sedimentos con partículas pequeñas de materia orgánica y cantidades elevadas de agua intersticial (áreas normales de depósitos)	SP7, 8, 9, 10
Velocidad actual	
Aguas bajas, casi estancadas	SP1, 7a, 11
Aguas con corriente	SP2, 3, 7b (primaveras)
Areas con alta diversidad de macrófitas acuáticas	SP1, 2, 3, 6, 7
Areas sin macrófitas acuáticas	SP4, 5, 8, 9, 10, 11

La cantidad de especies de fitoplancton halladas en cada punto de prueba fue similar (Apéndice 3), pero el punto con menor riqueza fue el SP10. Solamente dentro de esta área focal se registraron géneros tales como el *Tetraëdron* (SP2, SP3); *Kirchneriella* (SP2, SP3 y SP4); *Synura* (SP1, SP2, SP3 y SP6) y *Dinobryon* (SP3, SP4 y SP5).

Notas sobre otros invertebrados

Surgió una gran diversidad de macroinvertebrados asociados con *Salvinia auriculata*. Aproximadamente se registraron 44 taxones diferentes que representaban a los grupos de herbívoros, detritívoros, predadores y otros grupos tróficos (Apéndice 5). También se halló la presencia de mineros que utilizan sedimento así como tejido folicular vivo. Se recolectaron aproximadamente 10 morfoespecies de Odonatas adultos.

Un arrecife de moluscos

En una sección del Río San Pedro conocida localmente como los «rápidos» notamos la presencia de arrecifes de agua fresca compuestos de moluscos bivalvos. Este tipo de sistema acuático es único en la región y no es común generalmente en los Neotrópicos. El fondo del río se compone de una capa de 35 cms de profundidad de conchas bivalvos. No se encontraron sedimentos orgánicos. Los bivalvos vivas se encontraron en la superficie del arrecife. El pez *Cichlasoma sunsphilium* se asoció con el arrecife y se encontró que se alimentaba del perifiton que crecía en las conchas. Dentro de las conchas se encontraron diferentes ninfas de Ephemeróptera (Baetidae) y Trichóptera, así como otros bivalvos, gasterópodos y platelmintos. Concluímos entonces que estas conchas son microhábitats importantes para los invertebrados que habitan el fondo del río en esta área.

En una extensión del río con corriente baja, se observaron macrófitas acuáticas (*Nymphaea ampla* y *Valisneria americana*), que formaban otro microhábitat para los invertebrados. Dentro de esa extensión se encontró una acumulación de material orgánico fino. La prueba cualitativa para los coliformes fecales y *E. coli* fue positiva, que coincidía con los asentamientos humanos que se encontraban cerca de la playa. Adicionalmente, identificamos la mayor riqueza de fitoplancton en este punto (44 especies), siendo mayores las densidades en las clases de Bacillariophyceae, seguidas por la Cianophyceae. Este fue uno de los pocos puntos en que se encontró *Tetraëdron* (Apéndices 3, 4).

Area focal del Río Escondido

Clasificación de macrohábitats

En las 10 estaciones de muestreo a lo largo del Río Escondido (con prefijo E), se encontraron cuatro tipos de macrohábitats. (Tabla 2.4)

Características del hábitat, calidad del agua y diversidad del fitoplancton

El Río Escondido tiene bancos relativamente empinados y un margen bastante estrecho. Las playas se componen en su mayoría de bosque, bosque alterado y sibal. Una gran sección del sibal estaba siendo quemada durante nuestra investigación. El bosque de la estación que se encontraba más al noreste, Punto Icaco (E8) tenía un bosque de dosel relativamente cerrado, diferente en carácter que las partes restantes del Río Escondido.

El Río Escondido termina en un canal bajo y sistema de ciénagas, sobre las cuales se encuentran secciones relativamente profundas que se ensanchan y estrechan regularmente. El fondo es principalmente barro suave, pero en el Punto Icaco se encontraron muchos escombros, troncos y hojarasca. Las lagunas del recodo y ramificaciones muertas comprenden muchos hábitats que se encuentran a los lados del Río Escondido, con frecuencia a través de los hábitats sibales y no en el bosque. Estos hábitats en la actualidad no son profundos, tienen barro bastante suave y fondos de barro/marga. Las características del área del Río Escondido se resumen en la Tabla 2.4 e incluyen lo siguiente:

- Las profundidades menores se registraron en las lagunas y bahías, y las de mayor profundidad en el río, en E8.
- La transparencia alcanzó el fondo en todos los puntos de prueba.
- En general, las aguas de este lugar tenían temperaturas altas (desde 26° C hasta 29.8° C) y los valores de pH fueron desde neutrales hasta levemente alcalinos (6.99 a 7.76).
- Los sedimentos tienen altos niveles de materia orgánica y agua intersticial.
- La conductividad eléctrica en el canal principal del Río Escondido fue baja en 1 MS/cm. Por otro lado, las ensenadas tienen valores de conductividad muy altos.
- La presencia de coliformes totales se registró en E1, E2, E5 y E8 y la presencia de coliformes totales y fecales y *E. coli* se registró en E10 y E11. No se detectó el origen de la contaminación fecal en otros puntos.
- La mayor riqueza del fitoplancton se registró en los puntos E1 (38 spp.) y E5 (35 spp.). Llovió durante el muestreo de este río, incrementando su corriente, así que es posible que el muestreo del fitoplancton haya afectado la disolución. Esto pudo haber ocasionado una reducción general en la riqueza de especies cuando se comparó con las muestras tomadas del San Pedro. Algunos géneros de fitoplancton registrados aquí, como *Cosmarium* (E1, E2 y E6) y *Mallomonas* (E1, E2, E5, E7 y E8), no se encontraron en el San Pedro.

Area focal de Flor de Luna

Tabla 2.4. Descripción de puntos de muestreo a lo largo del Río Escondido

Tipo de macrohábitat	Puntos de muestra
Río permanente	E3, 8, 11
Ensenadas del río	E1, 2
Río efímero (no profundo)	E10
Lagunas oxbow	E5, 6, 7, 9
Tipos de fondos	
Sedimentos con partículas grandes de material orgánico	Ninguno
Sedimentos con partículas pequeñas de material orgánico y cantidades elevadas de agua intersticial (áreas típicas de deposición)	Todos excepto E11
Fondo rocoso	E11
Velocidad de la corriente	
Aguas lentas, casi estancadas	E1, 2, 7, 8, 10
Aguas con corriente	E3, 9, 11
Areas con una alta diversidad de macrofitos acuáticos	E3, 8, 10
Areas sin macrofitos acuáticos	El resto

Clasificación de los macrohábitats

Los cuerpos de agua en este sitio consisten exclusivamente de lagunas, tres de las cuales fueron muestreadas (F1, F2 y F5). Las características del hábitat, calidad del agua y las comunidades del plancton se discuten a continuación para cada laguna en forma individual.

Laguna Flor de Luna (F5)

- Esta laguna es poco profunda (1.05m). Esto favorece la mezcla de toda la columna de agua debido a la acción del viento (principalmente vientos del noroeste).
- La laguna era homotérmica (26.3° C a 26.8° C), la conductividad eléctrica fue menor que 1 MS/cm y el pH fue de neutral a levemente alcalino. El análisis cualitativo para las coliformes totales y fecales y *E. coli* fue negativo.
- Durante la recolección de muestras de sedimento, se detectó la presencia del gas de sulfuro de hidrógeno (H_2S). Este gas, junto con el gas metano (CH_4), causa la reducción química de nutrientes inorgánicos en el agua debido a que la reducción se ve favorecida en el entorno hipóxico o anóxico en la interfase de sedimento y agua. Por esta razón, los macroinvertebrados bénticos no se observaron en los sedimentos. Sin embargo, los Gerridae fueron abundantes en la superficie de las hojas y en las flores de *Nymphea ampla* en toda la laguna.
- Aquí se registraron treinta y cuatro especies de fitoplancton. Este fue uno de los pocos puntos en los cuales se registró *Tetraëdron*, *Micrasterias*, *Staurastrum* y *Spirogyra*.

Laguna Santa Amelia (F2)

- Esta laguna también era poco profunda (1 m), lo que favorece las mezclas generadas por el viento.
- Como en Flor de Luna, la laguna fue homotérmica (27.7° C a 29.3° C).
- La conductividad eléctrica osciló entre 1.85 y 2 MS/cm.
- La saturación de oxígeno oscilaba entre 8.6% a 46.5% en la superficie.
- El agua exhibió un color oscuro natural; sin embargo, la transparencia persistió en el fondo del lago.
- El muestro del sedimento reveló la presencia de sulfuro de hidrógeno y no se observaron macroinvertebrados bénticos. *Nymphea ampla* fue abundante en toda la laguna.
- Las pruebas para la presencia de coliformes totales y fecales y *E. coli* fueron positivas, lo que indicó la contaminación fecal por humanos y ganado. Las aguas de la laguna parecen ser usadas para consumo humano y los niveles de contaminación en la laguna deben ser monitoreados. Aquí se encontraron varios géneros de fitoplancton no comunes, incluyendo *Staurastrum*, *Spirogyra* y *Mallomonas*.

Laguna Buena Vista (F1)

- Esta laguna también era poco profunda(0.75 m) y homotérmica (29° C a 29.1° C).
- El pH fue alcalino (8.23).
- La laguna midió 200x600 m y fue extremadamente homogénea.
- Las poblaciones cercanas utilizan esta laguna para consumo, para lavar y bañarse. No es sorprendente que la prueba de coliformes totales y fecales y *E. coli* fuera positiva.
- Varios géneros de fitoplancton no comunes se identificaron aquí—aunque en algunos casos estos también se registraron en F1 y F5—tal como *Micrasterias*, *Staurastrum*, *Sorastrum*, *Spirogyra* y *Mallomonas*.

Área focal del Río Chocop

Clasificación de macrohábitats

Aquí se localizaron cinco puntos de muestreo (con prefijo C), los que representan tres tipos de macrohábitats. (Tabla 2.5).
La calidad del agua y las comunidades de fitoplancton/invertebrados se resume en la Tabla 2.5 y se discute en forma individual bajo cada punto de muestreo.

Río Chocop (C1)

- El Río Chocop y el Río Candelaria están clasificados como ríos poco profundos—es decir, como arroyos semi-permanentes porque durante la temporada seca, los niveles del río bajan dramáticamente y sólo un poco de agua fluye por el caudal. Ambos ríos tenían orillas muy inclinadas y bosques con un dosel cerrado. El Chocop se encuentra casi seco con áreas de agua estancada y posiblemente un «biofilme» de bacteria en la superficie de estos charcos.
- El análisis para la presencia total de coliformes fue positivo, pero fue negativo para coliformes fecales y *E. coli*
- Había muy poco oxígeno disuelto en la columna de agua. El fondo del río consistía en sedimentos compactos y hojarasca de la vegetación terrestre. Debido a la falta de oxígeno, había pocos invertebrados acuáticos (el grupo troficó fragmentado no estaba presente). Sin embargo, se observaron en las partes menos profundas un gran número de belostomátidos (Hemíptera). Este río seguramente es muy diferente durante la temporada de lluvia. Para el fitoplacton, este es uno de los puntos con menor riqueza de especies, con solo 28.

Tabla 2.5. Descripción de puntos de muestreo a lo largo del Río Chocop

Tipo de macrohábitat	Puntos de muestra
Ríos enfímeros (poco profundos) con fondo oscuro	C1, Ca1
Ríos permanentes con fondo rocoso	Ca2
Lagunas	C2, 3

Lagunas la Pista (C3) y Guayacán (C2)

• Estas lagunas son ecológicamente similares y por lo tanto se tratan juntas. Guayacán tenía algunas piedras calizas en su margen y agua clara (>2m visibilidad) sobre sedimentos suaves. Tenía un margen con guamil y márgenes de bosque. Habían muchas posas pequeñas en el bosque asociado con esta laguna que puede presentar buenas condiciones.

• El viento probablemente mezcla toda la columna de agua en ambas lagunas. Como consecuencia, las lagunas fueron homotérmicas, en el fondo se encontraba oxígeno disuelto (>86% saturado) y la conductividad fue igual en toda la columna de agua.

• La prueba de los coliformes totales, coliformes fecales y *E. coli* fue negativa.

• A pesar de la presencia de oxígeno disuelto en las columnas de agua de ambas lagunas, detectamos un fuerte olor de sulfuro de hidrógeno en las muestras de sedimento. Estas observaciones sugieren una resistencia de una microestratificación de oxígeno en la interfase de agua y sedimento. Como consecuencia, esperamos ausencia de macroinvertebrados en ambas lagunas.

• Guayacán albergó el menor número de género de fitoplancton en el estudio (19), posiblemente debido al estado oligotrófico de esta laguna. La laguna La Pista mostró una riqueza mayor de fitoplancton.

Río Candelaria (Ca1, 2)

• En este río los dos puntos de muestra fueron un poco diferentes. En Cal, el río fue profundo y tenía poca corriente. Generalmente, las condiciones fueron similares a aquellas encontradas en Río Chocop.

• Se encontró oxígeno disuelto en el fondo del río, aunque en bajas concentraciones. El material vegetal fue abundante en el fondo y se detectó la presencia de gas de hidrógeno de sulfato.

• La prueba para coliformes totales fue positiva y para coliformes fecales y *E. coli*, la prueba fue negativa. Aquí se registraron veintisiete especies de fitoplancton.

• El río en Ca2 fue menos profundo (35 cm), tenía una corriente más rápida, y había oxígeno en el fondo.

• Los valores de conductividad fueron similares a los de Cal.

• La prueba de coliformes totales y fecales y *E. coli* fue negativa.

• En este lugar se registraron veintinueve especies de fitoplancton, de tal manera que ambos sitios de Candelaria fueron intermedios en riqueza comparados con los otros lugares. Se observaron diferentes grupos tróficos de macroinvertebrados bénticos en ambos puntos, incluyendo gasterópodos, herbívoros, efemerópteros y carnívoros odonatos.

CONCLUSIONES Y RECOMENDACIONES

Respecto a la conservación y manejo de los ecosistemas acuáticos del parque, se ofrecen dos recomendaciones:

• Existe una urgente necesidad de conservar los rápidos del Río San Pedro. Estos rápidos poseen un sistema de arrecife de agua dulce único, que también podría realizar una importante función al filtrar las partículas suspendidas del agua.

• Muchas lagunas y corrientes dentro del parque se encuentran poco contaminadas, y se debería tomar acciones para asegurar su integridad.

Respecto de estudios futuros y el monitoreo de los sistemas acuáticos dentro de la región, se ofrecen cinco recomendaciones:

• Los estudios futuros se deben ejecutar a nivel de cuenca y se deben conducir por períodos más extensos, acompañando la variación entre las estaciones lluviosa y seca. Se deben tomar muestras en diferentes profundidades y se deben considerar nutrientes totales y disueltos.

• Los efectos de las actividades humanas sobre el paisaje, como la agricultura y el derrame de químicos y basura, debe evaluarse a nivel de la cuenca.

• Para el monitoreo del fitoplancton sería útil estandarizar las horas del muestreo entre diferentes puntos de muestra para evitar los efectos de la variación lumínica en el movimiento de plancton en la columna de agua. Esto reduciría la subjetividad en las comparaciones entre muestras.

• Las densidades de la bacteria coliforme debe ser monitoreada y cuantificada para rastrear exactamente el grado de contaminación fecal en el parque e identificar aquellas áreas que son fuente de contaminación en el parque.

- Necesita iniciarse un monitoreo más intenso de poblaciones de insectos acuáticos y las muestras deben ser tomadas en diferentes épocas del año para comprender los efectos de la variación temporal natural y cómo esto puede interacturar con la variación espacial ocasionada por los procesos naturales y antropogénicos.

LITERATURA CITADA

AMSA. 1998. *Listado de Géneros Determinados de Algas en el Lago de Amatitlan.* Secretaría de la República de Guatemala.

Basterrechea, M. 1986. Limnología del Lago de Amatitlán. Revista de Biología Brasileira 46: 461-468.

Basterrechea, M. 1988. Limnología del Lago Petén Itzá, Guatemala. Revista de Biología Tropical 35: 123-127.

Basterrechea, M. 1991. *Evaluación del Impacto Ambiental de la Exploración Sísmica en la Cuenca del Lago de Izabal.* Guatemala.

Basterrechea, M. y M. Torres. 1992. *Hidrología y Limnología de los Humedales del Biotopo Río Escondido-Laguna del Tigre, Petén-Guatemala.* Centro de Estudios Conservacionistas y Fundación Mario Dary, Guatemala.

Herrera, K. 1999. Indicadores de la calidad del agua del Río Polochic y de la integridad biológica del Lago de Izabal. Master's thesis. Departamento de Biología, Universidad del Valle de Guatemala.

McCafferty, P. 1998. *Aquatic Entomology.* Jones and Bartlett Publishers, Sudbury, MA, USA.

McCune, B. y M. J. Mefford. 1999. PC-ORD. *Multivariate Analysis of Ecological Data*, version 4.0. MjM Software Design, Gleneden Beach, OR, USA.

Merritt, R. W. y K. W. Cummins. 1984. *An Introduction to the Aquatic Insects of North America*, 2nd ed. Kendall-Hunt Publishing Company, IA, USA.

Sandoval, K. 1997. Fitoplancton como indicadores de contaminación en tres regiones del Lago Petén Iztá. Informe final, Programa E.D.C. Escuela de Biología, Universidad de San Carlos de Guatemala.

UNESCO. 1971. *Memorias del Seminario sobre Indicadores Biológicos del Plankton.* Oficina Regional de Ciencia y Tecnología para América Latina y el Caribe, Montevideo, Uruguay.

CAPITULO 3

LAS COMUNIDADES ACUATICAS DE MACROFITAS DEL PARQUE NACIONAL LAGUNA DEL TIGRE, PETEN, GUATEMALA

Blanca León y Julio Morales Can

RESUMEN

- Una flora vascular de más de 130 especies, 120 géneros y 67 familias se registró en y a un lado de los hábitats acuáticos del Parque Nacional Laguna del Tigre (PNLT). Esta flora es similar en composición a la de las áreas vecinas de la Península de Yucatán, pero será necesario expandir el estudio para proporcionar comparaciones más precisas.

- La vegetación del PNLT incluye una diversidad de comunidades como los bosques ribereños grandes o medianos, bosques bajos, la sabana de palma inundada y el resto del mangle.

- Una población de *Vallisneria americana* asociada con bivalvos de los rápidos del arrecife fue única. Otro sitio con depósitos menores de bivalvos del Río Escondido fue especialmente rico en especies. Las inundaciones y erosión ocasionan que las comunidades sean altamente dinámicas.

- Los disturbios antropogénicos como el fuego, están alterando el hábitat natural y la composición de la vegetación en el PNLT.

INTRODUCCION

Guatemala es uno de los pocos países de Centroamérica que tiene una flora ampliamente estudiada y un largo historial de colección botánica (e.g. Standley, 1924-1964; Williams, 1966-1975). Sin embargo, pocos estudios han analizado específicamente la flora vascular acuática (por ejemplo, de Poll, 1983).

Para El Petén guatemalteco, los primeros estudios botánicos se realizaron en la primera parte del siglo XX y estaban asociados con exploraciones arqueológicas (Lundell, 1937). Por esta razón, el interés en comprender los cambios culturales del pasado se ha ligado a interpretaciones en los cambios de la cubierta forestal (Deevey et al., 1979; Islebe et al., 1996). Estos estudios interpretativos de cambios históricos, sociales y ambientales se basan en estudios paleolimnológicos. Sin embargo, este trabajo ayuda un poco a ampliar nuestro conocimiento sobre la macroflora acuática actual en la región.

La heterogeneidad de la vegetación del bosque terrestre del Petén no se cuestiona (Deevey et al., 1979). Las inundaciones debido a la estación lluviosa, y las alteraciones que producen también crean una gran heterogeneidad en la expresión de elementos florísticos de las comunidades de plantas acuáticas.

El Parque Nacional Laguna del Tigre (PNLT) se asocia con la presencia de áreas extensas de ciénagas, ubicadas entre un amplio rango de tipos de hábitats acuáticos (Tabla 3.1) en las cuales las macrófitas constituyen los elementos más sobresalientes. Nuestro objetivo es describir las relaciones entre la flora acuática y estos macrohábitats y describir las comunidades formadas dentro de ellos.

METODOS

Las macrófitas acuáticas se describieron en este estudio como plantas vasculares que se encuentran permanentemente sumergidas en agua (verdaderas acuáticas) y plantas con raíces en suelos supersaturados o inundados, con el resto de la planta emergente (anfibias).

Las definiciones de los macrohábitats se basaron en la hidrología de los puntos de muestreo (ver Herrera et al., este volumen). Respecto a los macrófitas acuáticos, se reconocieron seis hábitats: ríos, corrientes, lagunas, lagunas de recodo, caños y rápidos (arrecifes de agua dulce). De éstos, el macrohábitat del río se subdividió en las aguas (profundas) de exhibición permanente, agua efímeras (no profunda) y rápidos, dependiendo del volumen de agua y tipo de sustrato. Estos rápidos constituyen una parte de hábitats permanentes de río.

Los tipos de vegetación acuática se clasificaron por fisionomía y se subdividieron según las asociaciones observadas dentro de los transectos. Se reconocieron dos tipos de vegetación en este estudio: bosques ribereños, mientras las plantas herbáceas caracterizan a las ciénagas. Los nombres utilizados para las asociaciones emplearon terminología local reconocida en estudios anteriores (Morales, sin publicar).

Establecimos transectos temporales entre los dos tipos de vegetación. A lo largo de cada transecto, registramos la composición de macrófitas encontradas en cuadrantes de 50x50 cms. En ambientes lóticos (es decir, agua corriente) los cuadrantes se colocaron aleatoriamente a lo largo del banco del río a intervalos de 10 metros. En ambientes lénticos (es decir, aguas estancadas), los cuadrantes se colocaron a lo largo del diámetro máximo del cuerpo del agua a intervalos similares. Por cada cuadrante, registramos la profundidad del sustrato, la distancia hasta la playa y las especies de plantas visibles dentro de dos metros del cuadrante. Para poder muestrear el bosque ribereño, se establecieron tres transectos (100 m de longitud) a lo largo del banco del río, cada uno separado por 100 m. A lo largo de los tres transectos se registraron todos individuos de plantas leñosas que tenían más de 50% de su masa de raíz dentro del agua.

Se recolectaron los especímenes de prueba de todas las plantas vasculares fértiles que registramos. Las plantas se prepararon en una prensa de plantas y posteriormente se colocaron en alcohol para prepararlas y finalmente secarlas. Se realizaron identificaciones utilizando colecciones hechas en el Field Museum de Chicago, USA.

RESULTADOS Y DISCUSION

Composición de fauna del PNLT

En las ciénagas, bosques ribereños y bosques inundados debido a la estación se registraron un total de 67 familias, 120 géneros y más de 130 especies de plantas vasculares (Apéndice 6). De éstas, 33 familias incluyen especies acuáticas y/o anfibias, y las otras 34 están representadas por especies terrestres, tanto de leñosas como herbáceas, asimismo, las epífitas pueden tolerar cortos períodos de inundaciones.

Entre las familias más diversas se encontraban Cyperaceae y Fabaceae, de las cuales ambas contienen especies terrestres y anfibias. Cyperaceae y las leguminosas también contienen los elementos más sobresalientes tanto de las ciénagas como de los bosques ribereños. Entre las Cyperaceae, *Cladium jamaicense* (sibal) es el elemento dominante del paisaje de las ciénagas. Esta planta forma colonias extensas y densas y su base forma un sustrato que favorece el establecimiento de semillas de plantas terrestres, como las de la familia de las Asclepiadaceae, Cucurbitaceae, y Passifloraceae. Entre las hortalizas, *Mimosa pigra* y *Haemotoxylum compechianum* (tinto) son componentes dominantes del bosque ribereño. Entre otras familias dominantes se incluyen las palmas (Arecaceae) y los llanos (Poaceae). Ambos son importantes en hábitats tanto terrestres como acuáticos.

La composición de las plantas acuáticas y anfibias refleja vínculos con la flora mexicana (Lot y Novelo, 1990; Martínez y Novelo, 1993); no hay taxa que sean endémicas en el área de estudio. Pocas especies se conocen solamente de Centroamérica, incluyendo la *Cabomba palaeformis* y la *Haematoxylum compechianum*. El género más vistoso de las plantas acuáticas fue Utricularia y el más vistoso de las plantas anfibias fue la *Elocharis*.

Cincuenta de las especies registradas en el parque son acuáticas o anfibias (Apéndice 6). Nueve especies poseen flores que son estrictamente acuáticas: *Brassenia schreberi*, *Cabomba palaeformis*, *Najas wrightiana*, *Nymphaea ampla*, *Pistia stratiotes*, *Utricularia foliosa*, *U. gibba*, *U.* sp., y *Vallisneria americana*. Estas especies representan a seis familias: Araceae, Cabombaceae, Hydrocharitaceae, Lentibulariaceae, Najadaceae y Nymphaeaceae. Dos de éstas eran plantas flotantes: *Salvinia auriculata* y *S. minima*. En todo el estado de Campeche, y en la Península de Yucatán, Lot et al. (1993) mencionó 33 plantas con flores estrictamente acuáticas. Es posible que no hayamos registrado varias familias de las que existen en el parque, incluyendo Ceratophyllaceae, Lemnaceae, Menyanthaceae y Sphenocleaceae.

Tipos de vegetación

Las planicies de la Península de Yucatán, particularmente los "humedales de bosque" tienen ricas ciénagas (Lot y Novelo, 1990). Estos humedales en bosque se dan como un mosaico con ciénagas (humedales herbáceos) tanto en el parque como en el biotopo. La abundancia de cada uno de estos tipos de vegetación varía, en parte, debido a procesos como la variación de las estaciones en cuanto a la precipitación que modifica la saturación del sustrato, así como la intervención humana. Las propiedades hidrológicas del área juegan, sin lugar a dudas un rol importante en el mantenimiento del mosaico. (Lugo et al., 1990).

Ciénaga

La ciénaga es sin duda el tipo de vegetación más dominada por la verdadera vegetación acuática. La ciénaga ocupa las partes más bajas del parque y probablemente esté asociada con sedimentos depositados por el río o en las márgenes erosionados de ambientes lénticos. Estas plantas también forman pequeñas comunidades a lo largo de los bancos que están dominadas por el bosque ribereño. La presencia de verdaderas plantas acuáticas, en parte, está relacionada con la profundidad y el tipo de sustrato, el nivel de protección de un sitio en cuanto a fuertes corrientes y la frecuencia de las molestias. Las especies anfibias dominan las ciénagas.

Las ciénagas en el PNLT se componen de diferentes asociaciones dominadas por *Cladium jamaicense* (la asociación llamada sibal). *C. jamaicense* y *Phragmites australis* (sibal-carrizo) existen abundantemente a lo largo del Río San Pedro hasta la desembocadura del Río Escondido. La asociación de *C. jamaicense* y *Typha domingensis* (sibal-tul) es rara aparentemente y sucede sólo en el Arroyo la Pista (SP3). La asociación de *C. jamaicense* y *Acoelorraphe wrightii* (sibal-tasiste) domina las márgenes del Río Escondido.

La ciénaga representa el tipo de vegetación más diverso respecto de la cantidad de especies acuáticas y anfibias, así como taxa terrestre como la Asclepiadaceae, Convolvulaceae, Cucurbitaceae y la Passifloraceae. La vegetación terrestre es favorecida por el sibal ya que se había quemado recientemente, usualmente durante la estación seca. Los incendios queman de manera heterogénea el sibal, dejando un mosaico de áreas gravemente consumidas, áreas mezcladas y áreas en las cuales solamente las partes superiores de las plantas se han quemado. La heterogeneidad producida por el fuego también se refleja en el nuevo crecimiento de especies de madera localizadas en los bancos de la ciénaga.

Otro factor que afecta la ciénaga y su biota es el cambio en los patrones de drenaje debido a la construcción de caminos. Se pueden observar cuerpos de agua a lo largo de las rutas de acceso a las lagunas, llenos con grama y guamiles.

Humedales en bosque

Es difícil delimitar este tipo de ciénaga durante la estación seca, ya que solamente se puede observar el límite inferior de su extremo durante este período. En México, este tipo de ciénaga se reconoce en diferentes clases, incluyendo bosques ribereños altos y medianos, bosques ribereños pequeños, maleza de palmas en humedales y mangles (Lot y Novelo, 1990). Estos cuatro tipos se presentan en el PNLT.

El bosque ribereño alto es dominado por *Pachira aquatica, Inga vera, Lonchocarpus hondurensis y Phitecellobium* sp. y éstas pueden ser vistas cerca de la Estación Biológica Guacamayas (EBLG) sobre el Río San Pedro. Estos se encuentran entre los bosques más altos en el área, pero los árboles raramente pasan de 30 metros, debido a la ausencia de suelos profundos. El bosque ribereño pequeño, comprendía la mayor parte de las áreas boscosas que examinamos. Representados por diversas asociaciones, dominadas por *Haematoxylum campechianum* (tintal), que se encuentra en la Laguna Flor de Luna (F1). Adicionalmente, la asociación de *Bucida buceras* y *Pachira aquatica* (pucte-zapote bobo) sucede en partes de los ríos San Pedro y Sacluc; la asociación de *B. buceras* y *Metopium* (pucte-chechen negro) sucede en la Laguna Guayacán (F5); y *B. buceras* y *Diospyros* sucede en la parte alta del Río Escondido, en donde también se encontraron las especies *Chrysobalanus icaco* y *Metopium*. La asociación más común fue *Bucida-Pachira*, especialmente a lo largo del Río San Pedro (SP6, SP9, y los rápidos) y el Río Sacluc (Tabla 3.1). En general, el bosque ribereño es menos rico en especies de macrófitas acuáticas que en la ciénaga (Tabla 3.1). Sin embargo, otras especies, algunas de las cuales tienen importancia desde el punto de vista económico para las poblaciones locales (por ejemplo *Calophyllum brasiliense*)— podrían aparecer en este bosque de manera cíclica.

La maleza de palmas inundadas se desarrolla en forma sobresaliente en la parte central y alta del Río Escondido. El "tasistal" describe una de sus asociaciones principales - aquélla en la cual *Acoelorraphe wrightii* es dominante y es acompañada por hierbas tales como *Cladium jamaicense, Osmunda regalis* en su variedad *spectabilis*, o bien árboles de *Pachira aquatica*.

El mangle no se desarrolló en el área de estudio. Solamente pocos individuos del *Rhizophora mangle* ocupan las márgenes del Río San Pedro y están rodeadas por un ´sibal´ alterado. Esta población aislada es la forma más continental conocida en la península (Nicholas Brokaw, pers. comm.). Se localiza fuera del parque y consecuentemente no recibe ninguna protección oficial.

Todos los tipos de vegetación están sujetos a ser quemados, ya sea directamente a través del uso de fuego para la caza de venados o indirectamente a través de incendios que escapan de los lotes adyacentes, en los que se practica el método corte y quema (la roza). Fue difícil evaluar los efectos del fuego en la ciénaga y su impacto sobre la biodiversidad en este estudio a corto plazo. El SP7 había sido quemado un año antes del estudio (Tabla 3.1). En el anterior, registramos 17 especies; en este último, registramos 19 especies. Ambas compartían la mayoría de sus especies. Sería útil registrar el impacto de estas alteraciones durante períodos más largos. Igualmente interesante sería un estudio en el que se documentara la probabilidad de supervivencia de ciertas especies de plantas cuando se pierden sus partes aéreas durante los incendios. Observamos que en segmentos parcialmente quemados de ramales de base, los pseudo-bulbos de las orquídeas *Eulophia* y *Habenaria* pueden sobrevivir.

Macrohábitats acuáticos y sus características florales

En base a las características del fondo del río y la forma de la corriente, los hábitats del río se pueden subdividir en ríos permanentes, ríos estacionales y macrohábitats rápidos (referirse a la Tabla 1.1).

Del río permanente (profundo) se obtuvieron muestras en seis localidades: cuatro en el Río San Pedro, una en el Río Sacluc y una en el Río Escondido (Tabla 3.1). Estas áreas se caracterizaban por fondo de roca calcárea, aguas profundas y flujo de agua durante todo el año. Este hábitat sostiene un mosaico de tipos de bosques ribereños y ciénagas. La mayor variación en ambientes acuáticos, tipos de vegetación y cantidad de especies de macrófitas acuáticas se encontró en el Río San Pedro. Nuestros registros sugieren que el macrohábitat del río es uno de los más diversos, aunque la mayoría de sus especies no son únicas en este macrohábitat.

Del poco profundo macrohábitat del río se obtuvieron muestras en tres ubicaciones de diferentes áreas focales (Tabla 3.1). De éstas, solamente E10 se encontraba en un área que había sido modificada extensivamente por la actividad humana. Este punto de muestreo fue rodeado por un bosque secundario depurado, dominado por *Bucida buceras* con arbustos de *Mimosa pigra*. Las plantas flotantes acuáticas se limitaban a *Salvinia minima*. En el caso de los ríos Chocop y Candelaria, cada uno tenía bosques ribereños, caracterizados por asociaciones distintas que aparentemente estaban intactas. El bosque ribereño en el Chocop albergaba *Bucida buceras* y *Matayba*, mientras que el bosque en Candelaria consistía de la asociación de *Bucida buceras* y *Diospyros*. En estos ríos, las plantas acuáticas sumergidas son muy escasas; solamente se encontró *Cabomba* en una curva del Candelaria.

Los rápidos del Río San Pedro están formados por depósitos de bivalvos y son florísticamente interesantes debido a la presencia de una población única de *Vallisneria americana*. Sin embargo, las especies presentes en el bosque ribereño, así como la ciénaga que rodea los rápidos se encuentran en otras áreas. Adicionalmente, en el E9 (Canal Manfredo) del Río Escondido, también se encontraron depósitos de bivalvos que fueron comparativamente ricos en taxa. Los depósitos de bivalvos pueden contribuir a la diversidad de microhábitats para las plantas.

Tabla 3.1. Macrohábitats, tipos de vegetación y número de especies de plantas registradas en los 25 puntos de muestreo.

Punto de muestreo	Macrohábitat	Tipo de vegetación	Número de taxa
SP3: Arroyo La Pista	Tributario	Ciénaga	14
SP4: El Sibalito	Río profundo	Ciénaga	12
SP6: Caracol	Tributario	Bosque ribereño	9
SP7: San Juan	Tributario	Ciénaga	17
SP8: La Caleta	caño	Ciénaga	10
SP9: Pato 1	Río profundo	Bosque ribereño	11
SP11: Río Saclúc	Río profundo	Bosque ribereño	12
SP12: Rápidos	Río profundo	Arrecife, bosque ribereño	6
SP13: Est. Guacamayas	Río profundo	Ciénaga	19
E5: Laguna Cocodrilo	Laguna oxbow	Ciénaga	4
E6: Laguna Jabirú	Laguna oxbow	Ciénaga	5
E7: Caleta Escondida	Laguna oxbow	Ciénaga, bosque ribereño	14
E8: Punto Icaco	Río profundo	Bosque ribereño	9
E9: Canal Manfredo	bivalvo/rápidos	Ciénaga	28
E10: Nuevo Amanecer	Río poco profundo	Bosque ribereño	5
F5: Flor de Luna	Laguna	Ciénaga, bosque ribereño	9
F6: Pozo Xan	Laguna	Ciénaga	6
C1: Chocop	Río poco profundo	Bosque ribereño	4
C2: Guayacán	Laguna	Ciénaga	7
C3: La Pista	Laguna	Ciénaga	3
C4: Tintal	Laguna	Bosque ribereño	5
C6: Laguna La Lámpara	Laguna	Ciénaga, bosque ribereño	8
C7: Laguna Poza Azul	Laguna	Bosque ribereño	7
Amelia 2	Laguna	Ciénaga	3
CA1: Candelaria	Río poco profundo	Bosque ribereño	6

El macrohábitat tributario se encontró en tres ubicaciones: dos dentro de la ciénaga y una en un bosque ribereño (Tabla 3.1). Este macrohábitat es relativamente rico en especies, aunque éstas pueden encontrarse en otras partes, exceptuando las *Najas* cf. *wrightiana* (SP3).

El macrohábitat de caño se caracteriza por aguas estables, turbias. Nuestras observaciones acerca de este hábitat se limitan a ciénagas compuestas de un sibal en el Río San Pedro que no fue quemado durante dos años, aunque condiciones similares suceden en las estaciones tributarias del Río Sacluc. En este punto de muestreo, descubrimos que el final del sibal incluía especies terrestres del género *Ipomoea* y *Passiflora* y una planta de la familia de las Longaniaceae.

El macrohábitat de la laguna incluye cuerpos de agua aislados y se representa en ocho de nuestros puntos de muestreo (Tabla 3.1). En la mayoría de las lagunas, está presente el bosque ribereño en los alrededores, y las macrófitas acuáticas prosperan, dependiendo del tipo de sustrato encontrado en una profundidad de un metro y las características del banco. La riqueza de este marcohábitat varía entre tres y nueve especies, la mayoría de las cuales también se encuentran en los macrohábitats. La flora en el Pozo Xan (F6) contribuyó con especies conocidas solamente en este macrohábitat, incluyendo *Brasenia schreberi* y *Xyris* sp. La Laguna Flor de Luna (F5) aparentemente representa una laguna en el proceso de eutroficación, con dos tipos de vegetación en bancos opuestos - uno dominado con ciénagas por *Cladium* en uno y *Eleocharis* en el otro—y un bosque ribereño dominado por *Haematoxylum campechianum*. La superficie de la laguna se encuentra casi completamente cubierta de *Nymphaea ampla*. La laguna Guayacán es un cuerpo de agua grande en el cual parte del banco está compuesto por roca calcificada, cubierta con bosque no inundado. La vegetación de ciénagas en esta laguna es limitada, pero puede encontrarse en interiores protegidos. Estos interiores aparentan ser corrientes que alimentan la laguna.

Las muestras de la laguna de recodo muestran una riqueza desde muy baja (3) hasta muy alta (28) en cuanto a riqueza de especies. La variación en riqueza puede interpretarse como resultado de la diferencia en edades de estas lagunas, duración del aislamiento del canal principal del río y características geológicas e hidrológicas. La muestra del punto E9 (Canal Manfredo) incluía las especies encontradas solamente en este lugar dentro de este estudio, como las dos especies de *Utricularia*, *Sagittaria lanciforme*, *Hydrocotyle* cf. *bonariensis* y una orquídea entre otras.

CONCLUSIONES Y RECOMENDACIONES

- Debería ser una prioridad comprender las características físicas e hidrológicas de las áreas dentro y alrededor del parque para poder entender de mejor manera los procesos que determinan la heterogeneidad de vegetales en ciénagas, así como para administrar y evitar la contaminación en los humedales.

- El impacto del desarrollo de actividades petroleras se debería evaluar, y los cambios generados por la construcción de caminos y sus efectos sobre el drenaje de las aguas.

- En el futuro los estudios biológicos deberían evaluar la riqueza y composición en cada humedal del área y debería relacionar estos patrones con patrones de amplia escala a lo largo de la Península de Yucatán.

- Debe ser una prioridad reconocer y distinguir los procesos naturales y antropogénicos que afectan los ecosistemas acuáticos del parque.

- El fuego constituye una de las perturbaciones más importantes en el parque y afecta a todos los tipos de ambientes acuáticos. Los estudios a mediano y largo plazo se deberían dirigir para determinar los efectos del fuego sobre las poblaciones de plantas en ciénagas así como en bosques ribereños.

- La vegetación ribereña que se encuentra a lo largo de ríos no profundos como el Chocop y el Candelaria difiere de la que se encuentre a lo largo de ríos con mayor longitud y profundidad, como el San Pedro. Por lo tanto, cada uno de estos tipos de bosque deberían ser conservados con un esfuerzo similar. La presencia de la especie *Arthrostylidium* a lo largo del Candelaria debería ser causa de preocupación debido a que la especie se esparce de manera agresiva en áreas sujetas a incendios.

- El Río Sacluc debería incorporarse dentro del PNLT de manera que el bosque de encino (*Quercus* sp.) de ese lugar fuera protegido, así como los bosques ribereños que conectan con el San Pedro.

- Las áreas con depósitos de bivalvos en el Río Escondido poseen especies ricas y vulnerables a la destrucción debido a la erosión ocasionada por el tránsito de lanchas de motor, la ampliación del canal y los incendios. Los rápidos del Río San Pedro albergan a poblaciones únicas por lo que esta área se debería conservar.

- Existe una potencial amenaza de contaminación por petróleo en el Río Escondido, ya que la tubería pasa cerca de las aguas principales del río. Tomando en cuenta el efecto devastador que un derrame de petróleo podría tener a lo largo del humedal, se debería crear un plan para evitar este tipo de accidentes.

LITERATURA CITADA

Deevey, E. S., D. S. Rice, P. M. Rice, H. H. Vaughan, M. Brenner y M. Flannery. 1979. Mayan urbanism: impact on a tropical karst environment. Science 206: 298-306.

Islebe, G. A., H. Hooghiemstra, M. Brenner, D. A. Hodell y J. Curtis. 1996. A Holocene vegetation history of lowland Guatemala (Lake Petén-Itza). Doctoral dissertation. Amsterdam, the Netherlands.

Lot, A. y A. Novelo. 1990. Forested wetlands of Mexico. Pp. 287-298 in *Ecosystems of the World 15: Forested Wetlands* (A. E. Lugo, M. Brinson, and S. Brown, eds.). Elsevier, Amsterdam, the Netherlands.

Lot, A., A. Novelo y P. Ramirez-García. 1993. Diversity of Mexican aquatic vascular plant flora. Pp. 577-591 in *Biological Diversity of Mexico: Origins and Distribution* (T. P. Ramamoorthy, R. Bye, A. Lot, and J. Fa, eds.). Oxford University Press, New York, NY, USA.

Lugo, A. E., M. Brinson y S. Brown. 1990. Synthesis and search for paradigms in wetland ecology. Pp. 447-460 in *Ecosystems of the World 15: Forested Wetlands* (A. E. Lugo, M. Brinson, and S. Brown, eds.). Elsevier, Amsterdam, the Netherlands.

Lundell, C. L. 1937. The vegetation of Petén. Publication 478. Carnegie Institution of Washington, Washington, DC, USA.

Martínez, M. y A. Novelo. 1993. La vegetación acuática del Estado de Tamaulipas, México. Anales Inst. Biol. Univ. Nac. Autón. México, Ser. Bot. 64: 59-86.

Morales Can, J. 1998. Tesis de bachillerato. Universidad Nacional de San Carlos, Guatemala.

de Poll, E. 1983. Plantas acuáticas de la región El Estor, Izabal. Universidad de San Carlos de Guatemala, Facultad de Ciencias Químicas y Farmacia. Escuela de Biología, Guatemala.

Standley, P. 1924-1964. Flora of Guatemala. Fieldiana Botany Vol. 24. Field Museum of Natural History, Chicago, IL, USA.

Williams, L. O. 1966-1975. Flora of Guatemala. Fieldiana Botany Vol. 24. Field Museum of Natural History, Chicago, IL, USA.

CAPITULO 4

UN ESTUDIO ICTIOLOGICO DEL PARQUE NACIONAL LAGUNA DEL TIGRE, PETEN, GUATEMALA

Philip W. Willink, Christian Barrientos, Herman A. Kihn y Barry Chernoff

RESUMEN

- Se recolectaron peces de agua dulce en 48 localidades de cinco puntos focales regionales del Parque Nacional Laguna del Tigre (PNLT) y el biotopo.

- Se capturaron cuarenta y un especies de peces de 55 conocidas o que se sospecha existen en la región. Varias de las especies capturadas representan nuevos registros para la Cuenca del Río San Pedro en Guatemala (incluyendo *Batrachoides goldmani*, *Mugil curema* y *Aplodinotus grunniens*). Se encontraron dos especies exóticas (*Oreochromis* sp.) y el césped (*Ctenopharyngodon idella*). Se capturaron dos especies no descritas, un pez gato y otro que pertenecía al género *Atherinella*.

- Nuestros resultados muestran que la mayoría de las especies conocidas del Río San Pedro hace 60 años aún se encuentran presentes.

- El punto focal del Río San Pedro es el más diverso, pero con pocas excepciones, la mayoría de las especies se encuentra distribuida homogéneamente entre las regiones dentro de la cuenca o entre los macrohábitats. Los hábitats de la laguna todos aparentemente tenían docenas de especies, sin importar las alteraciones a la vegetación marginal, el uso humano o la proximidad a los caminos.

INTRODUCCION

Los peces que viven en las aguas dulces de El Petén de Guatemala y México representan una ictiofauna relativamente bien conocida y bien descrita. En gran parte, y empezando con las publicaciones de Alfreid Gúnther (1868) que se basa en las recolecciones de Osberd Salvin en 1859, ha habido una serie de publicaciones que tratan con los peces de esta interesante región. C. Tate Regan (1906-08) fue el primero en reconocer y establecer provincias biogeográficas para los peces de México y Centroamérica. Carl L. Hubbs realizó exploraciones críticas del sur del Petén, el lago Petén Itzá y el Río San Pedro en 1935. Sus recolecciones forman una base importante de comparación para nuestra encuesta. Desde Hubbs, se han descrito muchas especies, y se han entablado debates sobre la naturaleza de las provincias biogeográficas (Miller, 1976).

El Río San Pedro es un tributario principal del Río Usumacinta. El Río Candelaria es un afluente independiente de la Laguna de Términos. Los peces de las tres cuencas se conocen porque se componen de 48 especies en base a especímenes registrados en el museo. Adicionalmente, las fuentes conocidas y las recolecciones del Río Pasíon en Sayaxché sugieren cinco especies adicionales, a las que nosotros agregamos dos exóticas. Esto eleva la ictiofauna conocida y potencial a 55 especies. Estas especies se enumeran en el Apéndice 7. Esta ictiofauna se compone de especies que viven principalmente en aguas dulces, además de las que libremente extienden sus rangos desde hábitas marinos y estuarinos hasta las aguas dulces, tales como las lisas (*Mugil* spp.).

La mayoría de los componentes de la ictiofauna encontrados en el Parque Nacional Laguna del Tigre (PNLT) y el biotopo pertenecen a un grupo ampliamente distribuido de peces encontrados entre el Río Coatzacoalcos al norte y el Río Polochic/Río Sarstún en el sur. Regan (1906-08) y Miller (1966, 1976) se han referido a esta distribución general como la Provincia del Río Usumacinta, aunque Bussing (1987) incluyó el Río Motaga en la Provincia del Río Usumacinta, esto no recibió mucha atención. Sin embargo, Chernoff (1986) descubrió que un grupo monofilético de *Atherinella* incluye la especie encontrada en el Motagua, Petén-Itzá y los drenajes del Usumacinta al Coatzacoalcos. Otros grupos presuntamente monofiléticos (por ejemlo *Phallichthys* y *Xiphophorus*) también incluyen al Motagua en las distribuciones de la Provincia del Usumacinta. Estas relaciones de grupo de hermanas, incluyendo el Río Motugua, crean la hipótesis de la presencia de ancestros comunes que una vez habitaron las regiones del Motagua y del Usumacinta, seguidos por algún grado de evolución local después de que las dos regiones se aislaron, probablemente durante el Terciario menor (Rosen y Bailey, 1959; Rosen, 1979; Weyl, 1980).

A excepción del Lago Petén Itzá y los lagos relacionados, hay poco o no hay endemismo de peces dentro de El Petén propiamente. Esto incluye al PNLT y al biotopo. Sin embargo, notamos que existe un gran grado de endemismo al nivel de la Provincia del Usumacinta *sensu* Regan (1906-08). Por lo tanto, las áreas protegidas dentro de la provincia mayor deben ser establecidas para poder conservar su interesante fauna. Uno de nuestros objetivos fue descubrir si el PNLT y el biotopo podrían servir para este propósito. Esto es particularmente importante ya que en los estados mexicanos vecinos de Campeche y Tabasco hábitats similares han sido totalmente o altamente degradados debido al ganado y exploración petrolera.

El propósito de esta encuesta fue triple: 1) documentar la biodiversidad de peces en el PNLT y el biotopo, 2) documentar, a través de protocolos de evaluación rápida, la diversidad de peces asociada con puntos focales y macrohábitats, y 3) identificar al mayor grado posible el estado de salud de los hábitats, las regiones críticas para mantenimiento de biodiversidad y las amenazas potenciales o reales así como sus impactos potenciales.

METODOS

Durante el curso de 20 días, se realizaron 48 recolecciones de peces (Apéndices 8-9). Los peces se capturaron utilizando los siguientes métodos:
• atarrayas (5 m x 2 m x 3.4 mm; 1.3 m x 1 m x 1.75 mm);
• redes (2 m, 1.5 m);
• redes experimentales de arroyo (48 m);
• pesca manual con redes (3 m) y
• redes sumergidas.
La pesca con redes barrederas fue bastante difícil porque no encontramos ninguna playa o áreas libres para halar las redes barrederas.

Todos los especímenes de peces recolectados se preservaron inicialmente en una solución de 10% de formalina, luego se transfirieron a una de 70% de etanol para su almacenamiento y archivo. Las identificaciones se hicieron en el laboratorio con la ayuda de descripciones originales y estudios de revisión más recientes (Regan, 1905; Rosen y

Bailey, 1959, 1963; Hubbs y Miller, 1960; Miller, 1960; Rivas, 1962; Suttkus, 1963; Deckert y Greenfield, 1987; Greenfield y Thomerson, 1997; y otros). Los especímenes se depositaron en el Field Museum de Chicago y en el Museo de Historia Natural, Universidad de San Carlos de Guatemala. Para ubicar la diversidad de peces, se muestrearon tantos hábitats como fue posible (referirse a la Tabla 1.1). Las áreas muestreadas se trataron jerárquicamente. En el nivel superior de la jerarquía se encuentran las áreas focales descritas en el "Itinerario de expedición y Diccionario de geográfico." Dentro de estas áreas focales, todo el equipo de ciencias acuáticas coordinó estudios a una escala más fina (muestra o puntos de georeferencia). Nuestras 48 muestras, en la mayoría correspondían a 40 puntos de georeferencia (Apéndice 1, 8). Pocas muestras se encontraban fuera de estos regímenes de muestras coordinadas. Los Indices de Similitud de Simpson (cantidad de taxa compartida entre dos grupos/número de taxa en el grupo menor) se utilizaron para realizar comparaciones entre los macrohábitats, áreas focales y tipo de equipo. Seleccionamos el Indice de Similitud de Simpson porque éste enfatiza la presencia de taxa utilizando el número de especies en el grupo menor del denominador (Chernoff et al., 1999). Consideramos la ausencia de taxa como ambigua debido a que una taxa determinada puede en realidad estar ausente o pudo haber estado presente pero no haber sido detectada.

RESULTADOS Y DISCUSION

Efectividad del muestreo

Aunque se cree que aproximadamente 55 especies de peces habitan la región estudiada (Apéndice 7), recolectamos 41 (Apéndice 8). Algunos de los peces no recolectados no son comunes en el área. Entre los ejemplos se incluye *Ictiobus meridionalis* y *Gobiomorus dormitor*. Los demás son peces grandes que probablemente nadan a través de los ríos largos pero no permanecen en un lugar determinado durante mucho tiempo. Estos son difíciles de recolectar durante muestreos cortas, pero a veces son atrapados por pescadores locales o bien personas que realizan estudios a largo plazo. Entre los ejemplos se incluye *Anguilla rostrata*, *Mugil cephalus* y *Centropomus undecimalis*. La curva de acumulación de especies demuestra que alcanzamos la asíntota después de ocho días y creemos que la taxa recolectada es una buena representación de las especies que se encontraban allí en ese momento (Figura 4.1).

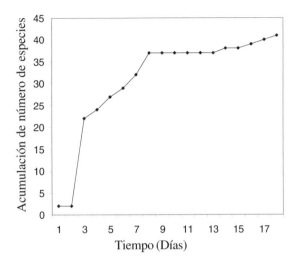

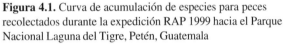

Figura 4.1. Curva de acumulación de especies para peces recolectados durante la expedición RAP 1999 hacia el Parque Nacional Laguna del Tigre, Petén, Guatemala

Puntos focales (ver Itinerario y boletín de la expedición).

Río San Pedro

Diversidad de peces

Del Río San Pedro, recolectamos 38 especies de peces durante el muestreo RAP (Apéndice 8) de un estimado de 55 especies para la región. Catorce de las especies recolectadas eran cíclidos. El segundo grupo más dominante fue el de poecílidos, con 8 especies. Se recolectaron diez especies a lo largo del punto focal y en una variedad de macrohábitats. Estos incluyen *Astyanax aeneus, Cichlasoma pasionis, Belonesox belizanus, Cichlasoma synspilum, Cichlasoma helleri, Gambusia sexradiata, Cichlasoma heterospilum, Petenia splendida, Cichlasoma meeki,* y *Poecilia mexicana.*

Batrachoides goldmani, Mugil curema y *Aplodinotus grunniens* son los primeros registros oficiales para estas especies en esta cuenca de Guatemala. En comparación a las recolecciones de Hubbs en 1935, la diversidad y los números aparentemente no han cambiado significativamente.

Diversos hábitats aquí son de especial atención para el mantenimiento de la diversidad de peces y se discuten a continuación.

1) Los bosques bajos, similares a varzea, inundados cerca de Paso Caballos y a lo largo del Río Sacluc son áreas críticas de crianza para peces jóvenes y probablemente funcionan como áreas en las cuales los peces, como el *Brycon* spp., comen y dispersan semillas. Este tipo de hábitat no es reconocido y no se describe en Mesoamérica.

2) Sibal (dominado por *Cladium jamaicense*) es probablemente un microhábitat importante para las cíclidos y el barbo. Probablemente también sirve de criadero para peces que depositan huevos de acuerdo con los niveles de fluidez del agua y que se alimentan de semillas. Aún durante la estación seca, el agua es una parte integral del sibal ya que el agua penetra en el sibal. La hojarasca en el fondo del río podría proporcionar una importante fuente de alimento para los peces, ya que la microflora constantemente crece en las hojas. Las hojas también pueden ser un escondite para los peces pequeños.

3) Los tributarios que entran al Río San Pedro desde la parte norte del río (Paso Caballos, SP2; Arroyo La Pista, SP3) fueron muy claros y poco profundos y de corriente rápida. El arroyo La Pista se encontraba sobre un sustrato rocoso, casi formando canales con troncos. Sin embargo, la diversidad de los peces fue relativamente baja, con solamente 12 especies en total. En el Arroyo La Pista, observamos enormes agregados de *Dorosoma petenense*, de aproximadamente 60 mm de longitud, que se reunían sobre las rocas. Estos hábitats de corrientes posiblemente se encuentran amenazados debido a la modificación y la contaminación del hábitat.

Observaciones

Los peces aguja y *Hyphessobrycon* fueron atrapados desde el fondo. Los peces aguja fueron abundantes y podrían ser los piscívoros dominantes. Muchos peces, incluyendo los aguja y los *Petenia*, estaban llenos de huevos o tenían testas grandes, lo cual indicaba que se estaban preparando para la reproducción. Se recolectaron los especímenes de *Atherinella alvarezi*. Aunque algunos especímenes fueron recolectados por Hubbs en 1935, las especies no se han recolectado en ocasiones más recientes del Río Pasión o cerca de Sayaxché.

Amenazas

Las aglomeraciones humanas son muy significativas debido a la quema de sibales, contaminación de fuentes y deforestación.

Fragilidad

La fragilidad es alta para el sibal y varzea debido a que se encuentran fragmentados.

Río Escondido

Diversidad de peces

Se encontraron solamente 27 especies de peces de las 55 encontradas en todos los macrohábitas dentro del punto focal del Río Escondido (Apéndice 8). Las especies del Río Escondido contienen las 10 especies comunes (Tabla 4.1) más *Atractosteus tropicus, Dorosoma petenense, Cichlasoma friedrichsthalii* y *Cathorops aguadulce*, entre otros.

Observaciones

Muchos de los peces se encontraban en condición reproductiva.

Amenazas

Las amenazas principales incluyen la quema de sibal y la destrucción del bosque. Fuimos testigos de la quema de sibal y la conversión activa de hábitats de bosques para pasto y usos agrícolas. Estas actividades tienden a incrementar la contaminación del río, y eliminan importantes áreas de criadero y microhábitats con sombra y más frescos en los cuales los peces puedan vivir.

Fragilidad

Los hábitats de sibal y de bosques son frágiles porque sus bases se están degradando y/o volviendo más pequeñas. Casi todas las márgenes boscosas han sido habitadas por personas ya que esta zona ofrece suelo sólido y se encuentra cerca del acceso al agua. La fragilidad del río desde la perspectiva de proporcionar hábitats adecuados para la supervivencia de los peces no se conoce en este momento.

Flor de Luna
Diversidad de peces

De las 55 especies, se encontraron 22 dentro del punto focal de Flor de Luna (Apéndice 8). La fauna encontrada dentro de las lagunas fue un subgrupo del Río San Pedro, incluyendo las diez especies más comunes.

Amenazas

En las instalaciones de petróleo Xan, existe una amenaza potencial de contaminación a causa de petróleo de un pozo en funciones que se encuentra adyacente a la laguna y la ciénaga. Se notó que los peces recolectados por toxicología tenían una alta incidencia de aletas podridas y que un cíclido tenía una masa fibrosa en su hígado.

Es difícil estimar las amenazas a otras lagunas debido a que encontramos esencialmente las mismas especies ya sea si la laguna tenía márgenes vírgenes, o se encontraba rodeada de bosques explorados y campos agrícolas.

Fragilidad

Es difícil estimar la fragilidad de los hábitats de la laguna debido a que aparentemente soportan contra los impactos humanos, modificaciones de bosque y construcciones de caminos. No existen datos para comparar nuestro muestreo con muestras previas a la modificación de muchas de estas lagunas.

Río Chocop
Diversidad de peces

Solamente se encontraron 15 especies dentro del punto focal (Apéndice 8). Sin embargo, *Xiphophorus hellerii* se encontró de manera única en este punto focal.

Amenazas

Entre las principales amenazas se incluye la deforestación y la quema. La deforestación y la quema aumentará la contaminación en estos sistemas de agua relativamente claros, lo que en última instancia podría afectar la diversidad de peces. Estas amenazas ya están sucediendo a una velocidad rápida aunque estas localidades se encuentren dentro del PNLT.

Fragilidad

Las corrientes de agua principales como las que encontramos en este punto focal tienden a ser más frágiles que las secciones de corriente hacia abajo. La contaminación debido a la deforestación y conversión de hábitat cambian el carácter de las corrientes de aguas de manera dramática.

Río Candelaria
Diversidad de peces

Solamente se obtuvieron muestras de una localidad dentro de este punto focal, de manera que es difícil hacer estimaciones de la diversidad. Recolectamos 13 especies (Apéndice 8). Solamente en este lugar recolectamos *Atherinella* c.f. *schultzi*. Debido a que el Río Candelaria es un drenaje independiente del Río San Pedro, existe una posibilidad de que existas taxa novedosas.

Amenazas

Al igual que en las cercanías del punto focal del Río Chocop, la deforestación y la quema son las amenazas principales. Ambas se observaron durante nuestro estudio. La deforestación y la quema aumentarán la contaminación en estos sistemas de agua relativamente claros, que en última instancia podría afectar la diversidad de peces. Estas amenazas ya están ocurriendo a una velocidad rápida, aunque estas localidades se encuentren dentro del PNLT.

Fragilidad

Las corrientes principales como las que encontramos en este punto focal tienden a ser más frágiles que las secciones río abajo. La alteración por deforestación y conversión de hábitat cambian el carácter de las corrientes principales de manera dramática.

Comparaciones entre macrohábitats, puntos focales y equipo

El río permanente (es decir, el Río San Pedro) fue el macrohábitat con la mayoría de especies (N=33) (Apéndice 9). Las lagunas, tributarios y ríos estacionales también tuvieron números relativamente altos de especies (23, 22 y 24 respectivamente). El río permanente es el tipo de macrohábitat acuático más común en el PNLT, aunque abundan las lagunas en las áreas focales de Flor de Luna y Chocop.

Tabla 4.1. Indices de Similitud de Simpson entre tipos de macrohábitats.
El promedio para todas las comparaciones es de 0.830, con una desviación estándar de 0.142

	Lagunas	Lagunas Oxbow	Caños	Tributarios	Ríos, estacionales	Ríos, permanentes	Ríos, bahías	Ríos, rápidos
Lagunas								
Lagunas Oxbow	1.000							
Caños	1.000	0.857						
Tributarios	0.818	0.882	1.000					
Ríos, estacionales	0.826	0.882	1.000	0.818				
Ríos, permanentes	0.870	0.882	1.000	0.864	0.833			
Ríos, bahías	0.923	0.692	0.571	0.846	0.846	1.000		
Ríos, rápidos	0.857	0.714	0.571	0.642	0.714	0.857	0.462	

En general, la similitud entre macrohábitats es alta (media en el Indice de Similitud de Simpson - 0.830; desviación estándar = 0.142; Tabla 4.1). Esto indica que existe un gran número de especies compartidas entre los macrohábitats. Ciertos macrohábitats difieren más fuertemente en la composición de especies - más notablemente en ríos/bahías comparado con caños, y todos los macrohábitats comparado con ríos/rápidos. Los bajos índices de similitud entre ríos/bahías, lagunas de recodo y caños fue una sorpresa debido a que frecuentemente se encuentra cerca y superficialmente parecen macrohábitats muy semejantes. Todas son áreas de aguas lentas a lo largo de las orillas de los ríos. Creemos que los bajos índices son mas bien debido al del bajo número de especies recolectadas en estos macrohábitats al compararlos con otros (Apéndice 9). Los índices de similitud tienden a ser bajos al compararlos con subgrupos de muestras pequeñas (Chernoff et al., 1999). Una explicación similar puede responder a la poca similitud entre los ríos/rápidos y otros macrohábitats. Un factor adicional que lleva a la diferencia de los rápidos es la presencia de especies recolectadas solamente o en gran parte en los rápidos, incluyendo *Batrachoides goldmani*, *Cichlasoma lentiginosum* y *Cichlasoma bifasciatum*. Los rápidos, con su suave corriente, sustrato rocoso y bivalvo, es un macrohábitat raro dentro del área estudiada.

El área local tuvo la mayor riqueza en cuanto a peces. Solamente se encontraron tres especies fuera de San Pedro (Apéndice 8). Basándose en los Indices de Similitud de Simpon, San Pedro, Escondido y Flor de Luna fueron bastante similares (Tabla 4.2) y por lo tanto representan la misma fauna. Las áreas focales Río Escondido y Flor de Luna son esencialmente subgrupos del San Pedro.

La media de similitud entre todas las áreas focales es de 0.873 (desviación estándar = 0.111). La mayoría de los valores inferiores provienen de comparaciones de otras áreas con Candelaria o Chocop. Atribuímos algunas de estas fuertes diferencias al hecho de que Candelaria se encuentra en un drenaje diferente y que alberga distintos tipos de macrohábitats. El hábitat de Candelaria es el más similar al de Chocop, pero su similitud es baja (Tabla 4.2). De nuevo, esto podría ser debido al bajo número de especies recolectadas en estas áreas (<16 especies en cada una) comparado al número de especies recolectadas en otras áreas (>21 especies cada una; Apéndice 8).

Utilizando los Indices de Similitud de Simpson para comparar tipos de equipo utilizado, se demuestra que las mismas especies fueron recolectadas utilizando la misma red barredera o atarraya (Tabla 4.3). Menos especies fueron recolectadas con la red barredera (Figura 4.2), pero éstas fueron un subgrupo de las recolectadas por la red y por la atarraya. La similitud fue más baja en contraste con la red de arroyos (Tabla 4.3). Esto se debe al número relativamente

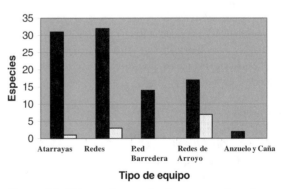

Figura 4.2. Número de especies de peces recolectados por cada equipo usado durante la expedición RAP en el 1999 en el PNLT.

grande de taxa recolectada exclusivamente con red de arroyos, (por ejemplo *Mugil curema*, la mayoría de los pez gato, *Aplodinotus grunniens*; (Figura 4.2). Solamente dos especies (*Brycon guatemalensis* y *Petenia splendida*) fueron atrapadas con anzuelo y caña, siendo un método poco frecuente, por lo que hay poco que decir sobre este tipo de utensilio. En general, es importante utilizar diversas técnicas de recolección para poder inventariar la totalidad de la fauna en un área determinada.

En resúmen, consideramos que cada una de las regiones focales contiene subgrupos de la ictiofauna encontrada dentro del Río San Pedro. Exceptuando los peces que prefieren hábitats de ríos grandes (- el ictalúrido y los peces gato, *Mugil*, *Aplodinotus*, etc.—)la mayoría de los peces están distribuidos en grupos más pequeños de manera homogénea entre macrohábitats y regiones focales. Las excepciones a este patrón incluyen *Xiphophorus hellerii*, *Heterandria bimaculata* y *Atherinella* cf. *schultzi*, las cuales econtramos en hábitats de aguas principales. El área de rápidos/arrecifes también tenía dos especies, *Cichlasoma lentiginosum* y *Batrachoides goldmani*, que no se encontraron en ningún otro lugar.

CONSIDERACIONES DE CONSERVACION

La homogeneidad de la distribución de especies entre macrohábitats y entre regiones focales sugiere que fácilmente se puede implementar un plan general para la conservación de la biodiversidad de peces en el PNLT. Recomendamos que

Tabla 4.2. Indices de Similitud de Simpson entre áreas focales. El promedio para todas las comparaciones es de 0.873, que es una desviación estándar de 0.111

	San Pedro	Escondido	Flor de Luna	Chocop	Candelaria
San Pedro					
Escondido	0.963				
Flor de Luna	1.000	1.000			
Chocop	0.867	0.933	0.733		
Candelaria	0.846	0.923	0.769	0.692	

Tabla 4.3. Indice de similitud de Simpson entre tipos de equipos. El promedio para todas las comparaciones es de 0.694, con una desviación estándar de 0.251.

	Atarrayas	Redes	Red barredera	Redes de arroyo	Anzuelo y Caña
Atarrayas					
Redes	0.935				
Red barredera	1.000	1.000			
Redes de arroyo	0.529	0.471	0.500		
Anzuelo y Caña	0.500	0.500	0.500	1.000	

se proteja una muestra de cada uno de los tipos de
macrohábitats como áreas de conservación, especialmente los
hábitats del arrecife y de aguas principales.

Taxones presentados

Se observaron dos especies no nativas: la especie
Oreochromis, que es un cíclido africano (no se guardaron
especímenes) y la *Ctenopharyngodon idella*, que es un pez
pequeño asiático. Estas especies han sido presentadas
alrededor del mundo para controlar la vegetación acuática y
sirven como fuente de alimento. Hasta donde sabemos, éstas
no han sido presentadas en el norte de Guatemala. Pero han
sido presentadas en algunos de los estados vecinos de
México. Es probable que estos peces hayan nadado a lo largo
del Usumacinta hasta el río San Pedro y luego al PNLT. Este
es un ejemplo gráfico de la interconexión entre los sistemas
acuáticos a través de las fronteras y resalta la necesidad de
estrategias de manejo ecoregionales. Cualquier iniciativa de
conservación o la introducción de especies exóticas, debe
tomar en cuenta la movilidad de la fauna.

Consideraciones biogeográficas

Si incluimos la especie *Oreochromis*, entonces registramos
42 especies durante la encuesta ictiológica en el PNLT. De
éstas, 34 son endémicas a la provincia del Usumacinta. Estos
taxa endémicos incluyen nueve cíclidos (*Cichlasoma
bifasciatum, C. helleri, C. heterospilum, C. lentiginosum, C.
meeki, C. pasionis, C. pearsei, C. synspilum* y *Petenia
splendida*), de las cuales todas son utilizadas como alimento.
Otras seis especies registradas (*Astyanax aeneus, Brycon
guatemalensis, Rhamdia guatemalensis, Belonesox
belizanus, Poecilia mexicana* y *Cichlasoma robertsoni*) se
encontraron en el Usumacinta, Motagua y posiblemente en
otros drenajes. Se integraron dos especies
(*Ctenopharyngodon idella* y *Oreochromis*). En resumen,
mucha de la taxa recolectada se encuentra solamente en el
drenaje del Usumacinta.

Se estima que existen doscientas o más especies en el
drenaje Usumacinta-Grijalva (Miller, 1976). Creemos que
existen 58 especies (56 nativas, 2 introducidas) en el norte de
las tierras bajas de Guatemala dentro de la cuenca del
Usumacinta. Estas 58 especies representan el 29% del total
de la fauna de peces del Usumacinta. El PNLT ocupa una
gran área del norte de las tierras bajas de Guatemala, y
relativamente no se ha modificado. Otras porciones del
drenaje del Usumacinta se encuentran en peligro de ser
desarrolladas. La preservación del PNLT ayudaría a proteger
a un cuarto de la fauna de peces del Usumacinta. Esto es
significativo debido al alto grado de endemismo global dentro
del drenaje del Usumacinta y el daño ambiental aparente
detrás de los límites del parque, incluyendo México.

RESUMEN DE HALLAZGOS Y CONCLUSIONES

Las siguientes son conclusiones generales que resultan de
nuestras interpretaciones de la distribución de peces en los
cinco puntos focales.

- Recolectamos 41 especies de peces de agua dulce de un
  potencial de 55 especies que se piensa que habitan esta
  región. Los peces que no recolectamos incluyen
  Centropomus undecimalis (reportado por pescadores
  de Sayaxché), *Rivulus tenuis* (reportado en estudios
  anteriores), *Tarpon atlanticus* y *Hyporhampus
  mexicanus* (reportado desde Sayaxché), *Ictiobus
  meridionalis* (conocido en el Río Pasión), y *Rhamdia
  laticauda* (conocido en el Río Pasión y en el Río
  Negro).
- Identificamos dos especies de peces introducidos, y
  recolectados por los pescadores: la tilapia *(especie
  Oreochromis*, no se guardaron especímenes) y la carpa
  (*Ctenopharyngodon idella*). Este es el primer registro
  oficial de esta especie en el norte de Guatemala.
- La lisa *(Mugil curema)* es el primer registro de la
  especie de la cuenca de San Pedro y Usumacinta en
  Guatemala.
- El tambor de agua dulce (*Aplodinotus grunniens*) y el
  pejesapo (*Batrachoides goldmani*) son los primeros
  registros en la cuenca de San Pedro.
- La fauna de peces está distribuida más o menos
  homogéneamente entre los cinco puntos focales
  regionales y entre los ocho macrohábitats. No pudimos
  indentificar algún patrón de especies entre los
  componentes principales o las comunidades de peces.
- Hubo un grupo de 10 especies que se encontró en la
  mayoría de puntos y macrohábitats. Este grupo incluye
  *Astyanax aeneus, Belonesox belizanus, Cichlasoma
  helleri, C. heterospilum, C. meeki, C. pasionis, C.
  synspilum, Gambusia sexradiata, Petenia splendida* y
  Poecilia mexicana.
- Ciertos grupos pequeños de especies se encontraron
  solamente en macrohábitats determinados. Los
  siguientes son algunos ejemplos: rápidos/arrecifes—
  Cichlasoma lentiginosum, Batrachoides goldmani, y en
  gran parte *Cichlasoma bifasciatum*; aguas principales
  con corriente—*Xiphophorus hellerii* y *Heterandria
  bimaculata*; canal principal del Río San Pedro—*Brycon
  guatemalensis*, pez gato e ictalurido, *Mugil curema,
  Aplodinotus grunniens*, y *Eugerres mexicanus*.
- Las faunas de las lagunas fueron marcadamente
  similares a pesar de las diferencias en cuanto a la
  proximidad a asentamientos humanos, caminos u otros
  disturbios principales.
- La heterogeneidad de los macrohábitats es baja. Los
  hábitats de los ríos y lagunas tenían márgenes bastante
  poco profundas.

- La fauna de peces es básicamente la misma que se encontró en muestreos cortos hace 60 años. No encontramos evidencia de extinciones. Sin embargo, debe notarse que se han introducido dos especies exóticas.

RECOMENDACIONES DE CONSERVACION

Estos resultados sugieren varias recomendaciones específicas para la conservación de poblaciones de peces y hábitats dentro del parque. Las recomendaciones no se presentan en orden de prioridad.

- Desarrollar un plan de conservación integrado para las áreas que contienen un número grande de lagunas, es necesario un estudio hidrológico para proporcionar información crítica sobre la conexión de lagunas, estanques y sibales durante las estaciones de inundación. ¿Cuáles son los roles o fuentes de las faunas de peces en éstos hábitats relativamente aislados? Grupos de lagunas y sibales se deben identificar como críticos para el mantenimiento de la biodiversidad.

- Los administradores de los parques deben reconocer la naturaleza crítica del bosque bajo, similar a varzea y los hábitats sibales para la reproducción de peces, criaderos, reproducción de larvas para peces y peces jóvenes. Estos hábitats deben considerarse frágiles y amenazados debido a la modificación humana, especialmente a través de la quema. Recomendamos que la franja de sibal no modificado que se encuentra al sur del Río San Pedro y la varzea a lo largo del Río Sacluc y debajo del Paso Caballos se debería incluir en el Parque.

- El impacto humano en el Río San Pedro, tanto en México como en Guatemala, se debería estudiar debido a que ha habido mucho deterioro de hábitats naturales a lo largo de las márgenes del río en la frontera sur del PNLT.

- Según los estudios preliminares realizados por hidroecólogos, los arrecifes de moluscos filtran el agua entre la Estación Biológica Las Guacamayas y el pueblo de El Naranjo. Estos tipos de arrecifes se encuentran raramente en aguas dulces y tienen una comunidad de peces única asociada. Esta región de macrohábitat debe ser protegida para poder asegurar la salud de la ictiofauna en la parte sureste del parque y también para asegurar la buena calidad del agua para las personas de El Naranjo.

- Los estudios sobre la dinámica de población de diferentes especies de peces para alimento, se deben implementar de inmediato, para proteger la ictiofauna de presiones generadas por los humanos debido a la subsistencia de pesca comercial. La presión proviene principalmente de la pesca de red no regulada, la pesca con anzuelo y caña, pesca con atarraya y pesca con red. Los objetivos principales son la *Petenia splendida*, y otros cíclidos, *Brycon guatemalensis* y los barbos grandes. Notamos que una de las fuentes principales de la presión de pesca sucede durante la Semana Santa, cuando los pescadores explotan los recursos fuertemente a lo largo de la cuenca.

- Se necesitan inmediatamente datos sobre el uso humano de peces dentro del parque y dentro del biotopo, para predecir las demandas futuras sobre poblaciones nativas de peces de agua dulce conforme la explosión demográfica en estas regiones.

- Un plan de conservación requerirá información sobre la ecología de la comunidad de peces durante las estaciones que no sean la estación seca.

- Los hábitats de las tierras bajas de los ríos y las lagunas dentro del PNLT y el biotopo se deben comparar con hábitats similares dentro de la provincia del Usumacinta, tanto en México como en Guatemala, y en otros lugares de la RBM. El PNLT y el biotopo pueden servir como un área de conservación adaptable para esta comunidad de peces ampliamente distribuida.

- La fuerte deforestación y destrucción de sibal deben evitarse, ya que podría afectar adversamente los hábitats acuáticos que se encuentran dentro de la cuenca, debido a una contaminación aumentada en el resto de las vías acuáticas.

LITERATURA CITADA

Bussing, W. L. 1987. *Peces de las Aguas Continentales de Costa Rica.* Editorial de la Universidad de Costa Rica, San Jose, Costa Rica.

Chernoff, B. 1986. Phylogenetic relationships and reclassification of menidiine silverside fishes, with emphasis on the tribe Membradini. Proceedings of the Academy of Natural Sciences of Philadelphia 138: 189-249.

Chernoff, B., P. W. Willink, J. Sarmiento, A. Machado-Allison, N. Menezes y H. Ortega. 1999. Geographic and macrohabitat partitioning of fishes in Tahuamanu-Manuripi region, upper Río Orthon basin, Bolivia. Pp. 51-67 in *A Biological Assessment of the Aquatic Ecosystems of the Upper Río Orthon Basin, Pando, Bolivia* (B. Chernoff and P. W. Willink, eds.). Bulletin of Biological Assessment 15. Conservation International, Washington, DC, USA.

Collette, B. B. y J. L. Russo. 1981. A revision of the scaly toadfishes, genus *Batrachoides*, with descriptions of two new species from the eastern Pacific. Bulletin of Marine Science 31: 197-233.

Deckert, G. D. y D. W. Greenfield. 1987. A review of the western Atlantic species of the genera *Diapterus* and *Eugerres* (Pisces: Gerreidae). Copeia 1987: 182-194.

Greenfield, D. W. y J. E. Thomerson. 1997. *Fishes of the Continental Waters of Belize.* The University Press of Florida, Gainesville, FL, USA.

Gúnther, A. 1868. An account of the fishes of the states of Central America, based on collections made by Capt. J. M. Dow, F. Goodman, Esq., and O. Salvin, Esq. Transactions of the Zoological Society of London 6: 377-494.

Hubbs, C. L. y R. R. Miller. 1960. *Potamarius*, a new genus of ariid catfishes from the fresh waters of Middle America. Copeia 1960: 101-112.

Miller, R. R. 1960. Systematics and biology of the gizzard shad (*Dorosoma cepedianum*) and related fishes. Fishery Bulletin 173: 371-392.

Miller, R. R. 1966. Geographical distribution of Central American fishes. Copeia 1966: 773-802.

CAPITULO 5

CONTAMINACION DE HIDROCARBONO Y DAÑO AL ADN EN LOS PECES DEL PARQUE NACIONAL DE LA LAGUNA DEL TIGRE, PETEN, GUATEMALA

Christopher W. Theodorakis y John W. Bickham

RESUMEN

- Las muestras de sedimento y de tejido de dos especies de peces (*Thorichthys meeki* y *Cichlasoma synspilum*) se recolectaron de una laguna que se encontraba inmediatamente adyacente a la instalación de petróleo Xan 3, así como de nuestros sitios de referencia (Laguna Flor de Luna, Laguna Buena Vista, Laguna Guayacán, Río Escondido y Río San Pedro). Se utilizó el sedimento para determinar los niveles de hidrocarbonos policíclicos aromáticos (PAH, por sus siglas en inglés), que son químicos indicativos de contaminación a causa de petróleo. Los tejidos se utilizaron para evaluar el nivel de daño en el ADN, lo que indica exposición a PAH y otros diversos contaminantes.

- El daño de ADN se analizó utilizando electroforesis de gel agarosa (para determinar la interrupción de la cadena de ADN) y el flujo de citometría (para determinar la ruptura de cromosomas).

- Las concentraciones de sedimentos de PAH de la laguna en Xan 3 no fueron significativamente más altas que los sitios de referencia, pero los de la Laguna Buena Vista fueron marcadamente mayores que en cualquier otro sitio.

- Para *T. meeki*, los niveles de la cadena de ADN de ruptura fueron mayores en las poblaciones Xan 3 que en los de Flor de Luna y el Río San Pedro. Para *C. synspilum*, el nivel de ruptura cromosomal fue mayor en el Xan 3 que en el sitio del Río San Pedro. En ambos casos, la cantidad de daño de ADN en Buena Vista fue intermedio entre el Xan 3 y otros sitios de referencia.

- Es necesaria la investigación adicional en las conexiones hidrológicas del parque y sus efectos en la redistribución contaminante, así como en los niveles de otros contaminantes.

INTRODUCCION

Una amenaza potencial a la biota del Parque Nacional Laguna del Tigre (PNLT) que no ha sido tratada, es el impacto de las actividades de extracción de petróleo en los campos de petróleo Xan. La preocupación mayor es que dichas actividades podrían ocasionar contaminación accidental del ambiente alrededor. La contaminación ambiental podría tener impactos adversos sobre las poblaciones y comunidades (Newman y Jagoe, 1996). Primero, podría resultar en una disminución en el éxito reproductivo o un aumento en la mortalidad de organismos afectados. Esto podría provocar reducciones en la densidad poblacional de algunas o todas las especies presentes. Algunas especies pueden ser más resistentes a los efectos de contaminación que otras. Esto podría ocasionar cambios en la estructura de la comunidad (composición de especies y abundancia relativa). Los ambientes expuestos al petróleo frecuentemente poseen menos especies que los hábitats no contaminados ya que relativamente pocas especies pueden tolerar la exposición a contaminantes asociados con el petróleo (Newman y Jagoe, 1996).

Los ecosistemas acuáticos son particularmente susceptibles a la contaminación por diversas razones. Primero, la contaminación terrestre y la deposición atmosférica tienden a eliminar los contaminantes de sistemas terrestres a acuáticos (Pritchard, 1993). En segundo lugar, debido a sus pieles y ampliamente permeables, los peces y otros organismos acuáticos son particularmente susceptibles a los efectos de contaminación (Pritchard, 1993).

Muchos hidrocarburos están ya degradados en un ambiente aeróbico, pero el ambiente acuático frecuentemente contiene poco o nada de oxígeno. Por lo tanto, los contaminantes pueden persistir en ecosistemas acuáticos por mucho más tiempo que en los sistemas terrestres (Ashok y Saxena, 1995). Finalmente, los organismos acuáticos contaminados implican un riesgo importante de salud para las personas que los consumen o el agua en el que habitan.

Aquí evaluamos los efectos de la contaminación de hidrocarburos en el estado fisiológico de los peces como consecuencia de las actividades de extracción de petróleo alrededor de los pozos petrolíferos de Xan en el PNLT. Hemos tratado esto al examinar a los peces en las lagunas localizadas a diferentes distancias de los pozos petrolíferos de Xan. Nuestra hipótesis es que las concentraciones de contaminantes de hidrocarburo sería más alta cerca de la ubicación de petróleo Xan 3, y que este patrón sería paralelo al aumentar el daño fisiológico a los peces. Existen muchas variables naturales que pueden contribuir a la variación en sistemas biológicos a lo largo del área. Por lo tanto, si se encontraran diferencias entre el Xan 3 y las áreas de referencia, sería difícil determinar si estas diferencias se debían a efectos contaminantes o a otras fuentes de variación ambiental. Los patrones paralelos de daño elevado en a los peces junto con las elevadas concentraciones de hidrocarburos entre los sitios (es decir, una evaluación toxicológica) podrían proporcionar fuerte evidencia de que la contaminación está afectando las poblaciones de peces.

Para determinar los niveles de contaminantes en el ambiente, se midieron los niveles de hidrocarburos policíclicos aromáticos (PAH, por sus siglas en inglés). Estos compuestos son componentes característicos de la contaminación del petróleo que se acumulan en el sedimento (Mille et al., 1998). El análisis incluye la determinación de concentraciones de PAH tanto en agua como en sedimentos.

La evaluación de los efectos de estos compuestos en los peces se enfocaron en las consecuencias fisiológicas por diversas razones. Primero, frecuentemente es difícil asumir las concentraciones del tejido corporal y los efectos tóxicos de los contaminantes de su concentración ambiental (McCarthy y Shugart, 1990). Esto se debe a que es frecuente que la disponibilidad de ciertos químicos que son absorbidos en los tejidos de los peces ("biodisponibilidad"), es decir, la cantidad de hidrocarburos, se dan naturalmente en el agua y sedimento. Asimismo, los contaminantes ambientales se encuentran presentes usualmente como mezclas complejas, y la toxicidad de estas mezclas no pueden predecirse con exactitud para las mediciones de concentraciones de químicos individuales (McCarthy y Shugart, 1990). Por lo tanto, las medidas de las respuestas fisiológicas de peces que viven en estos ambientes se necesitan para evaluar rigurosamente los impactos de contaminación de hidrocarburos.

La respuesta fisiológica determinada que se utilizó para esta evaluación fue daño en el ADN (ácido desoxiribonucléico). Se ha descubierto que este tipo de daño es característico de la exposición a PAH y contaminantes de petróleo(Bickham, 1990; Shugart, 1988). Las dos medidas utilizadas para evaluar el daño de ADN fueron ruptura de cadena de ADN y daño cromosomal, medido como en variaciones de célula a célula en el contenido de ADN. El ADN es una molécula doble, y la exposición a los PAH puede ocasionar rupturas en las cadenas de ADN (Shugart et al., 1992). Esto se debe a que los PAH se ligan químicamente al ADN, y las encimas de reparación del ADN eliminan porciones de la cadena que contienen esta modificación. Encimas posteriores llenan estas brechas, pero hasta que lo hacen, existen rupturas de cadenas temporales (Hoeijmakers, 1993). Si las rupturas se encuentran solamente en una cadena de ADN (rupturas de una cadena), el cromosoma permanece intacto, pero la integridad se pone en riesgo. Si las rupturas de cadenas suceden en dos cadenas adyacentes, entonces se podría fragmentar la cromosoma. Una manera de medir esta fragmentación es determinar la variación de célula a célula en el contenido de ADN (Bickham, 1990). Cuando los cromosomas se fragmentan, la división de una célula podría ocasionar una distribución no igual de cadenas de ADN entre las células hijas. Esto ocasiona un aumento en la variación de célula a célula en cuanto al contenido de ADN.

El daño en el ADN puede producir la formación de tumores en los peces (Bauman, 1998) o una vida más corta (Agarwal y Sohal, 1996). También podría tener un efecto deteriorante en la fertilidad y en la función inmunológica (Theodorakis et al., 1996; O'Connor et al., 1996). Esto se debe a que las células que se dividen rápidamente son presa fácil del daño al ADN, a los órganos, gónadas y embriones en desarrollo, que forman células sanguíneas blancas con altos índices de división celular. Por lo tanto, el daño al ADN podría indicar un aumento en la mortalidad en poblaciones de especies además de la presencia de contaminantes tóxicos.

METODOS

Recolección de muestras

Los peces y las muestras de sedimentos se recolectaron en una laguna adyacente a una instalación de bombeo de petróleo, que según nuestra hipótesis, es la fuente de contaminantes en el área de estudio, Xan 3, y en tres lagunas que se ubican a diferentes distancias de Xan 3 (Buena Vista, 5.63 km; Flor de Luna, 13.46 km y Guayacán, 39.51 km). Adicionalmente, un muestreo en el Río San Pedro (a 66.89 km de Xan 3) sirvió como sitio de control distante, y probablemente no contaminado. Los peces que se recolectaron utilizando una atarraya, fueron devueltos vivos al compo. Se midió la longitud de cada pez, siendo después anestesiados. Se recolectaron muestras de hígado, baso y sangre, se congelaron y se preservaron en nitrógeno líquido.

El agua y las muestras de sedimento también se tomaron en los sitios de recolección.

Análisis de ruptura en la cadena de DNA

Se extrajo ADN del tejido del hígado para determinar la cantidad de rupturas de cadenas individuales de ADN (Theodorakis y Shugart, 1993). Las muestras de ADN se analizaron por medio de electroforesis de gel agaroso para poder determinar el promedio de longitud de las moléculas de ADN. La longitud se relaciona de manera inversa al número de rupturas de cadenas individuales; si existen más rupturas, entonces la longitud promedio de las moléculas de ADN será menor.

Análisis de daño cromosomal

El daño cromosomal se evaluó al determinar la cantidad de variación de célula a célula en el contenido del ADN (Bickham, 1990). Los núcleos se aislaron de la sangre y de los tejidos del bazo y la variación en el contenido del ADN se evaluó utilizando un citómetro de flujo, que mide las cantidades relativas de ADN en núcleos individuales. Una descripción más detallada de los métodos para la electroforesis y la citometría de flujo se pueden encontrar en el Apéndice 10.

Análisis de contaminantes

Se extrajeron muestras de agua y sedimentos con solventes orgánicos (hexano/acetona o cloruro de metileno). Las muestras se concentraron por evaporación bajo gas de nitrógeno y se analizaron por medio de cromatografía líquida de alta presión o cromatografía gaseosa.

RESULTADOS Y DISCUSION

Condición fisiológica y análisis de daño al DNA

La cantidad de cada especie recolectada en cada sitio se reporta en la Tabla 5.1. El único pez que mostró enfermedad superficial (aletas rotas) fue el recolectado en la laguna Xan 3. El problema de aletas rotas es ocasionado por una infección bacterial que produce erosión en las aletas. Usualmente no se ve en poblaciones de peces silvestres excepto bajo condiciones de estrés. Los hidrocarburos de petróleo y otros contaminantes pueden aumentar la incidencia de aletas rotas y otras infecciones bacteriales al eliminar la respuesta inmunológica del pez. Esta condición es causa de preocupación porque la erosión en las aletas impide la capacidad del pez de nadar y maniobrar y usualmente es un síntoma de una infección bacterial más amplia. En el caso de peces activos, si no se trata, los individuos con aletas rotas usualmente mueren de infección sistémica.

Dos especies comunes *Thorichthys meeki* y *Cichlasoma synspilum*, se utilizaron para análisis de ruptura de cadenas de ADN y análisis de daño cromosomal. Los análisis estadísticos no paramétricos (Kruskal-Wallis) se utilizaron para probar diferencias significativas entre los sitios. Para *C. synspilum*, la variación en el contenido de ADN entre células (un reflejo de la cantidad de daño cromosomal) fue mayor peces de Xan 3 que en peces del Río San Pedro y esta diferencia fue significativa estadísticamente (Tabla 5.2; $P < 0.05$). La cantidad de daño cromosomal en peces de la Laguna Buena Vista no fue diferente a la de los provenientes de la laguna Xan 3. Al contrario, no hubo diferencias estadísticamente significativas en la cantidad de daño cromosomal entre cualquier población de *T. meeki*.

Tabla 5.1. Cantidad de cada especie recolectada del Pozo Xan y cuatro sitios de referencia.

Especie	F6: Xan 3	F1: Laguna Buena Vista	F5: Laguna Flor de Luna	C2: Laguna Guayacán	SP2: Río San Pedro
Thorichthys meeki	12	14	20	17	18
Cichlasoma synspilum	6	8	0	0	14
C. uropthalmus	7	0	0	0	0
Astianyx fasctiatum	0	14	0	11	0
Poecilia mexicana	0	0	0	10	0
Gambusia sexradiata	0	0	23	0	0
G. yucatana	0	28	0	0	0

No hubo diferencias estadísticamente significativas en la cantidad de ruptura de cadenas de DNA entre las poblaciones de *C. synspilum*. Sin embargo, la cantidad de ruptura de cadenas de DNA en la población *T. meeki* de Xan 3 fue mayor que la de las poblaciones en Flor de Luna y el Río San Pedro, aunque esta diferencia no fue estadísticamente significativa con las poblaciones que habitan las lagunas Buena Vista o Guayacán (Tabla 5.2). Ningún pez del Río Escondido se recolectó para análisis.

Los patrones diferentes de daño de ADN entre *C. synspilum* y *T. meeki* podrían originarse de diferentes causas. Primero, podría haber diferente exposición a diferentes compuestos genotóxicos como resultado de diferencias en el uso del hábitat o bien preferencias alimenticias. Desafortunadamente no hay suficiente información sobre este tipo de cíclidos neotrópicos. También, pueden haber diferencias en la fisiología molecular de las dos especies. De nuevo, no existe tal información. Es claro que se necesita más investigación en la historia natural y fisiología de peces neotrópicos antes de poder emitir conclusiones definitivas en cuanto a las diferencias entre ambas especies. Adicionalmente, la diferencias en las respuestas genotóxicas entre las especies podría ser mediada por diferencias en la citometría de flujo y daño en el ADN. Esto podría estar relacionado con diferencias en la sensibilidad de los ensayos o el tipo de daño que midieron. Este ensayo electroforético mide las rupturas de una sola cadena de ADN, lo cual se podría deber a un corte directo de la base de de fosfato de azúcar o a la conversión de ciertas modificaciones de base de ADN (que también indican una exposición contaminante) hasta rupturas de cadenas individuales *in vitro* en un pH alcalino (llamados "sitios lábiles alcalinos"). El análisis

citométrico de flujo, por otro lado, mide el daño cromosomal, como resultado de ruptura de dos cadenas. Es más difícil reparar rupturas de cadenas dobles (Ward 1988), para que haya un menor efecto transitorio que una ruptura de una cadena o sitios lábiles alcalinos. Estas diferencias también pueden deberse a la naturaleza de los ensayos en sí ya que el ensayo de citometría de flujo utilizó núcleos intactos y los ensayos electroforéticos utilizaron ADN extraído.

El daño de ADN es un proceso de estado estable; en otras palabras, constantemente se forma y se repara (Freidber, 1985). Por lo tanto, cualquier daño de ADN que sea aparente es el que no se había reparado al momento en que el tejido fue recolectado y congelado. Esto responde al hecho de que cualquier diferencia entre los sitios contaminados y de referencia es pequeña. Otro factor que contribuye a este patrón es que una gran parte del daño al ADN, especialmente el daño cromosomal, es letal para las células. Por lo tanto, cualquier célula que se encuentre afectada por una gran cantidad de daño de ADN morirá. Esto limita el rango de daño de ADN que pueda ser detectado en tejido vivo. El hecho de que cualquier daño pueda observarse indica que este estado constante ha sido alterado de tal manera que el equilibrio se ha convertido en acumulación de daño de ADN, ya sea a través de un incremento en el número de eventos que dañaron al ADN o una disminución en la reparación de ADN. Esto sería un argumento a favor de ver el daño de ADN como un biomarcador de efecto contaminante y no simplemente de exposición. Por lo tanto, las pequeñas diferencias en la cantidad de ADN dañado podría indicar efectos que deterioran la salud del pez.

Tabla 5.2. Variación media en el contenido de DNA de célula en célula (variación de coeficiente entre células) y número de particiones individuales//10^5 nucleótidos (SSB) en dos especies de peces del Parque Nacional de la Laguna del Tigre, Petén, Guatemala.

Los valores con el mismo sobreescrito indican que estadísticamente no son diferentes significativamente.

	Número de interrupciones individuales		Variación en contenido de DNA	
Lugar	Media	Rango	Media	Rango
A. *Cichlasoma synspilum*				
F6: Xan 3	5.53[a]	4.01-17.80	2.57[a]	2.45-2.83
F1: Buena Vista	3.70[a]	1.38-10.28	2.41[a,b]	1.92-3.44
SP2: Río San Pedro	5.47[a]	2.85- 8.56	2.13[b]	1.94-2.34
B. *Thorichthys meeki*				
F6: Xan 3	5.25[a]	1.65-17.40	2.36[a]	1.82-2.56
F1: Buena Vista	4.80[a]	1.24-28.27	2.07[a]	1.94-2.57
F5: Flor de Luna	2.53[b]	0.96-11.68	2.12[a]	1.68-2.48
C2: Guayacán	6.16[a]	0.48-25.12	2.06[a]	1.14-3.05
SP2: Río San Pedro	3.25[b]	0.02-11.52	2.01[a]	1.71-2.54

Tabla 5.3. Concentraciones de sedimento de hidrocarburos aromáticos policíclicos.

Compuesto	Concentración surrogada corregida (ng PAH/g sedimento seco)[a]				
	F6: Xan 3	F1: Buena Vista	F5: Flor de Luna	C2: Guayacán	SP2: Río San Pedro
Naftaleno	6.1	15.4	2.8	56.0	2.4
Naftalenos C1	9.7	25.2	3.9	44.1	3.0
Naftalenos C2	13.6	90.1	21.5	44.0	6.7
Naftalenos C3	17.1	54.4	11.2	23.4	7.2
Naftalenos C4	13.9	85.4	11.6	34.1	8.8
Bifenil	2.7	5.3	0.7	11.9	0.8
Acenaftileno	BMDL[b]	0.7	BMDL[b]	BMDL[b]	BMDL[b]
Acentafteleno	2.3	9.7	0.9	6.0	0.8
Fluoreno	5.5	29.2	1.8	17.0	1.8
Fluorenos C1	11.3	37.6	3.6	9.5	3.2
Fluorenos C2	27.5	190.0	17.0	75.7	15.2
Fluorenos C3	13.0	53.5	9.3	18.6	13.9
Fenantreno	15.4	69.3	12.2	23.7	12.7
Antraceno	1.4	5.8	0.5	4.6	0.8
Fenantrenos/Antracenos C1	18.3	94.8	29.0	34.8	16.8
Fenantrenos/Antracenos C2	15.1	72.7	13.0	17.7	14.4
Fenantrenos/Antracenos C3	4.6	29.8	6.9	7.1	4.7
Fenantrenos/Antracenos C4	3.0	17.8	5.4	3.8	1.9
Dibenzotiofen	1.1	12.8	1.7	5.5	1.3
Dibenzotiopenos C1	4.5	31.0	4.6	4.2	4.7
Dibenzotiopenos C2	6.9	52.4	8.5	7.5	7.5
Dibenzotiopenos C3	4.5	42.5	4.4	3.8	3.5
Fluoranteno	1.8	17.6	2.3	10.0	2.4
Pireno	1.5	13.5	2.0	8.4	2.4
Pirenos/Fluorantenos C1	1.5	4.2	BMDL[b]	6.0	0.7
Benz(a)ntraceno	0.3	1.8	0.2	3.3	0.2
Criseno	0.3	2.1	0.3	3.1	0.3
Crisenos C1	4.9	16.4	2.6	6.1	1.2
Crisenos C2	10.9	37.4	16.8	14.9	1.7
Crisenos C3	2.8	49.3	1.1	24.5	2.0
Crisenos C4	0.7	11.0	1.3	3.4	2.8
Benzo(b)fluoranteno	0.6	1.8	0.7	2.8	0.5
Benzo(k)fluoranteno	BMDL[b]	2.1	0.8	2.7	0.3
Continuación de tabla 10					
Benzo(e)pireno	0.4	1.6	0.4	1.3	0.4
Benzo(a)pireno	0.3	0.8	0.7	1.5	0.4
Perileno	BMDL[b]	4.4	BMDL[b]	64.2	BMDL[b]
Indeno(1,2,3-c,d)pireno	BMDL[b]	1.5	0.7	BMDL[b]	BMDL[b]
Dibenzo(a,h)antraceno	BMDL[b]	BMDL[b]	0.5	BMDL[b]	BMDL[b]
Benzo(g,h,i)perileno	BMDL[b]	1.1	BMDL[b]	BMDL[b]	BMDL[b]
Total de PAHs	225	1190	202	612	148
Recuperación de % de surrogado relacionado					
Surrogado	% de recuperación				
Naftaleno d8	88	75	74	72	83
Acentafteno d10	95	102	97	78	109
Fenantreno d10	108	71	103	96	117
Criseno d12	70	65	72	62	77
Perileno d12	81	51	92	62	87

[a] Cada muestra de sedimento se relacionó con un sedimento de PAH y las concentraciones se normalizaron por el % de recuperación del surrogado
[b] Inferior al límite mínimo de detección

Análisis de contaminantes

Se determinaron las concentraciones de sedimentos de PAH para Xan 3, así como para las lagunas Flor de Luna y Buena Vista, y los ríos Escondido y San Pedro. Las muestras de sedimento para Guayacán se perdieron debido a la ruptura de los contenedores que se encontraban en camino a Texas A&M University. El total de las concentraciones de sedimentos de PAH para Flor de Luna, Río Escondido, Río San Pedro, Buena Vista y Xan 3, fueron respectivamente 202, 148, 612, 1190 y 225 ng PAH/g de sedimento seco (partes por mil de millones). Las concentraciones de isómeros individuales de PAH se enumeran en la Tabla 5.3.

Estos datos muestran que la cantidad de PAH en la muestra de sedimento de la Xan 3 son menores o iguales que los de los sitios de referencia. Sin embargo, los datos de los análisis del daño de ADN y del daño cromosomal indican que los peces de Xan 3 están siendo expuestos a un agente que daña el ADN. Esto es según la observación de que los peces de Xan 3 y ningún pez de cualquier otro sitio - mostró signos de enfermedad. Esto siguiere que los peces están siendo presionados por contaminantes u otros PAH. Existen muchos otros compuestos además de los PAH que ocasionan daño en el ADN, por ejemplo metales pesados, hidrocarburos clorinados, arsénico, cromo, pesticidas y compuestos nitroaromáticos. Los procedimientos de evaluación de riesgo detallados podrían implementarse, lo que haría posible identificar la fuente de su genotoxicidad. Un hallazgo no esperado fue que la cantidad de PAH en el sedimento fue mucho mayor en la Laguna Buena Vista que en cualquier otro sitio. Eso podría explicar las elevadas cantidades de ruptura de cadenas de ADN en los peces de este sitio. También hay evidencia de que los peces en Guayacán están expuestos a agentes de daño.

CONCLUSIONES Y RECOMENDACIONES

La hipótesis de que la instalación de petróleo Xan 3 tiene un impacto en el ambiente que la rodea, no puede ser rechazada porque hubo evidencia de estrés y daño de ADN en los peces desde la laguna adyacente. Concluimos que la cantidad de ADN, daña a los peces en Xan 3 y fue mayor que en el Río San Pedro, con resultados intermedios para el apareamiento de peces en las cercanías de Xan 3. El hallazgo de concentraciones de PAH más altas en la laguna Buena Vista que en cualquier otra fue inesperado. Este resultado puede deberse a patrones complicados de intercambio de aguas entre lagunas durante la estación lluviosa y otras fuentes contaminantes. Las concentraciones de PAH individuales en el sedimento, sin embargo, fue mucho más bajas que los criterios del US EPA, para la protección de invertebrados bénticos, que se basa en pruebas de toxicidad a corto plazo de químicos sencillos (US EPA, 1991a, 1991b, 1991c). Sin embargo, los efectos a largo plazo de la exposición a múltiples químicos en la salud ambiental y humana son desconocidos. El hecho de que la población humana local utiliza bastante a Buena Vista para pescar, bañarse, lavar y recreación es de mucha preocupación. Aunque los datos no indican una amenaza extrema al ambiente y salud humana, sugieren que se debe iniciar un manejo mejorado.

Recomendamos que se inicie un programa de monitoreo ambiental y que se realice una evaluación de riesgos ecológicos y salud humana con detalle para controlar varios problemas que incluyen lo siguiente:

- una determinación de la fuente de estrés, enfermedad de aletas y genotoxicidad en los peces de las lagunas adyacentes a las instalaciones petroleras de Xan;

- una descripción más detallada de los posibles contaminantes en las instalaciones petroleras de Xan;

- una determinación del origen de PAH en la Laguna Buena Vista y, más general, estudios de las conexiones hidrológicas que varían de acuerdo con la temporada en las cercanías de las instalaciones petroleras de Xan, para que los patrones de redistribución de los contaminantes potenciales se pueda comprender; y

- una evaluación de cualquier efecto de salud posible por la exposición de PAH en la población local que se encuentra en las áreas al rededor de las instalaciones de Xan.

LITERATURA CITADA

Agarwal, S. y R. S. Sohal. 1994. DNA oxidative damage and life expectancy in houseflies. Proceedings of the National Academy of Sciences 91: 12,332-12,335.

Ashok, B. T. y S. Saxena. 1995. Biodegradation of polycyclic aromatic-hydrocarbons—a review. Journal of Scientific & Industrial Research 54: 443-451.

Bauman, P. C. 1998. Epizootics of cancer in fish associated with genotoxins in sediment and water. Mutation Research—Reviews in Mutation Research 411: 227-233.

Bickham, J. W. 1990. Flow cytometry as a technique to monitor the effects of environmental genotoxins on wildlife populations. Pp 97-108 in *In Situ Evaluation of Biological Hazard of Environmental Pollutants* (S. Sandhu, W. R. Lower, F. J. DeSerres, W. A. Suk, and R. R. Tice, eds.). Environmental Research Series Vol. 38. Plenum Press, New York, NY, USA.

Bickham, J. W., J. A. Mazet, J. Blake, M. J. Smolen, Y. Lou y B. E. Ballachey. 1998. Flow cytometric determination of genotoxic effects of exposure to petroleum in mink and sea otters. Ecotoxicology 7: 191-199.

Freeman, S. E. y B. D. Thompson. 1990. Quantitation of ultraviolet radiation-induced cyclobutyl pyrimidine dimers in DNA by video and photographic densitometry. Analytical Biochemistry 186: 222-228.

Freidberg, E. C. 1985. *DNA Repair*. Plenum Press, New York, NY, USA.

Hoeijmakers, J. H. J. 1993. Nucleotide excision repair II, from yeast to mammals. Trends in Genetics 9: 211-217.

McCarthy, J. F. y L. R. Shugart. 1990. Biomarkers of environmental contamination. Pp. 3-16 in *Biomarkers of Environmental Contamination* (J. F. McCarthy and L. R. Shugart, eds.). Lewis Publishers, Boca Raton, FL, USA.

Mille, G., D. Munoz, F. Jacquot, L. Rivet y J. C. Bertrand. 1998. The Amoco Cadiz oil spill: evolution of petroleum hydrocarbons in the Iie Grande salt marshes (Brittany) after a 13-year period. Estuarine Coastal and Shelf Science 47: 547-559.

Newman, M. C. y C. H. Jagoe (eds.). 1996. *Ecotoxicology: A Hierarchical Treatment*. Lewis Publishers, Boca Raton, FL, USA.

O'Connor, A., C. Nishigori, D. Yarosh, L. Alas, J. Kibitel, L. Burley, P. Cox, C. Bucana, S. Ullrich y M. 1996. DNA double strand breaks in epidermal cells cause immune suppression in vivo and cytokine production in vitro. Journal of Immunology 157: 271-278.

Pritchard, J. B. 1993. Aquatic toxicology—past, present, and prospects. Environmental Health Perspectives 100: 249-257.

Shugart, L. R. 1988. An alkaline unwinding assay for the detection of DNA damage in aquatic organisms.

Shugart, L. R., J. Bickham, G. Jackim, G. McMahon, W. Ridley, J. Stein y S. Steinert. 1992. DNA alterations. Pp. 125-154 in *Biomarkers: Biochemical, Physiological, and Histological Markers of Anthropogenic Stress* (R. J. Huggett, R. A. Kimerle, P. M. Mehrle Jr., H. L. Bergman, eds.). Lewis Publishers, Boca Raton, FL, USA.

Theodorakis, C. W., B. G. Blaylock y L. R. Shugart. 1996. Genetic ecotoxicology I: DNA integrity and reproduction in mosquitofish exposed *in situ* to radionuclides. Ecotoxicology 5: 1-14.

Theodorakis, C. W. y L. R. Shugart. 1993. Detection of genotoxic insult as DNA strand breaks in fish blood cells by agarose gel electrophoresis. Environmental Toxicology and Chemistry 13: 1023-1031.

Theodorakis, C. W., S. J. D'Surney, J. W. Bickham, T. B. Lyne, B. P. Bradley, W. E. Hawkins, W. L. Farkas, J. F. McCarthy y L. R. Shugart. 1992. Sequential expression of biomarkers in bluegill sunfish exposed to contaminated sediment. Ecotoxicology 1:45-73.

U.S. Environmental Protection Agency. 1991a. Proposed sediment quality criteria for the protection of benthic organisms: Acenapthene. U.S. EPA, Washington, DC, USA.

U.S. Environmental Protection Agency. 1991b. Proposed sediment quality criteria for the protection of benthic organisms: Fluoranthene. U.S. EPA, Washington, DC, USA.

U.S. Environmental Protection Agency. 1991c. Proposed sediment quality criteria for the protection of benthic organisms: Nanthrene. U.S. EPA, Washington, DC, USA.

Vindelov, L. L., I. J. Christianson, N. Keiding, M. Prang-Thomsen y N. I. Nissen. 1983. Long-term storage of samples for flow cytometric DNA analysis. Cytometry 3: 317-320.

Vindelov, L. L., and I. J. Christensen. 1990. A review of techniques and results obtained in one laboratory by an integrated system of methods designed for routine clinical flow cytometric DNA analysis. Cytometry 11: 753-770.

Ward, J.F. 1988. DNA damage produced by ionizing-radiation in mammalian-cells-identities, mechanisms of formation, and repairability. Prog. Nucl. Acid. Res. Mol. Biol. 35: 95-125.

CAPITULO 6

UNA EVALUACION RAPIDA DE DIVERSIDAD DE la AVIFAUNA EN HABITATS ACUATICOS DEL PARQUE NACIONAL LAGUNA DEL TIGRE, PETEN, GUATEMALA

Edgar Selvin Pérez y Miriam Lorena Castillo Villeda

RESUMEN

- Un total de 173 especies de aves se registraron en el Parque Nacional Laguna del Tigre (PNLT), aproximadamente 60.7% del número total de especies (N=285) reportadas del parque. De estas, 157 se reportaron en conteo de puntos; el resto en conteos no sistemáticos.

- Se realizaron once registros nuevos de aves para la lista oficial del parque, para un total de 296 especies de aves. Los registros nuevos incluyen *Aramus guarauma, Calidris minutilla, C. melanotus, C. fuscicollis, Cyanocorax yucatanicus, Cardinalis cardinalis,* e *Himantopus mexicanus.* Las extensiones de la distribución también se registraron para tres de los nuevos registros: *Tyrannus savana monachus, Laterallus exilis* y *Pyrocephalus rubinus.*

- En general, los ciconiformes (aves zancudas grandes) se distribuyeron ampliamente en el parque; eran más abundantes en las ciénagas del Río Escondido y el bosque de dosel abierto del Río Sacluc y Río San Pedro. Los especialistas forestales eran numerosos en colinas y bosques cerrados del este del área de San Pedro; estos incluyen *Crax rubra, Agriocharis ocellata, Automolus ochrolaemus,* dendrocoláptidos y trogloditidos.

- La fuente de variación principal en la composición de especies en el parque es el contraste entre las partes del oeste, incluyendo el biotopo y los humedales al rededor y las áreas boscosas de las partes centrales (Chocop/Candelaria) y orientales (Río San Pedro/Sacluc) del parque. El área oriental fue única y contenía muchas especies encontradas solamente allí, incluyendo *Calidris minutilla, C. fusciola, C.melanotus, Himantopus mexicanus, Laterallus exillis, Phorphyrula martinica, Jacana spinosa* (asociadas con playones, aguas estancadas poco profundas), *Tyrannus savana monachus* y *Pyrocephalus rubinus* (asociada con sabanas inundadas en La Lámpara).

INTRODUCCION

El Parque Nacional Laguna del Tigre representa un área de humedales muy importante internacionalmente para Guatemala. El parque se registró como un sitio Ramsar en 1990 y también se incluyó en el registro Montreaux 1993. El listado anterior fue necesario debido a la amenaza ocasionada por la explotación de petróleo en el parque (Carbonell et al., 1998). La meta de este estudio fue documentar y comparar la diversidad de las aves del humedal en diferentes áreas dentro del PNLT y contribuir al desarrollo de estudios de monitoreo de aves en esta área de humedales importante.

Tres tipos de hábitat de aves importantes se pueden distinguir en los humedales del PNLT: 1) bosques ribereños; 2) sabanas inundadas, pantanos o humedales; y 3) zonas de contaminación (operaciones petroleras, agricultura, colonizaciones, etc). Estos hábitats y los microhábitats que contienen, tienden a mantener diferentes tipos de aves, incluyendo especies que utilizan árboles ribereños y de humedales y arbustos; áreas con vegetación espesa; vegetación emergente alta y robusta; usualmente en áreas abiertas y árboles altos en áreas ribereñas. Así, la estructura de la vegetación se espera que sea un determinante principal de los patrones de composición de especies para aves de humedales (Zigüenza, 1994).

Un estudio base de aves para un programa de monitoreo en PNLT (Méndez et al., 1998) realizó 41 registros importantes para un total de 285 especies, incluyendo un nuevo registro para El Petén (*Zenaida asiatica*) y el índice de extensión del *Nyctibius grandis*. Nuestro objetivo fue complementar este esfuerzo al ampliar el número de asociaciones de vegetación y áreas muestreadas dentro del PNLT. También proporcionamos un foco en las relaciones entre aves y hábitats acuáticos que no se ha visto previamente en estudios.

METODOS

Las aves fueron muestreadas en tres hábitats diferentes. Las áreas ribereñas sirvieron para tomar muestras a lo largo de los ríos San Pedro, Sacluc, Candelaria, Chocop y partes del Río Escondido. Se consideraron tres tipos de macrohábitats dentro de las áreas ribereñas. Estudiamos el bosque ribereño abierto y cerrado, los pantanos en combinación con los bosques ribereños inundaban la sabana compuesta de sibales (sibal), árboles (por ejemplo, *Crescentia* sp.), tasiste (*Acoelorraphe wrightii*) y hierbas. Este último macrohábitat era más común en el área de Río Escondido y en determinadas lagunas como La Lámpara. En segundo lugar, se estudiaron las ciénagas/humedales, ejemplificados por la vegetación del Río Escondido dentro del biotopo y en su confluencia con el Río San Pedro. Esta categoría de hábitat fue la menos común de los tres tipos de hábitats. Finalmente, se examinaron las áreas de vegetación desequilibradas y regeneradas (guamiles) tal como aquellas que se encuentran cerca de los poblados, operaciones petroleras y campos agrícolas abandonados.

El muestreo se enfocó en los cuatro puntos principales utilizados en este estudio RAP (refiérase a "Itinerario de la Expedición y Diccionario Geográfico"). Para el muestreo de reptiles y anfibios (Castañeda Moya et al. este volumen), el uso de puntos discretos de muestreo se consideró inadecuado debido a que las aves utilizan hábitats sobre escalas relativamente amplias. Así, muestreamos aves en las orillas de los lugares con agua utilizando conteos de 20 puntos por ubicación. Las unidades de muestra individuales tratadas incluían el Río San Pedro, Río Sacluc, Río Escondido, Laguna Flor de Luna, Río Chocop, Río Candelaria y Laguna La Lámpara (Tabla 6.1, Apéndice 1). En el Río Chocop pudimos muestrar solamente 10 puntos. Para estudios sobre ríos, el conteo de puntos se distribuyeron a lo largo del río. Para estudios de lagunas, el conteo de puntos fue tomado mientras caminábamos a lo largo de la orilla de los mismos.

Para elegir los métodos más efectivos y convenientes para muestrear especies de aves acuáticas en el parque, comparamos inicialmente tres estrategias de muestreo. Conteos de puntos audiovisuales realizados al establecer 10 puntos establecidos en forma ecuánime en una distancia de tres kilómetros (es decir, conteo de un punto cada 300

metros) que fueron estudiados en una sóla sesión. Las especies observadas y escuchadas se registraron durante 10 minutos en cada punto. Las aves también fueron muestreadas mediante viajes a una velocidad constante (10 km/h) en una lancha de motor y grabando cada ave observada y escuchada en un área de tres kilómetros. Finalmente, muestreamos aves durante la noche en áreas de tres kilómetros para grabar sitios amplios durante la noche, especialmente para especies comunales. Para cada esquema de muestreo, registramos el tipo de hábitat en el cual se encontró cada especie y su comportamiento en ese hábitat (por ejemplo, alimentación, el momento de pernoctar).

Después de una evaluación inicial de estos métodos, decidimos utilizar solamente conteo de puntos audiovisuales. Este método probó ser el más efectivo para grabar el número más grande de especies en diferentes tipos de hábitats así como para identificar cada ave activa como pájaros carpinteros, pericos y crácidos. Nuestro censo fue apoyado por el uso de un micrófono direccional para registrar las llamadas que no fueron reconocidas inmediatamente. También se registraron las observaciones no sistemáticas pero no se incluyeron en el análisis de datos.

Tabla 6.1. Distribución de la intensidad del muestreo por punto focal, Parque Nacional Laguna del Tigre, RBM, Petén.

Áreas focales	Extensión del muestreo	Tipos de hábitats muestreados
Río San Pedro	De Paso Caballos(SP2) a La Caleta (SP8).	Bosque conectado con bosques altos en las montañas, áreas pantanosas.
Río Sacluc	Casi todo el río Sacluc.	Bosques abiertos y cerrados.
Río Escondido	Casi 2/3 del río (cuando se puede navegar).	Bosques abiertos y cerrados, ciénagas con sibal, caños, áreas quemadas.
Flor de Luna	Laguna Flor de Luna	Bosques abiertos, sabana inundada,, "guamiles" y ciénagas con sibal.
Río Chocop	Partes del río Chocop y Candelaria. Aparte, Laguna La Lámpara.	Bosques abiertos y cerrados, ciénagas, áreas con problemas (guamiles de operaciones petroleras, áreas quemadas), vegetación basta con *Nymphea*.

RESULTADOS Y DISCUSION

Registramos un total de 2103 individuos en 173 especies, 42 familias y 17 órdenes (Apéndice 11). La mayoría de los individuos fueron paseriformes. El número de especies contadas corresponde al 60.7% de la avifauna encontrada en el parque (Méndez et al., 1998).

Los registros de las nuevas especies para el PNLT incluyen *Aramus guarauma, Calidris minutilla, C. melanotus, C. fuscicollis, Cyanocorax yucatanicus, Cardinalis cardinalis,* e *Himantopus mexicanus.* Los rangos de distribución se ampliaron para *Tyrannus savana monachus, Laterallus exilis* y *Pyrocephalus rubinus* (en base a los mapas en Howell y Webb 1995). Se observó *Jabiru mycteria* que no se había registrado en esta área por 10 años (Herman A. Kihn, pers. com.).

Comparaciones de avifauna entre áreas focales

En general, las especies de aves fueron distribuidas homogéneamente en todo el parque. La mayor riqueza se encontró en los ríos San Pedro, Sacluc y Escondido, mientras que se observaron relativamente menos especies en los ríos Chocop y Candelaria (Figura 6.1). Esto último puede ser debido al área tan reducida de bosque cerrado. La riqueza de especies encontradas en los ríos de San Pedro y Sacluc puede ser debido a condiciones de clima favorables durante el muestreo y la proximidad de los grandes bosques altos. Muchas especies de aves registradas por los especialistas forestales en estas áreas (c.f. Whitacre, 1996), incluyendo pájaros carpinteros, trogones, motmots, pavos de Americanos, chachalacas, guacos y algunos mosqueritos. Algunas de estas especies se encontraron únicamente en estos lugares, como *Xiphorynchus flavigaster, Agriocharis ocellata, Crax rubra* y *Trogon massena.*

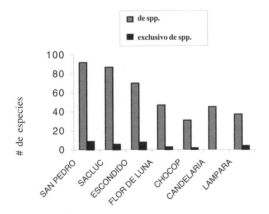

Figura 6.1 Patrones de riqueza de especies de aves y el número de especies exclusivas (no se encontró en otra muestra) entre las áreas de muestra en el PNLT.

El área del Río Escondido alberga diversos macrohábitantes ribereños, incluyendo sibales, bosques abiertos y cerrados (es decir, Punto Icaco, E8). Las zonas adyacentes en el lado oeste del parque, junto con la Laguna La Lámpara, representan un hábitat único compuesto de ciénagas y sabanas inundadas. Estos hábitats albergan especies de aves como *Rostramus sociabilis, Laterallus exillis, Aramus guarauma* y *Jabiru mycteria.*

El área central del parque, representado aquí por el área principal de Chocop/Candelaria, presentó especies de aves únicas como *Dendrocygna anabatina* (piche piquirrojo), que se observó en una laguna cerca del Pozo Xan 1. Esta zona tiene bosques cerrados y albergan especies principales boscosas como *Micrastur semitorquatus* y *Xenops minutus,* que también se registraron en los ríos Sacluc y San Pedro durante estudios anteriores (Pérez, 1998).

La jacana (*Jacana spinosa*) se asocia con vegetación artificial y se registró solamente en lagunas que poseen la ninfa *Nymphaea ampla* y en los humedales poco profundos del Río Escondido, en donde se encontró *Porphirula martinica, Laterallus ruber* y *L. exillis. Busarellus nigricollis* y *Rostramus sociabilis* se registraron en lagunas aisladas y se asociaron con vegetación emergente en ciénagas con sibales.

Los análisis de similitud revelaron que generalmente había un alto grado de similitud en la composición de especies entre las áreas de muestra (Tabla 6.2). Las áreas de San Pedro y Sacluc comparten 60 especies debido a su proximidad. La mayoría del avifauna de Río Escondido y Laguna La Lámpara se encontró en San Pedro y la mayoría de estas especies se asociaron con bosques abiertos y de humedales (consulte Léon y Morales Can, este volumen). Las especies que habitan los prados inundados, sin embargo, no se encontraron en el área de San Pedro; estos incluyen *Tyrannus savana monachus, Pyrocephalus rubinus* y *Aramus guarauma.* De los sitios visitados, solamente La Lámpara mostró prados inundados, compuestos principalmente de *Cladium jamaicense* y otras especies de Cyperaceae, pastos, árboles (*Crescentia alata*) y tasiste (*Acoelorraphe wrightii*) en determinadas áreas. Este tipo de hábitat también es llamado "campería" por los habitantes locales. Un número pequeño de familias se ha establecido en las orillas de esta laguna. Una combinación de estrés debido a la variación del clima (Kushlan et al., 1995) y una alteración atropogénica local pueden amenazar la existencia de un hábitat crítico en este lugar.

El tipo de bosque cerrado se comparte entre los ríos Chocop, Candelaria y Sacluc; como consecuencia, la avifauna de estos sitios es similar (Tabla 6.2). Las especies de aves acuáticas de este tipo de bosque son tipificadas por la presencia de guardarríos como *Chloroceryle americana* y *C. aenea,* que a menudo pernoctan en áreas cercanas al agua.

Tabla 6.2. Los números y porcentajes de especies compartidas entre los sitios de estudio. NS es el número de especies compartidas, % es el porcentaje de especies de la fauna menor compartida con la fauna mayor en cada comparación (es decir, el índice de similitud de Simpson).

	Río Candelaria		Río Chocop		Río Escondido		Flor de Luna		Laguna La Lámpara		Río San Pedro	
	NS	%	NS	%	NS	%	NS	%	NS	%	NS	%
Chocop	20	64.51										
Escondido	22	48.88	15	48.38								
Flor de Luna	20	44.44	11	35.48	31	66						
Lámpara	13	35.13	5	16.13	29	78.38	19	51.35				
San Pedro	28	62.22	16	51.61	45	64.28	26	55.31	28	75.67		
Sacluc	30	66.67	26	83.87	38	54.28	30	63.83	25	67.57	60	69

Distribución del avifauna entre los macrohábitats

Bosque ribereño

El mayor número de especies de aves (117) se registró en este tipo de hábitat. Existen dos tipos de bosques ribereños: cerrados y abiertos. En bosques cerrados, un número mayor de especies de mosqueritos está asociado con maleza. Otras especies que son normales de este hábitat incluyen *Thryothorus maculipectus,* varios pájaros carpinteros (*Dendrocinchla homochroa, D. autumnalis, Automolus ochrolaemus, Xiphorynchus flavigaster, Xenops minutus*), *Micrastur ruficollis, Platyrhynchus cancrominus,* dos guardarríos (*Chloroceryle americana, C. aenea), Geotrygon montana* y *Columba nigrirostris.* Los estudios del uso de aves como indicadores biológicos indican que las especies de aves que merodean en el medio y en la maleza de los bosques maduros (es decir, pájaros carpintero, ratonas y mosqueritos tiranos) son sensibles a los cambios de hábitat (Whitacre, 1996). Esto se debe al hecho de que estas especies sedentarias están expuestas a los efectos de cambios a un grado mayor que aquellas especies que habitan en las áreas superiores (Levey, 1990).

Los bosques abiertos (incluyendo humedales boscosos) es el tipo de bosque más común en esta área, que se encuentra en los ríos San Pedro, Sacluc y Escondido y cerca de todas las lagunas que observamos. Este hábitat alberga la mayoría de especies que registramos. La estructura compleja del bosque presenta oportunidades para una variedad de rincones con forraje, incluyendo insectívoros de pabellón, halconeros (mosqueritos y golondrinas), aves zancudas grandes (Ciconiiformes) que cazan en las orillas del río y predadores como halcones y guardarríos—la mayoría de estos requieren vegetación pequeña. Los habitantes de este tipo de bosque incluyen raptores (*Busarellus nigricollis, Pandion haliaetus* y *Rostramus sociabilis*); ciconiiformes (*Ardea herodias, Egretta thula, E. caerulescens, E. alba, Butorides striatus, Tigrisoma mexicanum, Cochlearius cochlearius, Nycticorax violaceus, Anhinga anhinga* y

Phalacrocorax olivaeus); mosqueritos (*Miozetetes similis, Pitangus sulphuratus, Megarhynchus pitanga* y *Legatus leucophaius*); y palomas (*Columba speciosa* y *C. cayenensis*).

Las especies en peligro como la guacamaya escarlata (*Ara macao*), se registraron en bosques abiertos cerca de una colina junto al área principal de Chocop. Estudios anteriores (Pérez, 1998) sugieren que muchas de las plantas de alimento favoritas de las guacamayas se encuentran en este hábitat, incluyendo el ramón (*Brosimium alicastrum*), miembros del Zapotaceae (*Pouteria* spp.), parras de la familia Bignoniaceae (malerio bayo, malerio blanco, guaya [*Talisia olivaeriformes*] y jocote jobo [*Spondias* spp.]), así como las tres especies en las que las guacamayas hacen sus nidos, como *Acacia glomerosa.* Estas tres están asociadas con tierras bajas inundadas, a menudo cerca de lagunas. La agricultura de corte y quema es la amenaza principal de los bosques ribereños del parque.

Sabana inundada

Este tipo de hábitat se registró únicamente en un área cercana a la Laguna La Lámpara. Las especies exclusivas de este sitio incluyen *Tyrannus savana monachus, Pyrocephalus rubinus* y *Aramus guarauma.* Otras especies comúnes en el área son *Tyrannus melancholicus, Myiozetetes similis, Megarynchus pitanga, Agelaius phoeniceus, Columba speciosa, C. flavirostris* y *C. cayenensis.* Este tipo de hábitat único puede ser amenazado por la colonización humana al rededor de la laguna—especialmente por incendios iniciados durante las temporadas secas para limpiar la tierra de la agricultura.

Prados inundados

Este hábitat también es conocido como sabana húmeda, ciénagas salítres, planos inundados o sibales. Es uno de los tipos de ecosistemas tropicales más extensos (Dugan 1992). Los prados inundados se encuentran exclusivamente en el Río Escondido. También fue uno de los más diferentes en términos de composición de especies de aves. Existe una gran diversidad de microhábitats en este hábitat debido a sus características hidrológicas. Por ejemplo, los bancos arenosos son los mejores hábitats para especies como *Calidris melanotus*, *C. minutilla*, *C. fuscicollis*, *Laterallus exilis*, *L. ruber* y *Tringa flavipes*. Registramos la mayoría de las especies con penacho observadas en el estudio de este hábitat, así como *Columba* spp. y el caracol (*Rostramus sociabilis*), el cual depende principalmente de caracoles acuáticos (*Pomacea* sp.). También se observó aquí el jabirú (*Jabiru mycteria*). Este hábitat está amenazado por el desarrollo debido a la proximidad de los caminos y las instalaciones petroleras; en particular, el ancho del canal del río para navegación se está deteriorando.

CONCLUSIONES Y RECOMENDACIONES

Consideramos que la salud en general de la biodiversidad de las aves en los bosques del parque es intermedia. La abundancia de especies en general que son indicadoras de problemas no es tan grande en estudios similares de sitios de guamil en la jungla Lacandona cercanas a Chiapas, México (Warkentin et al., 1995). Sin embargo, los cambios y la pérdida forestal continua, es inminente, si la actividad agrícola de los colonos continúa sin resolverse. Con relación a este problema, el área central del PNLT (es decir, las áreas principales de Flor de Luna y Chocop) es más fácil que se vea más afectada por la colonización debido a la presencia de caminos en esta zona y su proximidad a los hábitats de humedales. Debido a que la fauna de aves es única en esta zona, los impactos de la destrucción del hábitat tendrán consecuencias desastrosas para la fauna de aves del parque (Méndez et al., 1998).

Nuestras observaciones sugieren que los hábitats más importantes para las aves acuáticas en el PNLT, incluyendo tanto humedales como bosques altos/ribereños, son raros en las áreas cubiertas de la Reserva de la Biosfera Maya (RBM; Carbonell et al., 1998). La conservación estricta de dichos hábitats en estas áreas deberían ser una consideración importante en el diseño y manejo de la RBM (Méndez et al., 1998). Hasta este momento, es muy importante que estudios futuros en el parque proporcionen una comprensión más detallada de lo que constituye los diferentes hábitats de aves y sus distribuciones (en lugar de confiar en simples divisiones zonales). Ofrecemos varias recomendaciones específicas para estudios y conservación de aves dentro del PNLT; las recomendaciones no están en orden de importancia:

• Investigar los efectos sobre la población de aves de la ampliación del canal en Río Escondido.

• Investigar la asociación entre *Tyrannus savana monachus* y *Pyrocephalus rubinus* con hábitats de la sabana inundada de La Lámpara.

• Implementar un programa de monitoreo para documentar los cambios en la diversidad de las aves durante los cambios en los diferentes tipos de hábitats y usar estos datos para establecer los límites de tolerancia para especies determinadas de aves.

• Proporcionar una perspectiva unificada regional para el manejo de la avifauna de los humedales del oeste de la RBM.

• Proporcionar una clasificación detallada de los hábitats de aves para las áreas en toda la RBM.

LITERATURA CITADA

Carbonell, M., O. Lara, J. Fernández-Porto, and others. 1998. Convention on Wetlands. Proceedings of the Orientation of Management. Ramsar Site: Laguna del Tigre, Guatemala. Gland, Switzerland.

Dugan, P. J. (ed.) 1992. Conservation of Wetlands: An Analysis of Actual Issues and the Necessary Actions. IUCN, Switzerland.

Howell, S. y S. Webb. 1995. *A Guide to Birds of México and Northern Central America.* Oxford University Press, New York, NY, USA.

Kushlan J. A., G. Morales y P. Frohing. 1985. Foraging niche relations of wading birds in tropical west savannas. Pp. 663-682 in *Neotropical Ornithology.* Ornithological Monographs 36.

Levey, D. 1990. Habitat-dependent fruiting behaviour of an understorey tree, *Miconia centrodesma*, and tropical treefall gaps as keystone habitats for frugivores in Costa Rica. Journal of Tropical Ecology 6: 409-420.

Méndez, C., C. Barrientos, F. Castañeda y R. Rodas. 1998. Programa de Monitoreo, Unidad de Manejo Laguna del Tigre. Los Estudios Base para su Establecimiento. CI/ProPetén, Conservation International, Guatemala.

Pérez E. S. 1998. Evaluation of available habitat for the scarlet macaw (*Ara macao*) in Petén, Guatemala. Tésis de Licenciatura. Universidad de San Carlos de Guatemala, Facultad de Ciencias, Químicas y Farmacia, Escuela de Biología.

Warkentin I. G., T. Russell y J. Salgado. 1995. Songbird use of gallery woodlands in recently cleared and older settled landscapes of the Selva Lacandona, Chiapas, México. Conservation Biology 9: 1095-1106.

Whitacre, D. 1996. *Ecological Monitoring Program for the Mayan Biosphere Reserve*. Draft. The Peregrine Foundation.

Zigûenza R. 1994. Evaluación de las fluctuaciones poblacionales de 31 especeis de aves en el área de protección especial Manchón-huamuchal. Tésis de Licenciatura. Universidad de San Carlos de Guatemala, Facultad de Ciencias, Químicas y Farmacia, Escuela de Biología.

CAPITULO 7

LA HERPETOFAUNA DEL PARQUE NACIONAL LAGUNA DEL TIGRE, PETEN, GUATEMALA, CON ENFASIS EN LAS POBLACIONES DEL COCODRILO DE MORELET (*CROCODYLUS MORELETII*)

Francisco Castañeda Moya, Oscar Lara y Alejandro Queral-Regil

RESUMEN

Estudio de la población del *Crocodylus moreletii*

- Los índices del *C. moreletii* registrada en el camino de la Xan-Flor de Luna y en Laguna la Pista son los más altos registrados hasta ahora en Guatemala.

- Laguna la Pista, Laguna el Perú y el Río Sacluc son sitios ideales en los cuales se puede implementar un programa de manejo para *Crocodylus moreletii* en el Parque Nacional Laguna del Tigre (PNLT) debido a los altos porcentajes de adultos en estos sitios así como la infraestructura disponible para los investigadores.

- El hábitat sibal (de *Cladium jamaicense*) se utilizó más frecuentemente por *C. moreletii* en el PNLT. Este hábitat ofrece aislamiento adecuado, disponibilidad de alimentos y protección para estas especies.

- Áreas de 500 mts. muestreados en 150 min parece ser una buena estrategia de muestreo para *C. moreletti* en áreas dentro del PNLT.

Estudio de herpetofauna en la vegetación ribereña

- Existe una disminución de la abundancia de la herpeofauna en los límites de los cuerpos de agua en el interior del bosque durante la época seca en el PNLT.

- Las especies de reptiles y anfibios registrados en el PNLT tiene distribuciones amplias dentro de la Península de Yucatán y muestran poca especificación en el área de Petén. *Dermatemys mawii*, sin embargo, es un endémico regional y debería dársele énfasos por medio de los administradores del parque.

INTRODUCCION

La herpetofauna de Guatemala es extremadamente rica. Campbell y Vannini (1989) enumeraron 326 especies para Guatemala y 160 de estas se encuentran en la región de Petén (Campbell, 1998). Sin embargo, durante los últimos diez años, han habido varias revisiones taxonómicas y los análisis filogenéticos que han revelado una herpetofauna más rica que la que se pensaba previamente (Campbell y Mendelson, 1998). Actualmente, conocemos de 145 especies de anfibios y 242 especies de reptiles para un total de 387 especies en Guatemala, aunque cálculos recientes sugieren que pueden haber más de 400 especies (Campbell y Mendelson, 1998).

A pesar de estos estudios, la herpetofauna de El Petén y específicamente en el Parque Nacional Laguna del Tigre, es poco conocida. Recientemente, Méndez et al. (1998) condujo un estudio de las plantas, aves y anfibios y reptiles del parque. Este estudio proporcionó datos base sobre la herpetofauna, aunque la cobertura geográfica de este estudio fue muy limitada.

Un componente importante de la herpetofauna de El Petén es el cocodrilo pantanero (*Crocodylus moreletii*). *C. moreletii* es una especie endémica de la Península de Yucatán y está registrada en el Libro Rojo de UICN (1996) con una deficiencia de datos y en el Apéndice I de CITES. Estudios previos de población del *C. moreletii* en Guatemala han mostrado que la persistencia de las especies en el área está amenazada por la caza ilegal y por la destrucción incrementada del hábitat debido a la invasión del hombre (Lara, 1990; Castañeda, 1997).

METODOS

Estudio de la población del *Crocodylus moreletii*

El muestreo de los cocodrilos se realizó en los siguientes lugares:

1) Area principal de San Pedro: Río San Pedro (ca. 38 km junto al río, desde la comunidad de Paso Caballos hasta la comunidad de El Buen Samaritano) y Río Sacluc (ca. 7 km desde la confluencia con el Río San Pedro hasta el final del Río Sacluc).

2) Area principal de Río Escondido (ca. 29 km cubriendo la mayor parte de la sección navegable del río).

3) Área principal de Flor de Luna (ca. 1.3 km de las orillas de la Laguna Flor de Luna y ca. 1 km de las orillas de la Laguna Vista Hermosa).

4) Area principal del Río Chocop: Laguna Tintal (C4, 1.4 km de las orillas); La Pista (C3; 2 km de las orillas); Laguna Guayacán (C2; 1.3 km de las orillas); y varias otras aguadas (pequeños estanques) y lagunas a lo largo de la carretera de Xan-Flor de Luna (refiérase a "Itinerario de la expedición y Diccionario Geográfico").

5) Además, se tomó una muestra en Laguna El Perú (5 km de las orillas), lo que se encuentra cerca del sitio arqueológico El Perú (un sendero del sitio empieza cerca de SP10).

Realizamos un censo de población del *Crocodylus moreletii* utilizando linternas en las orillas de los ríos y lagunas desde un bote inflable con motor fuera de borda. Contamos cualquier cocodrilo visible. Intentamos capturar cada animal que veíamos, aunque casi siempre estaban fuera del alcance del bote. Ordenamos por categorías tanto animales como neonatos observados y capturados, juveniles, subadultos o adultos en base al tamaño aproximado del cuerpo (Castañeda, 1998). Calculamos el tamaño de cada individuo utilizando la relación entre la distancia de donde nos encontrábamos y la distancia hacia el punto enfocado (Lara, 1990).

Para determinar los tipos de hábitats utilizados, también registramos el hábitat en el cual un animal se observó por primera vez. Los tipos de hábitats considerados fueron 1) agua: aquellos registros de individuos que no flotaban en la orilla o en la vegetación; 2) bosque ribereño: aquellos registros en vegetación arborescente en las orillas de los cuerpos de agua; 3) guamil: vegetación de crecimiento secundario sin árboles; 4) sibal: guamiles (*Cladium jamaicense*) en vegetación de ciénagas; y 5) vegetación emergente.

Estudio de herpetofauna en vegetación ribereña

La herpetofauna que ocupa los bancos de los ríos se examinó en puntos de muestreo terrestres a lo largo del Río San Pedro (SPT1), Río Sacluc (SPT5) y Río Escondido (ET2, ET3; refiérase a "Itinerario de la Expedición y Diccionario Geográfico").

Realizamos transectos de 500 mts. de largo y 5 de ancho en las orillas de estos ríos. Cada corte se dividió en cinco transectos más pequeños de 100 mts. Estandarizamos los esfuerzos de muestreo al limitar la búsqueda a 30 minutos cada 100 metros. Además, se registraron las observaciones puntuales en áreas muestreadas de cocodrilos.

RESULTADOS

Estudio de la población de *Crocodylus moreletii*

En general, se registraron 130 individuos de *C. moreletii* en 87.14 km de estudios en cuerpos de agua del PNLT. La densidad promedio en los puntos de muestreo fue de 4.3 cocodrilos/km. Los sistemas acuáticos cerrados tenían una densidad promedio mayor comparada con sistemas abiertos (ríos; Tabla 7.1). Entre sistemas acuáticos abiertos, el Río Sacluc poseía una densidad mayor que la del Río San Pedro o Río Escondido. Las lagunas del área principal de Flor de Luna y la Laguna la Pista (SP3) tienen las densidades mayores entre los sistemas acuáticos cerrados. La estructura de la población del *C. moreletii* en el PNLT se basa en clases de edades menores (Figura 7.1).

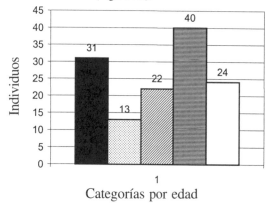

■Adult □Subadult ■Neonate ■Juvenile □Unknown

Figura 7.1. El número de individuos de *C. moreletti*, por clase de edad, observados durante estudios dentro del LTNP.

La Laguna el Perú tuvo el porcentaje más alto de adultos, constituyendo un 32.26% de todos los adultos en el estudio, seguido por el Río Sacluc (Tabla 7.2). La Laguna la Pista también contenía un mayor número de adultos y mostró el porcentaje más alto de subadultos en el estudio (ca. 46% de todos los subadultos). En el Río Escondido, como en el San Pedro, observamos un porcentaje alto de neonatos así como de jóvenes. Las muestras de muchas lagunas en las áreas principales de Flor de Luna y Río Chocop mostraron números bajos; en estas áreas, la mayoría de los individuos eran jóvenes.

El porcentaje más alto de registros de hábitats de *C. moreletii* se encontraba en sibal (39%), seguido por el agua lejos de la orilla (31%), bosque ribereño (25%), vegetación emergente (4%) y guamil (1%; Tabla 7.3).

Los adultos tenían patrones similares de abundancia entre hábitats con valores relativamente altos en agua, bosque ribereño y sibal (Tabla 7.2). Por otro lado, los neonatos y los jóvenes están más asociados con el sibal.

Estudio de herpetofauna en la vegetación ribereña

Encontramos un total del 14 especies anfibias y 22 especies de reptiles (incluyendo *Crocodylus moreletii*; Apéndice 12) en los transectos muestreados. Registramos 240 reptiles y anfibios individuales, sin tomar en cuenta el *C. moreletii*. Las especies anfibias más abundantes fueron las fringe-toed foamfrog (*Leptodactylus melanonotus*; 119 individuos), *Hyla microcephala* (25) y *Rana berlandieri* (24). Estas especies se encontraron a lo largo de los transectos muestreados. Las especies más abundantes de reptiles fueron los lagartos rayados (*Basiliscus vittatus*; 12) y *Norops bourgeai* (7). Entre las tortugas, la *Kinosternon leucostomum* fue la más abundante, aunque la mayoría de los individuos de estas especies fueron registradas como observaciones anécdotas. Encontramos la riqueza más amplia de especies en el Río San Pedro (17 especies), seguido por el Río Sacluc (13) y el Río Escondido (11; Apéndice 12). Con relación a la abundancia, sin embargo, los valores más altos se observaron en el Río Sacluc (123), seguido por el Río San Pedro (68) y Río Escondido (40).

DISCUSION

Estudio de población de *Crocodylus moreletii*

C. moreletii es llamado comúnmente el lagarto pantanero debido a su preferencia en estos hábitats (Alvarez del Toro, 1974). Esto explica las densidades relativamente altas de estas especies en lagunas y otros sistemas acuáticos cerrados en comparación con sistemas acuáticos abiertos como los ríos.

Dentro de la zona que rodea la Estación Biológica Las Guacamayas, el Río Sacluc es uno de los sitios más importantes para la reproducción de estas especies (Castañeda, 1998). Las densidades registradas en este estudio (2 individuos/km), en comparación con los valores en otros ríos como el San Pedro (0.60) y Escondido (0.76), confirman la importancia del Río Sacluc en la región. Las densidades registradas en las lagunas combinadas a lo largo de Xan-Flor de Luna (12.28) y Laguna la Pista (11) son las densidades más altas registradas en Guatemala. Los otros estudios disponibles para la comparación—los de Lara (1990) y Castañeda (1998)—reportaron un índice de densidades de 0 - 5.9.

Las lagunas El Perú, La Pista y el Río Sacluc son sitios ideales de los cuales se puede iniciar programas de manejo para *C. moreletii* dentro del PNLT por las siguientes razones: 1) La Pista es de fácil acceso y está relativamente cerca al puesto de control del CONAP; 2) El Perú se encuentra adyacente al puesto de monitoreo de CI-ProPetén, tiene servicios y tiene un gran número de individuos adultos y 3) Río Sacluc se encuentra cerca de la estación biológica. El hecho de que cada uno de estos sitios tenga una proporción grande de individuos adultos es un factor importante para considerar en el manejo del éxito reproductivo del *C. moreletii*.

Tabla 7.1. Densidades calculadas para *Crocodylus moreletii* en el Parque Nacional Laguna del Tigre, Petén, Guatemala

Sistemas acuáticos abiertos			
Punto de muestreo	**Longitud (km)**	**# de individuos**	**Individuos/km**
Río San Pedro	38	23	0.61
Río Sacluc	7	14	2.00
Río Escondido	29	22	0.76
TOTAL	*74*	*59*	*Promedio= 1.12*
Sistemas acuáticos cerrados			
Laguna Flor de Luna	1.3	5	3.85
Laguna Vista Hermosa	1	3	3.00
Laguna Tintal	1.4	5	3.57
Laguna la Pista	2	22	11.00
Laguna Guayacán	1.3	2	1.54
Laguna el Perú	5	22	4.40
Otras lagunas a lo largo del camino de Flor de Luna	1.14	14	12.28
TOTAL	*13.14*	*73*	*Promedio= 5.66*
GRAN TOTAL	*87.14*	*130*	*Promedio= 4.30*

Tabla 7.2. Estructura de la población de *Crocodylus moreletii* por muestra.

Data: Total # de individuos
        % de muestra
        % de individuos

Clase de edad → / Lugar de Muestreo ↓	Adulto	Neonato	Desconocido	Jóvenes	Subadultos	Total Individuos/ Muestra
Aguada 1	0 0.00 0.00	0 0.00 0.00	0 0.00 0.00	1 100.00 2.50	0 0.00 0.00	1
Aguada 2	0 0.00 0.00	0 0.00 0.00	0 0.00 0.00	1 100.00 2.50	0 0.00 0.00	1
El Perú	10 45.45 32.26	0 0.00 0.00	5 22.73 20.83	5 22.73 12.50	2 9.09 15.38	22
Escondido	2 9.09 6.45	10 45.45 45.45	1 4.55 4.17	8 36.36 20.00	1 4.55 7.69	22
Establo	0 0.00 0.00	0 0.00 0.00	0 0.00 0.00	6 100.00 15.00	0 0.00 0.00	6
Flor de Luna	1 20.00 3.23	3 60.00 13.64	0 0.00 0.00	0 0.00 0.00	1 20.00 7.69	5
Garza 1	0 0.00 0.00	0 0.00 0.00	1 33.33 4.17	1 33.33 2.50	1 33.33 7.69	3
Guayacán	0 0.00 0.00	1 50.00 4.55	0 0.00 0.00	1 50.00 2.50	0 0.00 0.00	2
La Pista	6 27.27 19.35	4 18.18 18.18	5 22.73 20.83	1 4.55 2.50	6 27.27 46.15	22
Román	0 0.00 0.00	0 0.00 0.00	0 0.00 0.00	3 100.00 7.50	0 0.00 0.00	3
Sacluc	5 35.71 16.13	0 0.00 0.00	7 50.00 29.17	2 14.29 5.00	0 0.00 0.00	14
San Pedro	4 19.05 12.90	4 19.05 18.18	3 14.29 12.50	8 38.10 20.00	2 9.52 15.38	21
Tintal	2 40.00 6.45	0 0.00 0.00	1 20.00 4.17	2 40.00 5.00	0 0.00 0.00	5
Vista Hermosa	1 33.33 3.23	0 0.00 0.00	1 33.33 4.17	1 33.33 2.50	0 0.00 0.00	3
TOTAL de Individuos	31	22	24	40	13	130

Las lagunas y estanques a lo largo de la carretera de Xan-Flor de Luna son de gran interés no sólo por el número de animales, sino también porque son sitios excelentes en los cuales se pueden estudiar los patrones de migración de estas especies. La mayoría de los animales registrados son jóvenes que posiblemente migraron de otros sitios, lo que coincide con la edad de los animales.

Al comparar los valores de densidad obtenidos por Castañeda (1998) en los ríos Sacluc (4.35 individuos/km) y San Pedro (2.10) con aquellos registrados en este estudio en las mismas ubicaciones (2.00 y 0.61 respectivamente), observamos que los valores pueden variar pero que el Sacluc mantiene una densidad mayor en ambas muestras. Esta variación puede ser debido al uso de un diseño de muestreo más intenso en un área más reducida por Castañeda (1998; dos transectos y ocho estudios), mientras que en este estudio se muestrearon diez transectos solamente una vez cada uno, permitiendo una cobertura mayor del parque.

La estructura de la población registrada en este estudio es atípico en el que los subadultos fueron la clase menor. Las poblaciones estables normalmente tienen adultos como la clase de edades menores (ver las referencias en Castañeda, 1998). Esto se puede explicar por medio del comportamiento de subadultos o por la posibilidad de que los adultos migraron de sus áreas natales a otros sitios en busca de territorios.

La similitud de los patrones de abundancia entre los adultos y la clase "desconocida" (Tabla 7.2, 7.3) con relación a los puntos muestra y los hábitats sugiere que en muchos casos los desconocidos fueron adultos. Es posible que estos individuos hayan tenido experiencias negativas debido a la presencia humana en el área, lo que incluye caza (Castañeda, 1998) y, de esa forma, tienden a ocultarse antes de que puedan ser marcados. Las lagunas El Perú y La Pista y el Río Sacluc pueden ser considerados áreas de conservación importantes para *C. moreletii* dentro del PNLT debido a la alta proporción de adultos que poseen—especialmente debido a que el éxito reproductivo de la población de cocodrilos depende de un gran número de adultos reproductores. Con respecto a esto, también es importante notar que la Laguna La Pista también tenía un gran número de subadultos que contribuirán a la población reproductora si sobreviven a la madurez. Las proporciones relativamente altas de jóvenes encontradas en el Río Escondido, Río San Pedro y Laguna La Pista puede ser debido a la eclosión temprana de los huevos en estos sitios o al éxito reproductivo variable entre la población de cocodrilos. La influencia de la actividad humana en la reproducción de cocodrilos es un tema importante para el estudio en el PNLT.

El tipo de vegetación sibal (ciénaga) es relativamente común en el PNLT y es un hábitat importante para especies animales como el pescado y tortugas, así como para cocodrilos que hacen sus nidos en esta vegetación. La importancia de este hábitat puede reflejarse en el uso relativamente alto de sibal por cocodrilos registrados en este estudio (Tabla 7.3). Es imposible que los cocodrilos usen el sibal por otras razones, incluyendo sus niveles relativamente altos de aislamiento durante la mayor parte del día, la gran

Tabla 7.3. Uso del hábitat de *Crocodylus moreletii* por clase de edad.

Hábitat	Cuerpo de agua	Clase de edad					Totales de Hábitat
		Adultos	Neonatos	Desconocido	Joven	Subadultos	
Aguas abiertas	Laguna	13	0	7	11	3	34
	Río	4	1	1	0	0	6
Total		*17*	*1*	*8*	*11*	*3*	40
Bosque ribereño	Laguna	1	1	3	4	1	10
	Río	5	3	7	7	1	23
Total		*6*	*4*	*10*	*11*	*2*	33
Emergente Vegetación	Laguna	0	0	0	4	0	4
	Río	0	0	0	1	0	1
Total		*0*	*0*	*0*	*5*	*0*	5
Guamil	Laguna	0	0	0	1	0	1
	Río	0	0	0	0	0	0
Total		*0*	*0*	*0*	*1*	*0*	1
Sibal	Laguna	6	7	3	3	6	25
	Río	2	10	3	9	2	26
Total		*8*	*17*	*6*	*12*	*8*	51
Total general		**31**	**22**	**24**	**40**	**13**	**130**

Programa de Evaluación Rápida

disponibilidad de alimentos debido al uso de sibal por los peces, el poco acceso para los humanos y la tranquilidad del ambiente.

A pesar de la importancia de los sibales dentro del parque, este hábitat se quema mucho cada año y este proceso constituye la amenaza principal para las poblaciones de *C. moreletii* en la región (Castañeda, 1998).

Los adultos a menudo se asociaron con el agua abierta, posiblemente porque nuestro estudio correspondió a la época de apareamiento de los cocodrilos que empieza en marzo en México (Alvarez del Toro, 1982). En el caso del PNLT, este período puede empezar en marzo o abril (Castañeda, 1998). Los adultos se encontraron a menudo en lagunas, posiblemente debido al poco acceso de los humanos hacia las lagunas en comparación con los ríos.

Estudio de herpetofauna en vegetación ribereña

La abundancia general de especímenes registrados a lo largo del Río Sacluc se debe a la gran abundancia de fringe-toed foamfrog (*Leptodactylus melanonotus*), la cual es una especie común con una distribución amplia en el PNLT. La mayoría de los animales se detectaron en las orillas de los cuerpos de agua y otros pocos se encontraron en más de cinco metros tierra adentro. Esto se debe a la dependencia de la herpetofauna en áreas cercanas al agua durante la época seca (Méndez et al., 1998).

Es aparente que modificar el diseño del transecto utilizado por Méndez et al. (1998) de 200 m muestreados en 120 min a una estrategia de muestreo de 500 m en 150 min fue más eficiente. Sin embargo, se necesitan esfuerzos de estudio más intensos en el PNLT. Los nuevos registros para el área, como la serpiente barriga manchada con rayas (*Coniophanes quinquevittatus*) que se encontraron en este estudio, surgen de dichos esfuerzos.

Otros reptiles amenazados de El Petén que se observaron durante estos estudios incluyen la tortuga de río de América Central (*Dermatemys mawii*), cuya (Campbell, 1998) y así como boa (*Boa constrictor*). *D. mawii* se encuentran enumerados por la UICN como especies en peligro y ambos reptiles se encuentran enumerados en el Apéndice II de CITES. Junto con el cocodrilo pantanero, estas especies deberán ser un foco de futuros estudios y planes de manejo para especies dentro del parque.

RECOMENDACIONES

- Monitorear al *Crocodylus moreletii* en el PNLT, haciendo énfasis en el Río Sacluc y las lagunas El Perú y La Pista.

- Iniciar un programa de manejo cautivo para el *C. moreletti* en el que se recolecten los huevos y se incuben bajo condiciones controladas. Esto ayudará a reducir el impacto de pérdidas naturales y el superhábit puede ser utilizado como medida alterna para las poblaciones humanas locales.

- Usar la estrategia de muestreo de transectos de 500 m/ 150 min, utilizando tanto señales visuales como auditivas para inventarios y monitoreos de herpetología en el PNLT.

- Incluir la tortuga de río de Centro América (*Dermatemys mawii*) y la boa (*Boa constrictor*) en programas de monitoreo.

LITERATURA CITADA

Alvarez del Toro, M. 1974. *Estudio Comparativo de los Crocodylia de México*. Ed. Inst. Instituto Mexicano de Recursos Naturales Renovables, A.C. Mexico.

Alvarez del Toro, M. 1982. *Los Reptiles de Chiapas*. Instituto de Historia Natural del Estado, Departamento de Zoología.

Campbell, J. A. 1998. *Amphibians and Reptiles of Northern Guatemala, the Yucatán, and Belize*. The University of Oklahoma Press, Norman, OK, USA.

Campbell, J. A. y J. R. Mendelson III. 1998. Documentación de los anfibios y reptiles de Guatemala. Mesoamericana 3:21.

Campbell, J. A. y J. P. Vannini. 1989. Distribution of amphibians and reptiles in Guatemala and Belize. Proceedings of the Western Foundation of Vertebrate Zoology 4:1-21.

Castañeda, F. 1997. Estatus y manejo propuesto de *Crocodylus moreletii* en el Departamento de El Petén, Guatemala. Pp. 52-57 in *Memorias de la 4ta. Reunión Regional del Grupo de Especialistas de Cocodrilos de América Latina y el Caribe*. Centro Regional de Innovación Agroindustrial, S.C. Villahermosa, Tabasco.

Castañeda, F. 1998. *Situación actual y plan de manejo para Crocodylus moreletii en el área de influencia de la Estación Biológica Las Guacamayas, Departamento de Petén, Guatemala*. Tésis de Licenciatura en Biología, Facultad de Ciencias Químicas y Farmacia, Universidad de San Carlos, Guatemala.

Lara, O. 1990. *Estimación del tamaño y estructura de la población de Crocodylus moreletii en los lagos Petén-Itza, Sal-Petén, Peténchel y Yaxhá, El Petén, Guate-mala*. Tésis de Maestría, Universidad Nacional, Heredia, Costa Rica.

Méndez, C., C. Barrientos, F. Castañeda y R. Rodas. 1998. *Programa de Monitoreo, Unidad de Manejo Laguna del Tigre. Los Estudios Base para su Establecimiento*. CI-ProPetén, Conservation International, Washington, DC, USA.

CAPITULO 8

LA FAUNA MAMIFERA DEL PARQUE NACIONAL LAGUNA DEL TIGRE, PETEN, GUATEMALA, CON ENFASIS EN MAMÍFEROS PEQUEÑOS

Heliot Zarza y Sergio G. Pérez

RESUMEN

- La fauna mamífera ha sobrevivido en cuatro áreas principales del Parque Nacional Laguna del Tigre (PNLT). El estudio se enfocó en mamíferos pequeños.

- Los murciélagos fueron capturados con redes y los roedores con trampas. Otros mamíferos fueron registrados con rastreos y señales.

- Se identificó un total de 40 especies de mamíferos de una lista potencial de 130 para el parque. No hubo diferencias significativas en la diversidad de los murciélagos entre las cuatro áreas principales. Las especies roedoras mostraron asociaciones de hábitat específicas tal y como se han reportado previamente en otros estudios.

- Reportamos un nuevo registro de especies para Guatemala: el ratón *Peromyscus yucatanicus*. También se observó otra especie endémica regional, el ratón coludo *Heteromys guameri*. Varios mamíferos raros y/o en peligro de extinción, incluyendo el tapir y el ciervo cola roja se observaron cerca de la Estación Biológica Las Guacamayas. Esta área puede ser de gran valor de conservación para los mamíferos.

INTRODUCCION

La deforestación produce pérdida y fragmentación de hábitats de mamíferos y es una de las peores amenazas para la biodiversidad mamífera en todo el mundo. Produce un cambio significativo en la composición y abundancia relativa de especies mamíferas, reflejado en la extinción local de especies raras (Williams y Marsh, 1998) e incrementa la abundancia de especies generales u oportunistas (Malcolm, 1997). Por ejemplo, los cambios a menudo conllevan la invasión de especies que generalmente se benefician con las actividades humanas, como los conejos (*Sylvilagus* spp.) y el coyote (*Canis latrans*; Garrott et al., 1993).

Entre los mamíferos, los murciélagos son componentes importantes de la biodiversidad en bosques tropicales y pueden representar hasta el 60% de la fauna mamífera local (Eisenberg, 1989). Son diversos tropicalmente, incluyendo especies que se alimentan de invertebrados y vertebrados, polen, néctar, frutas y sangre (Wilson, 1989). Los bosques neotropicales tienen la diversidad más grande de murciélagos en el mundo. De los 75 géneros presentes en los neotrópicos, 48 pertenecen a la familia Phyllostomidae (Wilson, 1989). La diversidad de los murciélagos filostómidos se refleja en el número de especies y su diversidad de dietas (Gardner, 1977) y estas características los hacen buenos indicadores biológicos (Fenton et al., 1992).

La composición y la abundancia relativa de murciélagos se determina por la complejidad y heterogeneidad del hábitat (LaVal y Fitch, 1977). Si el hábitat mantiene una complejidad y heterogeneidad alta, puede proporcionar un gran número de nichos por unidad de espacio (August, 1983). Así, los patrones de composición y diversidad de los murciélagos tienen gran potencial como detectores de cambios en el hábitat en los bosques tropicales (Amín y Medellín, *en imprenta*). Un estudio en la Sierra Lacandona, México (Medellín et al., *en imprenta*), encontró valores significativamente altos de diversidad de murciélagos en lugares con poca perturbación y alta abundancia de subfamilias Phyllostominae en hábitats con cambios. Patrones similares se reportaron en el este de Quintana Roo, México (Fenton et al., 1992). Se ha propuesto que la identidad de las especies más sencillas de murciélagos abundantes también pueden ser un indicador de cambios. Por asocian con los árboles altos como los amates (*Ficus* spp.) y jobos (*Spondias* spp.). En condiciones con cambios moderados, el género *Carollia* es dominante; en hábitats altamente con cambios, el género sencillo más abundante es *Sturnira* (Amín y Medellín,*en imprenta*).

Los mamíferos pequeños no voladores como los roedores, son grupos funcionalmente importantes en los bosques tropicales y constituyen entre el 15% y 25% de la fauna mamífera en los bosques neotropicales (Voss y Emmons, 1996). Los roedores ejercen una influencia sobre las plantas al dispersar las semillas, son fuentes de alimento importantes para los predadores (Dirzo y Miranda, 1990) y son sensibles al cambio de hábitat. Los mamíferos de bosques neotropicales como el tapir, el jaguar y los monos, pueden ser especialmente sensibles al cambio de hábitat y a la actividad humana debido a su rareza, susceptibilidad a la caza y timidez (Reid, 1997).

METODOS

Se reunió una lista de especies que podrían encontrarse potencialmente en el PNLT utilizando mapas de índices y descripciones de distribución en Hall (1981), Emmons y Feer (1997) y Reid (1997); éstos se compararon con los datos que obtuvimos.

Muestreo de roedores

Establecimos transectos de 540 metros cerca del punto focal de cada área principal y muestreamos roedores utilizando 90 trampas (Victor Rat Trap y Museum Special Mammal Trap) distribuidos en pares en 45 estaciones localizadas cada 12 metros a lo largo del área del transecto. Por las tardes, en las trampas se colocaron carnadas con una mezcla de mantequilla de maní, pasas, nuez molida y tocino y se revisaron al día siguiente por la mañana (Jones et al., 1996). Las trampas fueron operados dos o tres noches por transecto. Cada roedor capturado fue identificado y preparado en el campo como una piel de estudio y se colocaron pruebas en formol líquido. Todas las especies se almacenaron en Colecciones Zoológicas del Museo de Historia Natural de la Universidad de San Carlos.

Muestreo de murciélagos

Utilizamos cuatro redes que median 2.6x12 m, colocadas en forma de "T" para optimizar el esfuerzo de captura (Kunz et al., 1996). Las redes fueron colocadas en diferentes ubicaciones cada noche, se abrieron al anochecer (ca. 18:30 hrs) y se cerraron cuatro horas más tarde (ca. 22:30 hrs). Este es el período durante el cual los murciélagos son más activos (Fleming et al., 1972; Willig, 1986). Cada murciélago capturado se identificó con especies utilizando Medellín et

al. (1997) y se liberaron en el mismo lugar. En algunos casos, los murciélagos fueron preparados como muestras de estudio y depositados en las Colecciones Zoológicas del Museo de Historia Natural de la Universidad de San Carlos.

El esfuerzo de captura y la abundancia relativa se calculó con el método propuesto por Medellín (1993), en donde las capturas por metros totales de red/hora se utilizaron para calcular la abundancia relativa. Para calcular la riqueza de las especies de murciélagos, creamos curvas de acumulación de especies utilizando el software EstimateS 5 (Colwell, 1999), utilizando el calculador de riqueza de especies ACE (Calculador de cobertura en base a abundancia de riqueza de especies) esto se basa en aquellas especies con 10 o menos individuos en la muestra. Calculamos la diversidad de las ubicaciones utilizando el ídice de diversidad Shannon-Wiener y estos valores se compararon entre las áreas principales utilizando ANOVA (Zar, 1984). También consideramos la comunidad de murciélagos, de acuerdo con el hábito alimenticio. Cada especie fue clasificada en diferentes categorías de dietas de acuerdo con Medellín (1994).

Otros mamíferos

La mayoría de los mamíferos grandes son difíciles de observar. Utilizamos, huellas, piel, heces, cadáveres o restos óseos y madrigueras (Aranda, 1981) para proporcionar datos sobre la presencia y ausencia de mamíferos grandes. Realizamos estudios visuales diarios durante las pruebas para observar los mamíferos.

RESULTADOS Y DISCUSION

Murciélagos

Registramos un 39% de las especies potenciales (en base a mapas) de murciélagos (23 especies) para el PNLT (Apéndice 13). Doscientos veintiún individuos y capturamos 20 especies con un esfuerzo de muestreo de 684 metros de red/57 horas (14 noches). El esfuerzo promedio por noche fue de 3.7 horas con 44 metros de red (Tabla 8.2). Capturamos e identificamos tres especies sin redes, que no se consideraron en el análisis.

El índice de captura más grande se observó en el área focal del Río San Pedro, lo cual contrasta con los valores más bajos registrados en las áreas focales de Flor de Luna y Río Chocop (Tabla 8.1). Estos valores bajos probablemente representan la "fobia lunar", es decir una influencia negativa en la actividad de los murciélagos debido a que la luna estaba llena durante el muestreo en ambas localizaciones (Morrison, 1978). La curva de acumulación de especies de murciélagos (Figura 8.1) muestra que durante las primeras ocho noches, se recolectaron 19 especies. Esto representó un 95% del total de especies capturadas en el estudio. En las siguientes seis noches, solamente una especie se agregó. El calculador ACE de la riqueza de especies fue de 26.5.

No hubo diferencia significativa en la diversidad de las comunidades de murciélagos de las cuatro localizaciones ($F = 1.116$, $p>0.05$, $n = 12$). La diversidad total fue H' = 1.471 para las cuatro localizaciones combinadas. Entre las localizaciones, Río Escondido tenía el valor más alto de diversidad (Tabla 8.2). La paridad más alta se encontró en Río Chocop ($E = 0.94$), posiblemente debido al bajo número de especies ($S = 8$) y a la baja abundancia (dos de tres individuos/especies).

Las 20 especies de murciélagos capturados pertenecían a cuatro familias. La familia Phyllostomidae fue la más abundante, con 198 individuos de 17 especies y se encontró en todas las localidades. Cuatro especies (tres phyllostomidos y una mormoópido) fueron comunes en todas las localizaciones. La abundancia relativa (Tabla 8.3) muestra que las especies dominantes de la comunidad son *Artibeus intermedius, Carollia brevicauda* y *A. lituratus*. Estas especies comprenden el 54% del total de murciélagos (consulte también el Apéndice 13). Un segundo grupo menos abundante consta de 11 especies, algunas de las cuales pueden considerarse comunes (por ejemplo, *Sturnira lilium, Dermanura phaeotis*) y otras que son menos capturadas (por ejemplo, *Uroderma bilobatum*). Un tercer grupo de especies raras consta de seis especies capturadas solamente una vez, juntas representan el 2.7% de los murciélagos capturados (por ejemplo, *Chrotopterus auritus, Mimon crenulatum*).

Entre las especies más abundantes (*A. intermedius, C. brevicauda* y *A. lituratus*), dos son dominantes en el hábitat sin cambios: *A. intermedius* y *A. lituratus*; la otra, *C. brevicauda*, es dominante en condiciones de cambios moderados. La abundancia de cada una de estas especies es < 25% del total de todos los murciélagos capturados, lo que sugiere que el hábitat se encuentra sin cambios de acuerdo a los criterios de Amín y Medellín (*en imprenta*).

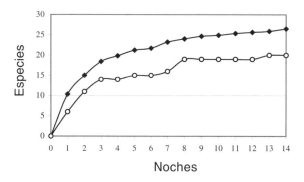

Figura 8.1. La curva de acumulación de especies para los murciélagos del PNLT. La curva más baja (círculos abiertos) indica la acumulación observada y la curva superior (diamantes sólidos) indica la acumulación calculada del calculador de cobertura en base a la abundancia (ACE).

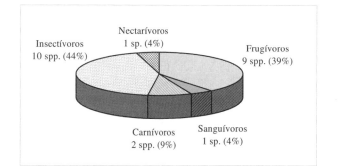

Figura 8.2. Porcentaje de especies de murciélagos que pertenecen a diferentes grupos trópicos en el PNLT.

Tabla 8.1. Esfuerzo de captura y éxito utilizando redes en el Parque Nacional Laguna del Tigre, Petén, Guatemala.

	San Pedro	Sacluc	Escondido	Flor de Luna	Chocop	Total	Promedio	Prom/ Noche
Redes-m	144	144	96	132	144	660	132	44.00
Noches	3	3	2	3	3	14	2.8	0.93
Horas	12	12	8	12	12	56	11.2	3.73
# de murciélagos	100	38	38	19	25	220	44	14.67
# especies	14	10	12	8	8	20	10.4	
Redes-m/h	1728	1728	768	1584	1728	7536		
#muerciélagos/N-m/h	0.05787	0.02199	0.049479	0.01199	0.01446	0.029193		
#especies/N-m/h	0.00810	0.00578	0.015625	0.00505	0.00463	0.002654		

Los murciélagos capturados en este estudio representaron una variedad de grupos tróficos (Figura 8.2). Los insectívoros comprendieron el 43% de las especies, seguido por frutívoros en un 39%, carnívoros en un 8.7%, nectarívoros en un 4.3% y sanguívoros en un 4.3%. En sus estudios de murciélagos de bosques tropicales, Wilson et al. (1997) y Amín (1996) reportaron que los frutívoros eran el grupo tropical más grande (50%), seguido por insectívoros (23%). Nuestros resultados pueden diferir de los suyos debido a que nuestros puntos de muestreo se encontraban cerca de cuerpos de agua, en donde los insectívoros generalmente son más abundantes (Kunz y Kurta, 1988).

Análisis de la diversidad de murciélagos por sitios (Tabla 8.3)

Area principal de Río San Pedro
En esta localidad registramos el número más alto de especies (15 especies) y capturas (138 individuos). Esta localidad se subdividió en dos puntos para muestreo—el Río San Pedro y el Río Sacluc—debido al contraste de la vegetación entre ellos. En el Río San Pedro, el muestreo se realizó en el bosque ribereño (SPT1-3) y el bosque alto (SPT6; refiérase a la Tabla 1.2, Apéndice 2); aquí capturamos 100 individuos de 14 especies en tres noches. Las especies más abundantes fueron *Artibeus intermedius*, *Carollia brevicauda*, *A. lituratus* y *Sturnira lilium*. En Río Sacluc, las redes se colocaron en encinales (SPT4-5) y se capturaron 38

individuos de 10 especies. Las especies más abundantes fueron *A. lituratus*, *C. brevicauda* y *A. intermedius*. El éxito de la captura fue en parte por el hábitat del bosque alto (Tabla 8.1). No hubo diferencia estadísticamente importante entre las áreas principales (San Pedro vs. Sacluc: t =1.933, t =1.352, P >0.005), pero los valores de diversidad fueron un poco mayores en Río San Pedro (Tabla 11). El único murciélago cara manchada (*Centurio senex*) observado fue capturado en el sitio del Río Sacluc. Un murciélago negro (*Molussus ater*) se encontró bajo una tienda de campaña en la Estación Biológica Las Guacamayas (SP13) y un grupo de murciélagos proboscis (*Rhynchonycteris naso*; cinco de seis individuos) se observó en el tronco de un árbol directamente sobre el agua. En la Laguna El Perú, también observamos un murciélago pescador más grande (*Noctilio leporinus*).

Río Escondido
Aquí muestreamos en bosque alto con algunas áreas naturales abiertas (ET 1-2). El bosque se quemó hace aproximadamente 20 años. Capturamos 39 individuos y 13 especies. Las especies más abundantes fueron *Artibeus intermedius* y *Sturnira lilium*. Los valores de riqueza fueron similares a los del Río San Pedro, aunque la abundancia fue menor. El éxito de la captura fue un poco más alta que la del Río Sacluc (Tabla 8.1). Esta localización mostró valores de riqueza más altos (Tabla 8.2). Observamos un grupo de murciélagos con líneas blancas mayores (*Saccopteryx bilineata*) que utilizaba un tronco hueco de matapalos (*Ficus* sp.) como lugar de descanso.

Tabla 8.2. Abundancia y dominio relativos de murciélagos en el Parque Nacional Laguna del Tigre, Petén, Guatemala.

Rango	Especies	No. de individuos	Ind/N-m/h
1	*Artibeus intermedius*	48	0.006369
2	*Carollia brevicauda*	41	0.005441
3	*Artibeus lituratus*	30	0.003981
4	*Sturnira lilium*	20	0.002654
5	*Dermanura phaeotis*	17	0.002256
6	*Pteronotus parnellii*	16	0.002123
7	*Artibeus jamaicensis*	10	0.001327
8	*Glossophaga commissarisi*	8	0.001062
9	*Dermanura watsoni*	7	0.000929
10	*Rhynchonycteris naso*	5	0.000663
11	*Tonatia evotis*	4	0.000531
12	*Desmodus rotundus*	3	0.000398
13	*Mimon bennettii*	3	0.000398
14	*Uroderma bilobatum*	2	0.000265
15	*Mormoops megalophylla*	1	0.000133
16	*Micronycteris schmidtorum*	1	0.000133
17	*Chrotopterus auritus*	1	0.000133
18	*Centurio senex*	1	0.000133
19	*Mimon crenulatum*	1	0.000133
20	*Tonatia brasiliense*	1	0.000133

Tabla 8.3. Riqueza y diversidad de murciélagos por área principal en el Parque Nacional Laguna del Tigre, Petén, Guatemala.

H' = el índice Shannon-Wiener Index

Area principal	# Ind.	# ESPECIE	Diversidad (*H'*)	Pareamiento
Río San Pedro	100	14	1.6897	0.8270
Río Sacluc	38	10	1.3514	0.9050
Río Escondido	38	12	1.8475	0.9173
Laguna Flor de Luna	19	8	1.1769	0.8993
Río Chocop	25	8	1.2899	0.9404
Valores generales	220	20	1.4711	0.8978

Laguna Flor de Luna

Bosque bajo que consta de tintales que se muestreó aquí (FT 1-3). Capturamos a 19 individuos de ocho especies. Durante la primera noche de muestreo, obtuvimos el número más alto de individuos; las especies más abundantes fueron *Pteronotus parnellii* y *Sturnira lilium*. El éxito de la captura fue la más baja de todas las áreas principales (Tabla 8.1). El éxito de la riqueza, abundancia y captura disminuyó después de la segunda noche de muestreo, probablemente debido a la influencia de la luna llena (Morrison, 1978).

Río Chocop

Bosque alto fue muestreado y las especies de plantas dominantes fueron *Mataiba* sp., *Chrysophila* sp. y algunas sapoteceas (*Manilkara sapota* y *Pouteria* sp.). Capturamos 25 individuos y ocho especies. Las especies dominantes fueron *Artibeus intermedius* y *Pteronotus parnellii*. Estos resultados muestran una disminución clara en el índice de capturas de individuos mientras continuaba el muestreo. El éxito de la captura fue bajo, similar a la de Flor de Luna.

Roedores

Este fue el segundo grupo más frecuentemente registrado de mamíferos en este estudio (Apéndice 13). Identificamos ocho especies de roedores en siete géneros. Dos especies se registraron por observación directa y no se recolectaron. Los roedores capturados constituyen el 47% de las especies de roedores potenciales para el PNLT (en base a registros previos). Las especies capturadas fueron *Oryzomys couesi*, *Sigmodon hispidus*, *Ototylomys phyllotis*, *Peromyscus yucatanicus*, *Heteromys desmarestianus* y *H. gaumeri*). Dos de estos son endémicos a la Península de Yucatán (*Peromyscus yucatanicus* y *H. gaumeri*).

El Sigmodontinae (Muridae) consta de 21 géneros y 61 especies en el sureste de México y Centro América (Reid, 1997). Encontramos que la mayoría de las especies capturadas durante el muestreo pertenecen a esta subfamilia. La familia Heteromyidae está representada por dos géneros en Centro América y registramos un género (*Heteromys*) y dos de sus especies (todas las que podían ocurrir potencialmente) en el PNLT. El otro género (*Liomys*) está distribuido principalmente en las costas pacíficas en bosques secos del norte de México hasta Panamá (Reid, 1997).

En un total de 1149 trampas nocturnas, obtuvimos un total de 10 individuos y cinco especies (Tabla 8.4). La riqueza de especies más alta se encontró en el área focal del Río San Pedro/Río Sacluc (3 sp.), en comparación con las otras ubicaciones, en donde solamente se registró una especie. El éxito de las trampas fue constante para la mayoría de las ubicaciones, con la excepción de Río Sacluc (encinal), en donde se observó el valor más bajo. Nuestros datos y otras fuentes sugieren que las diferentes especies de roedores tienden a asociarse con diferentes tipos de vegetación en la región. *Sigmodon hispidus* está asociado con los guamiles (Río Escondido); *Heteromys desmarestianus* (Río San Pedro), *H. gaumeri* (Río Chocop) *y Ototylomys phyllotis* (Río San Pedro) con bosque alto; y *Oryzomys couesi* (Río Sacluc) y *Peromyscus yucatanicus* (Flor de Luna) con bosque bajo y tintales (Medellín y Redford, 1992; Young y Jones, 1983).

Obtuvimos un nuevo registro para Guatemala: el ratón ciervo de Yucatán (*Peromyscus yucatanicus*; Muridae). Este ratón es de tamaño moderado, más largo que el *P. leucopus* (Young y Jones, 1983). Este es un endémico de la Península de Yucatán, aunque su limite de distribución del sur no se ha determinado aún (Huckaby, 1980). Las localidades del sur, en donde se ha reportado esta especie en la literatura se encuentra 7.5 km al oeste de Escárcega, Campeche, México (Dowler y Engstrom, 1988). Cuatro especies de *P. y. badius* se recolectaron de Calakmul, Campeche y se depositan en la colección del Instituto de Biología, Universidad Autónoma de México (IBUNAM 37362, 37364, 37365, 37366), con 115 km ESE de acuerdo con el último registro. Nuestro registro constituye una extensión de 116 kms al sur de Escárcega y 134 km al suroeste de Calakmul.

Otros mamíferos

Se registraron otras once especies de mamíferos grandes en el PNLT (Apéndice 13). Confirmamos la presencia de varias especies amenazadas y en peligro, incluyendo el tapir de Baird *(Tapirus bairdii)*, coche de monte *(Tayassu tajacu)*, cabrito guitsisil *(Mazama americana)*, mono araña de Centro América *(Ateles geoffroyi)* y el mono aullador negro de Yucatán *(Alouatta pigra)*.

La presencia del mono araña en Centro América y el mono aullador negro de Yucatán en las cuatro áreas principales sugieren que todavía existen hábitats relativamente sin cambios en estas áreas. En Veracruz, México, Silva-López et al. (1993) reportó una reducción en la densidad de la población para ambas especies en condiciones de cambio y una abundancia relativamente mayor de monos araña con relación a los monos aulladores en los tramos con cambios.

El río San Pedro fue el único lugar en el cual se observó un tapir y un cabrito guitsisil. Ambas especies viven en hábitats sin cambios, aunque los tapires pueden tolerar las condiciones de cambios moderadamente (Fragoso, 1991). Sin embargo, los tapires se registraron como amenazados en el Apéndice I de CITES y el cabrito guitsisil se registró en el Apéndice III de CITES (Reid, 1997). La caza es la amenaza principal para el tapir y ha ocasionado extinciones en algunas partes de Centro América (March, 1994). Para preservar estas especies en Guatemala, será necesario conservar las poblaciones en la Reserva de la Biósfera Maya (Matola et al., 1997) y en particular en el PNLT.

Aunque no registramos el jaguar *(Panthera onca)* en el PNLT, una evaluación rápida reciente identificó áreas que contenían poblaciones viables de jaguares en Guatemala. La población más importante en Guatemala se considerará que reside dentro de la RBM, específicamente en el PNLT y el Parque Nacional Sierra del Lacandon (McNab y Polisar, *en imprenta*).

CONCLUSIONES Y RECOMENDACIONES

De este trabajo y otros estudios realizados en la región, surgen tres conclusiones:

* El PNLT es un refugio importante de alta diversidad mamífera y poblaciones de varias especies amenazadas y en peligro debido a su gran área, heterogeneidad de hábitat y estado primitivo.

* Dentro del PNLT, el área principal del río San Pedro es posiblemente, el área de conservación más importante debido a su abundancia y riqueza de murciélagos comparativamente alta y la presencia de especies amenazadas y en peligro.

* El PNLT representa uno de los lugares en la región en donde se pueden mantener las poblaciones viables a largo plazo de mamíferos grandes, como el tapir y el jaguar.

* Los factores principales que amenazan las poblaciones de mamíferos salvajes del PNLT son la agricultura de quema y rosa, caza y la expansión de las poblaciones humanas, aunque los impactos de estos factores en diferentes partes del parque son poco comprendidos. Para mejorar nuestra comprensión de estos procesos, se debe iniciar un programa de monitoreo que evalúe el impacto de los humanos sobre las poblaciones mamíferas en el PNLT. Para organizar dicho programa de monitoreo y manejar poblaciones mamíferas a largo plazo, será necesario desarrollar colaboraciones regionales con México y Belice con el objetivo de mantener la existencia de los corredores biológicos y evitar el aislamiento de las poblaciones. Un foco de dicho programa deberá enfatizar en las especies carismáticas como el tapir y el jaguar para garantizar su permanencia en Guatemala.

Tabla 8.4. Esfuerzo y éxito de captura para roedores en el Parque Nacional Laguna del Tigre, Petén, Guatemala.

Area principal	# esp	# Ind	Esfuerzo de captura (trampas-noches)	Éxito de captura (% de trampas)
Río San Pedro	2	2	243	0.82
Río Sacluc	1	1	276	0.36
Río Escondido	1	2	180	1.10
Laguna Flor de Luna	1	4	270	1.15
Río Chocop	1	1	180	1.10
Valores generales	5	10	1149	0.90

LITERATURA CITADA

Amín, M. 1996. *Ecología de comunidades de murciélagos en bosque tropical y hábitats modificados en la Selva Lacandona, Chiapas.* Tesis de Licenciatura. Universidad Nacional Autónoma de México.

Amín, M. y R. A. Medellín. *In press.* Bats as indicators of habitat disturbance. In *Single Species Approaches to Conservation: What Works, What Doesn't and Why* Island Press, Washington, DC, USA.

Aranda, J. M. 1981. Rastros de los mamíferos silvestres de México. Manual de campo. Instituto Nacional de Investigaciones sobre Recursos Bióticos, México.

August, P. V. 1983. The role of habitat complexity and heterogeneity structuring tropical mammal communities. Ecology 64:1495-1507.

Colwell, R. K. 1999. *EstimateS 5. Statistical Estimation of Species Richness and Shared Species from Samples.* Web site: viceroy.eeb.uconn.edu/estimates.

Dirzo, R. y A. Miranda. 1990. Contemporary Neotropical defaunation and forest structure, function, and diversity: A sequel to John Terborgh. Conservation Biology 4: 444-447.

Eisenberg, J. F. 1989. *Mammals of the Neotropics: The Northern Neotropics: Panama, Colombia, Venezuela, Suriname, French Guiana. Vol 1.* University of Chicago Press, Chicago, IL, USA.

Emmons, L. H. y F. Feer. 1997. *Neotropical Rainforest Mammals: A Field Guide.* Second edition. The University of Chicago Press, Chicago, IL, USA.

Fenton, M. B., L. Acharya, D. Audet, M. B. C. Hickey, C. Merriman, M. K. Obrist, D. M. Syme y B. Adkins. 1992. Phyllostomid bats (Chiroptera: Phyllostomidae) as indicators of habitat disruption in the Neotropics. Biotropica 24: 440-446.

Fleming, T. H., E. T. Hooper y D. E. Wilson. 1972. Three Central American bat communities: structure, reproductive cycles, and movement patterns. Ecology 53: 555-569.

Fragoso, J. M. 1991. The effect of selective logging on Baird's tapir. Pp. 295-304 in *Latin America Mammalogy: History, Biodiversity, and Conservation* (M. A. Mares and D. J. Schmidly, eds.). University of Oklahoma Press, Norman, OK, USA.

Gardner, A. L. 1977. Feeding habits. Pp. 351-364 in *Biology of Bats of the New World: Family Phyllostomidae Part II* (R. J. Baker, J. K. Jones, and D. C. Carter, eds.). Special Publications, Museum, Texas Tech University, Lubbock, TX, USA.

Garrott, R. A., P. J. White y C. A. Vanderbilt White. 1993. Overabundance: an issue for conservation biologists? Conservation Biology 7: 946-949.

Hall, E. R. 1981. *The Mammals of North America, Vol. 1.* John Wiley and Sons.

Huckaby, D. C. 1980. Species limits in the *Peromyscus mexicanus* group (Mammalian: Rodentia: Muridea). Contributions Sciences National History Museum, Los Angeles 326: 1-24.

Jones, C., W. J. McShea, M. J. Conroy y T. H. Kunz. 1996. Capturing mammals. Pp. 115-155 in *Measuring and Monitoring Biological Diversity: Standard Methods for Mammals* (D. Wilson, J. R. Cole, J. D. Nichols, R. Rudran, and M. S. Foster, eds.). Smithsonian Institution Press, Washington, DC, USA.

Kunz, T. H. y A. Kurta. 1988. Capture methods and holding devices. Pp. 1-29 in *Ecological and Behavior Methods for Study of Bats* (T. H. Kunz, ed.). Smithsonian Institution Press, Washington, DC, USA.

Kunz, T. H., D. W. Thomas, G. C. Richards, C. R. Tidemann, E. D. Pierson y P. A. Racey. 1996. Observational techniques for bats. Pp. 105-114 in *Measuring and Monitoring Biological Diversity: Standard Methods for Mammals* (D. Wilson, J. R. Cole, J. D. Nichols, R. Rudran, and M. S. Foster, eds.). Smithsonian Institution Press, Washington, DC, USA.

LaVal, R. K. y H. S. Fitch. 1977. Structure, movement, and reproduction in three Costa Rican bat communities. Occasional Papers, Museum of Natural History, University of Kansas 69:1-28.

Malcolm, J. R. 1997. Biomass and diversity of small mammals in Amazonian Forest fragments. Pp. 207-221 in *Tropical Forest Remnants: Ecology, Management, and Conservation of Fragmented Communities* (W. F. Laurace and R. O. Bierregaard, eds.). The University of Chicago Press, Chicago, IL, USA.

March, I. J. 1994. Situación actual del tapir en México. Centro de Investigaciones Ecológicas del Sureste (Serie Monográfica No. 1).

Matola, S., A. D. Cuarón y H. Rubio-Torgler. 1997. Status and action plan of Baird's tapir (*Tapirus bairdi*). In *Tapir: Status and Conservation Action Plan* (D. M. Brooks, R. E. Bodmer, and S. Matola, eds.). The IUCN/SSC Tapir Specialist Group Report. Web site: http://www.tapirback.com/tapirgal/iucn-ssc/actions97/cover.htm.

McNab, R. B. y J. Polisar. *In press*. A participatory methodology for a rapid assessment of jaguar (*Panthera onca*) distributions in Guatemala. Wildlife Conservation Society.

Medellín, R. A. 1993. Estructura y diversidad de una comunidad de murciélagos en el trópico húmedo mexicano. Pp. 333-354 in *Avances en el Estudio de los Mamíferos Méxicanos* (R. A. Medellín and G. Ceballos, eds.). Publicaciones Especiales, Asociación Mexicana de Mastozoología, A. C. 1:1-464.

Medellín, R. A. 1994. Mammals diversity and conservation in the Selva Lacandona, Chiapas, México. Conservation Biology 8:788-799.

Medellín, R. A., H. Arita, and O. Sánchez. 1997. Identificación de los murciélagos de México. clave de campo. Asociación Mexicana de Mastozoología, A.C. Publicaciones Especiales No 2. México.

Medellín, R. A., M. Equihua, and M. A. Amín. *In press*. Bat diversity and abundance as indicators of disturbance in Neotropical rainforest. Conservation Biology.

Medellín, R. A., and K. H. Redford. 1992. The role of mammals in Neotropical forest-savanna boundaries. Pp. 519-548 in *Nature and Dynamics of Forest-Savanna Boundaries* (P.A. Furley, J. Procyor, and J.A. Ratter, eds.). Chapman & Hall.

Morrison, D. W. 1978. Lunar phobia in a tropical fruit bat, *Artibeus jamaicensis* (Chiroptera: Phyllostomidae). Animal Behaviour 26:852-855.

Reid, F. A. 1997. *A Field Guide to the Mammals of Central America and Southeast Mexico*. Oxford University Press, New York, NY, USA.

Silva-López, G., J. Benítez-Rodríguez, and J. Jimenéz-Huerta. 1993. Uso del hábitat por monos araña (*Ateles geoffroyi*) y aullador (*Alouatta palliata*) en áreas perturbadas. Pp. 421-435 in *Avances en el Estudio de los Mamíferos Méxicanos* (R. A. Medellín and G. Ceballos, eds.). Publicaciones Especiales, Asociación Mexicana de Mastozoología, A.C. 1:1-464.

Voss, R. S., and L. H. Emmons. 1996. Mammalian diversity in Neotropical lowland rainforest: a preliminary assessment. Bulletin American Museum of Natural History 230:1-115.

Williams, S. E., and H. Marsh. 1998. Changes in small mammal assemblage structure across a rain forest/open forest ecotone. Journal of Tropical Ecology 14:187-198.

Willig, M. R. 1986. Bat community in South America: a tenacious chimera. Revista Chilena de Historia Natural 59: 151-168.

Wilson, D. E. 1989. Bats. Pp. 365-382 in *Tropical Rain Forest Ecosystems: Biogeographical and Ecological Studies* (H. Lieth and M. J. A. Werger, eds.). Elsevier Science Publications, Amsterdam, the Netherlands.

Wilson, D. E., R. Baker, S. Solari, and J. J. Rodríguez. 1997. Bats: Biodiversity assessment in the Lower Urubamba Region. Pp. 293-301 in *Biodiversity Assessment and Monitoring of the Lower Urubamba Region, Peru: San Martin-3 and Cashiriari-2 Well Sites* (F. Dallmeier and A. Alonso, eds.). SI/MAB Series # 1. Smithsonian Institution/MAB Biodiversity Program, Washington, DC, USA.

Wilson, D. E., and D. M. Reeder. 1993. *Mammal species of the world: a taxonomic and geographic reference*. 2nd. ed. Smithsonian Institution Press, Washington, DC, USA.

Young, C. J., and J. K. Jones. 1983. *Peromyscus yucatanicus*. Mammalian Species 196: 1-3.

Zar, J. H. 1984. Biostatistical Analysis. 2nd. Ed. Prentice Hall. Englewood Cliffs, NJ, USA.

CAPITULO 9

LAS HORMIGAS (HYMENOPTERA: FORMICIDAE) DEL PARQUE NACIONAL LAGUNA DEL TIGRE, PETEN, GUATEMALA

Brandon T. Bestelmeyer, Leeanne E. Alonso y Roy R. Snelling

RESUMEN

- Registramos 112 especies y 39 géneros de hormigas en el Parque Nacional Laguna del Tigre (PNLT). Estos valores son mayores que aquéllos registrados en varios otros estudios en la región.

- El género *Pheidole* contenía el número más grande de especies, pero otros géneros arborícolas fueron también altamente diversos.

- Se registraron tres especies endémicas de hormigas en la región del Bosque Maya, incluyendo *Sericomyrmex aztecus*, *Xenomyrmex skwarrae* y *Odontomachus yucatecus*. Un género extremadamente raro, *Thaumatomyrmex,* se descubrió por primera vez en Guatemala.

- La riqueza de las especies fue más alta en hábitats relativamente abiertos y con cambios, contrario a nuestras expectativas. Esto se debió a la alta diversidad de microhábitats de estos sitios. Un hábitat modificado fue similar relativamente a un sitio boscoso primitivo cercano, lo que coincide con la sugerencia de que la recuperación de la fauna de hormigas neotrópicas de los cambios es rápida.

- Los índices de cambio en la composición de las hormigas entre los sitios es muy alto, lo que impide una evaluación de asociaciones de especie-hábitat. La riqueza general de la fauna de hormigas del PNLT es mucho más alto que el registrado.

INTRODUCCION

Se cree que los insectos son el grupo más vistoso de los organismos sobre la tierra (Mayo, 1988). Este grupo alcanza su mayor diversidad en bosques tropicales (Erwin y Scott, 1980), así que el índice incrementado de deforestación tropical puede resultar en reducciones severas de diversidad de insectos en todo el mundo. Por esta razón, los estudios que documentan los patrones de diversidad de insectos entre los bosques tropicales son necesarios para priorizar la conservación de las regiones tropicales y los hábitats dentro de las regiones (Haila y Margules, 1996). Otra consecuencia de la gran diversidad de insectos es que es difícil logísticamente hablando considerar todos los insectos simultáneamente en dichos estudios. Así, los ecologistas confían en La información proporcionada por uno o más grupos de insectos para caracterizar los patrones de diversidad de insectos en las regiones.

Las hormigas (Hymenoptera: Formicidae) son un grupo que puede ser útil para representar las respuestas de insectos terrestres. Las hormigas son extremadamente llamativas, abundantes, activas y elementos diversos de los bosques neotropicales. Debido a que las hormigas son insectos sociales, son tomadas en grupos y son muy flexibles en sus respuestas a las diferencias en su entorno en tiempo y espacio (Wilson, 1987). Como consecuencia de esta flexibilidad, las hormigas son exitosas en aún los hábitats más difíciles. Las hormigas pueden encontrarse anidando en casi todas partes de los ecosistemas terrestres, desde las profundidades en columnas de tierra hasta las profundidades de hojas secas y tierra y en arbustos epifitos y troncos de árboles.

Sin embargo, la composición y riqueza de las especies de las comunidades de hormigas, revelaron respuestas a la variación del hábitat natural y a los cambios de hábitat (Andersen, 1991; Bestelmeyer y Wiens, 1996; Vasconcelos, 1999) debido a que las diferentes especies pueden estar asociadas con condiciones determinadas del suelo (Johnson, 1992), microclimas (Perfecto y Vandermeer, 1996) y aún la presencia de especies asociadas con plantas particulares (Alonso, 1998), todos estos pueden verse alterados por las actividades económicas humanas. En los bosques tropicales en particular, el dosel es parte importante de las características ambientales que determinan la distribución y abundancia de las especies de hormigas (MacKay, 1991; Perfecto y Snelling, 1995). Las respuestas de las hormigas para dicha variación puede utilizarse para evaluar tanto el valor de la conservación de determinadas áreas como los efectos del uso de tierra por el hombre en la biodiversidad.

Todavía no se han reportado estudios de comunidades de hormigas para los bosques de El Petén en Guatemala y nuestro objetivo principal en esta evaluación rápida fue proporcionar información preliminar de la riqueza y composición de la fauna de hormigas en el Parque Nacional Laguna del Tigre (PNLT) en un contexto biogeográfico. También deseamos proporcionar una evaluación preliminar de las relaciones entre la variación de hábitat y la variación en la composición y riqueza de la comunidad de hormigas. Específicamente, nos enfocamos en el contraste entre 1) hábitats de doseles cerrados que caracterizan el bosque alto, 2) los hábitats de doseles relativamente abiertos incluyendo encinales y tintales que se producen por medio de procesos naturales y 3) los guamiles que se producen por medio de sucesiones secundarias después de la deforestación. Con la hipótesis de que los hábitats de doseles abiertos tienen menos riqueza de especies y la composición de la comunidad alterada en comparación con los hábitats de doseles cerrados en el contexto de este bosque tropical, de acuerdo a lo que encontró MacKay (1991) y Vasconcelos (1999). Sin embargo, no muestreamos intensamente en hábitats de pastos recientemente cambiados. Dicho esfuerzo debería ser redundante ya que varios estudios han demostrado las consecuencias de dichos cambios drásticos en la vegetación de la diversidad de hormigas (Greenslade y Greenslade, 1977; Quiroz-Robledo y Valenzuela-Gonzáles, 1995; Vasconcelos, 1999). Observamos los patrones del dominio de las especies de hormigas en los hábitats de los guamiles, para comparación con otros estudios en la región. Utilizamos esta información para resaltar la importancia regional del parque e intentar identificar algunos procesos que pueden apoyar o amenazar a las hormigas y otras poblaciones artrópodas en el parque. Nuestras observaciones intentan sugerir rutas fructíferas para estudios más intensos en el parque así como proporcionar información tentativa sobre patrones de riqueza de especies de hormigas en la ausencia de dichos estudios.

AREAS Y METODOS DE ESTUDIO

Se muestrearon hormigas y variables ambientales en los transectos en uno o más tipos de hábitats en cada área focal (refiérase a "Itinerario de la expedición y Diccionario Geográfico"). No se hizo ningún intento de caracterizar la fauna de hormigas en áreas principales debido a que cada área principal contenía una variedad de distintos tipos de hábitats que no podían muestrearse adecuadamente en el tiempo pensado. Las especies de hormigas corresponden a la variación del hábitat que ocurre en cada escala, así que nuestro objetivo principal fue examinar las asociaciones de hábitats de hormigas en distintos tipos de hábitats. Las trampas se utilizaron para reunir muestras estandarizadas para comparaciones cuantitativas, de acuerdo a los procedimientos de Bestelmeyer et al. (*en imprenta*). Se colocó una trampa oculta (75 mm de diámetro) en diez puntos en un área con intervalos de 10 metros (consulte la Tabla 9.4). Las trampas se abrieron durante 48 horas. Se realizaron trampas ocultas en cinco puntos de muestras terrestres: SPT1 (hábitats de guamil; consulte la Tabla 1.2); SPT4 (hábitat de encinal); SPT6 (hábitat de bosque alto, ramonal); ET1 (recientemente ramonal quemado) y FT1 (hábitat de bosque bajo, tintal).

Además, se realizó la extracción de desechos Winkler en SPT1 y SPT6. La extracción de desechos se realizó utilizando el diseño "mini-Winkler", siguiendo los métodos descritos por Bestelmeyer et al. (*en imprenta*). Se muestreó toda la basura de hojas y ramas en un área de 1m^2 en 10 estaciones por área. Se tomaron muestras de desechos en por lo menos 2 metros de trampas ocultas.

Se utilizó carnada de atún para muestrear las especies de hormigas excavadoras/omnívoras y para proporcionar información sobre el comportamiento merodeador de algunas especies en ET1, ET2 (solamente hábitat de guamiles), FT1, CT1 (bosque alto) y CT4 (bosque alto). Se colocó una carnada del tamaño de la mitad de una cucharada de atún enlatado (ca. 3 g) en 10 puntos (espacios de 10 metros) a lo largo de un área de 90 metros. Esta actividad ocurrió al final de la tarde o en la noche durante los períodos de actividades pico de las hormigas durante el estudio. El número de individuos por especies de hormigas se registro cada 45 minutos después de colocar la carnada o varias veces en un período de dos horas (10, 30, 60, 90 y 120 minutos después de colocarlas), dependiendo de los objetivos de una sesión.

Finalmente, se utilizó el muestreo directo para registrar la presencia de especies de hormigas en cada área focal. Se realizó un muestreo directo en áreas de 90 metros en todos los puntos muestreados y se realizó un muestreo oportunista en toda el área de estudio. El muestreo directo incluyó la búsqueda de nidos en el suelo y basura al escarbar las hojas en un área de 1m^2 y la búsqueda visual de hormigas; rompiendo ramitas y troncos; buscando bajo las piedras; y en las hojas, sobre y bajo los árboles y en áreas de hasta 2 metros. Estos datos no se utilizaron para comparación debido a que los hábitats y transectos eran diferentes en la composición de elementos de microhábitat y debido a los diferentes lapsos de tiempo que estaban disponibles para la búsqueda en diferentes sitios.

Las especies y géneros fueron ordenados por categoría de acuerdo a grupos funcionales en base a observaciones de campo y reportes en la literatura. Grupos funcionales de Andersen (1995) y Bestelmeyer y Wiens (1996). Además, las especies fueron clasificadas de acuerdo con el tipo de distribución geográfica que exhibieron en base a la información de la distribución en Kempf (1972), Brandão (1991), Longino (en línea) y Ward (en línea). Se reconocieron seis tipos de índices aquí: Bosque Maya (bosque en la Península de Yucatán de México, el norte de Guatemala y noreste de Belice), el norte de Centro América (sur de México–Honduras), Centro América (sur de México–Colombia), en toda Centro América (sur de Estados Unidos–Centro América), Neotrópico (sur de México–norte de Argentina) y toda el área Neotropical (sur de Estados Unidos–centro de Sur América).

Se utilizó un densiómetro para cuantificar la extensión del bosque en cada punto a lo largo de los transectos en los cuales se realizó el muestreo.

Se comparó tanto la riqueza como la composición de las especies entre los puntos de muestra que representan distintos tipos o variantes de macrohábitats. La riqueza de especies medida, así como la riqueza real calculada se utilizó para comparar los puntos de muestra, considerando que los hábitats pueden variar en cuanto a lo completo de las especies muestreadas. La riqueza se calculó utilizando las técnicas de extrapoblación de riqueza (Colwell y Coddington, 1994) y utilizando el software EstimateS (versión 5.0.1; Colwell, 1997). Solamente se consideraron los cálculos en base en las incidencias ya que la abundancia de hormigas a menudo se ve atestado debido a su hábito de anidar. En primer y segundo lugar los calculadores de insectívoros y el calculador de cobertura basado en la incidencia (ICE) se reportan acá.

Los patrones de composición de especies se evaluaron en dos formas. Primero, comparamos cada punto de muestreo en una forma equitativa utilizando índices de similitud. Se usaron los índices de cálculo de Sorensen, los cuales se basaron en el número esperado de especies compartidas entre muestras de un calculador de cobertura (EstimateS 5.0.1, Colwell, 1997). Adicionalmente, se proporcionan valores de similitud no modificados en el formato de índices de Morisita-Horn. En segundo lugar, utilizamos técnicas de agrupación por aglomeración jerárquica para establecer los patrones de similitud en la composición de especies entre grupos de puntos de muestreo considerados como uno solo. Se utilizó el procedimiento de agrupación central llamado método de

variación mínima de Ward (PC-ORD 4.0; McCune y Mefford, 1999) después de las discusiones de Legendre y Legendre (1998).

Finalmente, los patrones de abundancia de hormigas en las carnadas se compararon entre dos transectos, uno en los guamiles (ET2) y el otro en bosque alto (FT1) para comparar los patrones de forraje y dominancia de las hormigas entre estos diferentes tipos de hábitats. La abundancia de hormigas en carnadas individuales se transformó de acuerdo con cinco escalas de abundancia (0-5) de acuerdo con Andersen (1991).

RESULTADOS

Sobre todos los puntos de muestreo, se registraron 112 especies y morfoespecies de 39 géneros de hormigas (Apéndice 14). De estas, 71 especies se reunieron utilizando métodos sistemáticos (extracción de trampas ocultas y basura) y las restantes 41 especies se reunieron solamente con la mano o con carnadas de atún. La recolección manual registró 79 especies. *Pheidole* fué el género más vistoso (20 especies y OTU), seguidos por *Pseudomyrmex* (9), *Camponotus* y *Cephalotes* (7), *Pachycondyla* y *Solenopsis* (6), *Azteca* (5) y *Odontomachus* (4). Se encontraron siete especies en las cuatro áreas focales que se pueden considerar comunes; éstas incluyen: *Camponotus atriceps*, *C. novogranadensis*, *C. planatus*, *Cyphomyrmex minutus*, *Dolichoderus bispinosus*, *Solenopsis geminata* y *Wasmannia auropunctata*.

Se conocen treinta y tres especies o se sospecha que son exclusivamente terrestres y 29 exclusivamente arbóreas. El resto de las especies pueden ocurrir en microhábitats o en sus afinidades de hábitat son desconocidas. Se buscó exhaustivamente una bromeliacea epífita en la cual se registró, *Aechmea tillandsia. Camponotus planatus*, *Cephalotes minutus*, *C. scutulatus*, *Crematogaster brevispinosus* y *Pseudomyrmex oculatus*. Las hormigas se recolectaron en 20 metros de longitud de un árbol *Piscidia piscipula* que cayó durante una tormenta. *Camponotus atriceps*, *Ectatomma tuberculatum*, *Hypoponera* sp. B (nidos en cortezas), *Pachycondyla stigma*, *Pheidole* sp. H (nidos en epífitos, *Epidendrum* sp.), *Pheidole punctatissima* y *Tapinoma inrectum* se recolectaron de este árbol.

Las especies registradas en la muestra que se restringen al norte de América Central (así se conocen hasta ahora) incluyen *Azteca foreli*, *Belonopelta deletrix*, *Cephalotes basalis*, *C. biguttatus*, *Hypoponera nitidula*, *Pseudomyrmex ferrugineus*, *P. pepperi*, *Rogeria cornuta* y *Trachymyrmex saussurei. Odontomachus yucatecus*, *Sericomyrmex aztecus*, *Xenomyrmex skwarrae* pueden clasificarse como endémicas del Bosque Maya. *O. yucatecus* se recolectó en trampas ocultas en ramas quemadas y sitios de guamiles, *S. aztecus* de las trampas ocultas en el guamil y *X. skwarrae* se recolectó manualmente en el sitio arqueológico El Perú. Se recolectó la hormiga rara *Thaumatomyrmex ferox* en el hábitat de ramonales utilizando la clasificación de deshechos Winkler. Otras especies raras incluyen *Discothyrea cf. horni*, el cual se encontró en el sitio del guamil utilizando un dispositivo de Winkler y *Belonopelta cf. deletrix*, el cual se recolectó manualmente en bosque alto cerca de Laguna Flor de Luna. Una forma no descrita de *Pseudomyrmex* (PSW-13) se capturó en una trampa oculta en el sitio de encinales. Solamente se reunieron 930 individuos con trampas ocultas y Winklers, promediando 13 individuos por trampa o cuadrante. La actividad de las hormigas en la basura o en la superficie del suelo fue muy baja. Las observaciones de carnadas sugirieron que los nidos en el suelo en algunas áreas se encontraban en el suelo. Muchas otras especies se anidan o refugian en árboles o en epífitos (Apéndice 14) y por lo tanto no se registraron en la superficie del suelo o en muestras de basura.

Las mediciones de las coberturas del dosel en franjas de trampas ocultas indican que el hábitat del guamil y tintal solamente han disminuido levemente la cobertura del dosel, en hábitats de ramonales y de ramonales quemados (Figura 9.1). Los valores del dosel oscilaron de 92 a 97% en estos hábitats. El encinal tenía una cobertura menor de 67%.

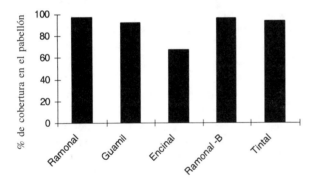

Figura 9.1. El porcentaje de cobertura de dosel medido en las estaciones de trampas ocultas para hormigas en cada punto de muestra.

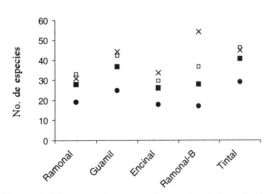

Figura 9.2. Patrones de riqueza de especies de hormigas entre los puntos de muestra. Círculo sólido = riqueza medida, cuadro sólido = calculador de primer orden insectívoro, cuadro abierto = calculador de segundo orden insectívoro, x = calculador de cobertura en base a incidencias.

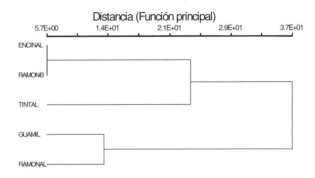

Figura 9.3. Análisis de los puntos de muestra, en base a la composición de especies de hormigas, utilizando el procedimiento de Ward.

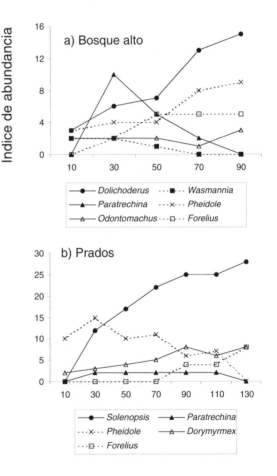

Minutos después de colocar la carnada

Figura 9.4. Patrones de dominación en hormigas con el tiempo transcurrido después de colocar la carnada en el suelo, es transectos con carnada de atún en hábitats bosque alto y prados. Abundancia de hormigas fue determinada como la suma de las abundancias transformadas en las 10 carnadas. (ver el texto).

La riqueza medida fue mayor en el hábitat de tintal seguido por el hábitat de guamil (Figura 9.2). Los calculadores de riqueza de insectívoros también sugieren que la riqueza fue mayor en estos hábitats. El calculador ICE indicó que el hábitat ramonal fue más rico, seguido por el hábitat del guamil y tintal.

La similitud esperada entre los hábitats fue generalmente más bajo, oscilando de 0.195 a 0.510 (Tabla 9.1). La comparación de hábitat encinal-tintal mostró el mayor valor de similitud, por lo tanto el ramonal y el ramonal quemado fueron la pareja de hábitat más diferentes. La pareja de ramonal-guamil y la pareja de encinal-ramonal quemado también tenían valores altos similares. El índice Morisita-Horn, que incluye una medición de abundancia, produjo un patrón diferente. Utilizando esta métrica, el hábitat del encinal y en ramonal quemado fueron los más parecidos, seguido por la pareja de ramonal y guamil y la de encinal y tintal, por lo tanto las otras parejas mostraron

similitud generalmente baja. El análisis de agrupación reveló dos grupos diferentes, uno que contenía el encinal, ramonal quemado y tintal y el otro guamil y ramonal (Figura 9.3). La similitud relativamente alta entre el encinal y el ramonal quemado observados en los índices de Morisita-Horn es aparente en el diagrama de grupos.

Los datos de carnadas del bosque alto y los guamiles revelaron diferencias grandes en el dominio de especies de hormigas (Figura 9.4). En el área del bosque alto, *Dolichoderus bispinosus* y *Pheidole* spp. incrementaron en abundancia en carnadas con tiempo, mientras que *Paratrechina* sp. B alcanzó un pico en 30 minutos después de colocarlas y declino posteriormente. En el área de los guamiles, *Solenopsis geminata* dominó extremadamene y la abundancia de *Pheidole* sp. Que disminuyó gradualmente con el tiempo. La abundancia de *Forelius* sp. y *Dorymyrmex* sp. incrementó con el tiempo mientras incrementaban las temperaturas de la superficie de la tierra durante la sesión de carnadas.

Tabla 9.1. Patrones de similitud de comunidades de hormigas entre los tipos de hábitats con base en los datos de trampas ocultas. Los valores en las celdas son índices de Sorensen, con base en el número esperado de especies compartidas entre muestras de un calculador con base en la cobertura. Los valores en paréntesis son índices de Morisita-Horn con base en los datos observados.

	Ramonal	Guamil	Encinal	Ramonal-B
Ramonal	--			
Guamil	0.460 (0.215)	--		
Encinal	0.212 (0.020)	0.285 (0.178)	--	
Ramonal-B	0.195 (0.130)	0.280 (0.116)	0.409 (0.635)	--
Tintal	0.237 (0.054)	0.320 (0.072)	0.510 (0.201)	0.286 (0.114)

Tabla 9.2. Una comparación de riqueza de especies y riqueza genérica entre las regiones neotrópicas con base en los estudios publicados.

Estos estudios en particular, fueron seleccionados para representar regiones debido a que eran bastante similares al estudio presentado aquí (es decir, una evaluación de heterogeneidad a nivel de paisajes). La escala representa la distancia máxima entre muestras. Cuando no se reportaron los valores, proporciono un cálculo de escala (seguido por un signo "?"), basados en la información en el texto, cuando era posible. Los valores se encuentran en orden descendiente de latitud.

Riqueza de especies	Riqueza genérica	Escala/# habitantes	Ubicación
103	46*	Dentro de10 km?/2	Veracruz, México[1]
87	32	Dentro 2 km?/3	Chiapas, México[2]
112	39	ca. 80 km/5	LTNP, Guatemala[3]
109	31	Dentro 35 km/4	Sarapaqui, Costa Rica[4]
127	49	?/elevación	BCI, Panamá[5]
156	49	?/3	Trombetas, Brasil[6]
184	45	?/3	Amazonas, Brasil[7]
104	30	ca. 10 km/4	Salta, Argentina[8]

*Los valores de la riqueza genérica se ajustaron si los sinónimos genéricos fueron reconocidos en listas de especies.
[1]Quiróz-Robledo y Valenzuela-Gonzáles 1995 [2]MacKay et al. 1991
[3] este estudio [4]Roth et al. 1994
[5]Levings 1983 [6]Vasconcelos 1999
[7]Majer y Delabie 1994 [8]Bestelmeyer y Wiens 1996

DISCUSION

La actividad de las hormigas durante el período de muestreo fue bajo, así que es normal que los valores de riqueza de especies sea menospreciado. Muchas de las especies de hormigas pueden haber estado completamente inactivas sobre la tierra debido a las condiciones ásperas e improductivas que caracterizan la época seca (consulte también Janzen y Schoener, 1968; Levings, 1983; Bestelmeyer y Wiens, 1996). Durante las temporadas secas tropicales, algunas especies de hormigas pueden descender del estrato superior a la columna de suelo en donde no pueden muestrearse utilizando sus trampas ocultas o extracción de basura de hojas (Levings y Windsor, 1982; Harada y Bandeira, 1994). De esa forma, el muestreo directo proporcionado por la carnada de atún parece haber sido la combinación más eficiente de los procedimientos de muestreo bajo las condiciones presentes durante el estudio, especialmente cuando se consideran el tiempo de preparación posterior al procedimiento.

A pesar de estas limitaciones, el número de hormigas registrado en el PNLT se compara favorablemente con los resultados de otros estudios (Tabla 9.2). Aunque el número de tipos de hábitats muestreados y las escalas sobre las cuales se tomaron las muestras difieren entre estos estudios, se pueden hacer algunas comparaciones con precaución. La riqueza de especies y el valor de género en este estudio son similares o mayores a los registrados en estudios realizados en bosques tropicales de México y Costa Rica, pero son menores a aquellos de estudios en bosques panameños y de la amazona brasileña. En general, este patrón sugiere una curva de riqueza en forma de campana en todos los bosques neotropicales, con un pico desde el área de Centro América hasta la amazona brasileña. Se necesitarán estudios adicionales que empleen técnicas de recolección estandarizadas y diseños de muestreo para determinar la validez de este patrón. Si se propicia una relación de una forma específica, podría servir como base desde la cual anticipara y evaluar la biodiversidad de las hormigas en ubicaciones tropicales.

Pocas especies parecen ser endémicas a los bosques mayas, aunque hacen falta datos que se puedan comparar con otros estudios. Comúnmente se encontraron varios géneros en otros estudios de hormigas en los bosques centroamericanos pero aquí no se encontraron, incluyendo *Adelomyrmex*, *Leptogenys* y *Oligomyrmex* y a una extensión menor *Megalomyrmex* y *Proceratium*. Por otro lado, se registraron varios géneros en este estudio que a menudo no se han registrado en otros estudios regionales, incluyendo *Belonopelta*, *Discothyrea*, *Thaumatomyrmex* y *Xenomyrmex*. El registro de *Thaumatomyrmex ferox* es importante. Se hizo referencia a esta hormiga como "hormiga milagrosa" por

Wilson y Hölldobler (1995) por su rareza y belleza. Se sabe que existen menos de 100 especímenes en museos y esto representa el primer registro que conocemos de Guatemala.

Contrario a nuestras expectativas, la riqueza de las hormigas no fue mayor en hábitats cerrados, pero fue mayor en hábitats que se encontraban un poco más abiertos y en el caso del guamil, más distribuidos (Figura 9.2). La riqueza del hábitat relativamente abierto del encinal fue similar a la del hábitat del ramonal de doseles cerrados. Además, la composición de la comunidad de las hormigas de los hábitats abiertos o con cambios no difieren sistemáticamente de aquellos de más doseles cerrados o hábitats "primitivos" (es decir, ramonal). Por el hecho de estudios previos de los efectos de cambios en el bosque, incluyendo los realizados por MacKay (1991) y Vasconcelos (1999), este estudio puede ser considerado hábitat que recientemente sufrió más cambios severos que aquellos ya considerados con anterioridad. MacKay (1991), por ejemplo, observó una reducción del 50% en la riqueza de especies en bosques de la región de Chiapas que sufren roza y quema y la pérdida se atribuye a la pérdida de especies arbóreas. En contraste, los sitios con cambios en este estudio tienen cobertura abundante de doseles (Figura 9.1). La recuperación de bosques maduros y comunidades puede tomar de 13 a 24 años después de los cambios (Roth et al., 1994; Vasconcelos, 1999) y el hábitat del guamil estudiado aquí se calculó que empezó a sufrir cambios hace más de veinte años atrás (Francisco Bosoc, pers. comm.). Además, nuestro estudio también consideró las comunidades de hormigas que se estaban realizando en tipos de bosques abiertos naturalmente en la región. Las poblaciones de muchas especies de hormigas ahora presentes en los bosques de El Petén pueden ser insensibles a las diferencias moderadas en la estructura de la vegetación debido a que estas hormigas han estado expuestas a incrementos históricos en las tierras áridas de la región, así como a los cambios antropogénicos de la misma (Méndez, 1999). Por razones similares, la riqueza más grande de hormigas observado en los hábitats de guamil y tintal puede deberse al hecho de que las hormigas que habitan estos sitios pueden acoplarse a las condiciones de la época seca. La riqueza observada del hábitat ramonal podría ser mayor al de los sitios con cambios si se muestreara en una época del año o si se comparara con sitios con cambios recientes. Desafortunadamente, el poder estadístico de este estudio es demasiado bajo para identificar especies determinadas que pueden disminuir cuando el bosque alto se convierte en guamil. Es posible que las especies crípticas y algunos depredadores especialistas en hábitats más pequeños y especies arbóreas (Apéndice 14) sería el más afectado (consulte Bestelmeyer y Wiens, 1996).

La similitud relativamente alta entre los hábitats ramonal y guamil puede estar dominada por su proximidad en lugar de su estructura de hábitat o historial de cambios.

Dicho efecto fue notado por Feener y Schupp (1998) y Bestelmeyer (2000). La similitud de los hábitats de ramonal quemado y de encino no se pueden atribuir a la proximidad y no está claro todavía qué es lo que puede causar este patrón.

El dominio relativo de *Solenopsis geminata* en el hábitat de guamiles (Figura 9.4) se ha observado en otros estudios realizados en Centro América (Perfecto, 1991; Perfecto y Vandermeer, 1996). Esta hormiga es favorecida por las condiciones climáticas cálidas bien aisladas en hábitats deforestados. La presencia de *Forelius* y *Dorymyrmex*, que es un género dominante en la zona árida, en los guamiles y otros hábitats cambiados fue sorprendente. No pudimos realizar registros de estos géneros en hábitats similares en estudios realizados más al sur de Centro América. La presencia de estos géneros en los bosques de Petén pueden relacionarse con la aridez relativa de muchos microhábitats naturales en la región y a su proximidad a áreas áridas del norte. *Dorymyrmex*, *Forelius* y *Solenopsis* son característicos de los desiertos neárticos y su dominio combinado en los guamiles disminuye el inmenso cambio en las condiciones ambientales en comparación con el bosque (también consulte Quiróz-Robledo y Valenzuela-Gonzáles, 1995).

CONCLUSIONES Y RECOMENDACIONES

Es muy útil ver los patrones de la diversidad de hormigas con relación a las clases de microhábitats disponibles para que los ocupen diferentes especies de hormigas. Generalmente, la riqueza de las hormigas parece estar a su máximo cuando varios tipos de microhábitats—basura de hojas y ramas, espacios de tierra suelta, ramas y troncos caídos, epífitos, arbustos y árboles de varias especies, por ejemplo—son abundantes en el área. La reducción o eliminación de estos hábitats de hormigas en un área resulta en disminuciones en la riqueza de hormigas y dominio incrementado por especies altamente competitivas de hormigas. Además, los cambios naturales frecuentes o antropogénicos eliminarán las especies de hormigas de las áreas aún cuando se encuentren disponibles los hábitats adecuados.

La evidencia presentada aquí y en estudios citados anteriormente sugieren que la recuperación de microhábitats adecuados para hormigas de bosques tropicales después de los cambios puede ser rápida. La recuperación tanto del bosque como de sus microhábitats es contingente en el grado del uso agrícola y erosión del suelo. La erosión del suelo puede ser importante en los bosques de Petén debido a los suelos delgados sobre las capas de las montañas (Beach, 1998). Además, la recolonización de hormigas en los microhábitats necesariamente depende de la presencia de la proximidad de las poblaciones de origen. La relación entre la distribución de la población de origen y los índices de

colonización no se han considerado en detalle, pero es posible que existan cantidades límite de deforestación más allá de las colonizaciones que se limiten por las especies de hormigas (Roth et al., 1994), pueden existir las extinciones en toda la región (Hanski y Gilpin, 1996). Estudios de evaluación rápida y los estudios de corta duración de los efectos de la alteración del hábitat en las comunidades de hormigas no pueden esperar detectar dichos límites o efectos regionales. Lo mejor que podemos hacer en este caso es estar conscientes de la potencia de estos efectos. A pesar de las habilidades de la relativamente buena colonización de hormigas y algunas otras especies de insectos, la deforestación amenaza la extinción de sus poblaciones. En este momento, sin embargo, la diversidad de las hormigas dentro del parque es alta, y debería realizarse cualquier esfuerzo para mantenerla.

El manejo de los bosques dentro del PNLT para la conservación de hormigas debería enfatizar la preservación de una variedad de hábitats. Una forma de lograrlo es conservar la extensión de los declives topográficos, que generalmente son reconocidos como determinantes la diversidad de las hormigas en territorios tropicales (por ejemplo, Olson, 1994) y de la biodiversidad en el PNLT en particular (Méndez et al., 1998). Los epífitos y otros microhábitats de doseles disponibles en los árboles altos puede contener muchas especies de hormigas únicas (Longino y Nadkarni, 1990), así los bosques altos deberían ser objeto de estudios adicionales y conservación.

La composición de las comunidades de hormigas es muy irregular y los índices de cambios son muy altos (Tabla 9.1), así que se necesitarán estudios más intensos para aclarar los efectos del patrón de espacio y los factores ambientales en la diversidad de las hormigas. Las hormigas responden en formas importantes e informativas a la variación del terreno y proporcionan un punto importante y útil en más estudios extensivos y monitoreos. Las hormigas responden a los procesos como la variación topográfica, patrones de inundaciones y de esa forma indicar la irregularidad ambiental y la dinámica que son importantes para los que manejan las tierras. El comprender las relaciones entre los cambios animales, drenajes e incendios, será importante para identificar las áreas de conservación prioritarias dentro de los hábitats terrestres del parque. Las recomendaciones específicas son las siguientes:

- Los bosques altos altamente complejos en áreas como El Perú y Laguna Guayacán que parecen haber sufrido relativamente pequeños cambios previos debería protegerse y examinarse en mayor detalle—especialmente el dosel y las comunidades.

- La conservación de una variedad de tipos de hábitats—en donde los tipos de hábitats se definen de acuerdo a la vegetación, topografía y características del suelo—es posible tener una buena estrategia para conservar las hormigas y otros artrópodos terrestres. En el PNLT, las tecnologías de censos remotos puede ayudar en este proceso.

- La fauna de hormigas de bosques altos y otros sitios debería investigarse en alguna época del año, como al principio de la época húmeda. Las trampas Malaise deberían utilizarse para capturar formas sexuales voladoras durante este período, además de las técnicas tradicionales. Esto posiblemente nos llevaría a un registro más completo de la fauna de las hormigas del parque (e insectos).

- Los esfuerzos adicionales de estudio y monitoreo deberían examinar el papel de la topografía, variación del suelo y el historial de incendios en las hormigas y otras comunidades de animales en el parque.

- Se necesitan los estudios repetidos en tipos de hábitats naturales como encinales y selva baja para proporcionar evaluaciones más detalladas del complemento de las faunas de hormigas entre los tipos de hábitat en el parque.

LITERATURA CITADA

Alonso, L. E. 1998. Spatial and temporal variation in the ant occupants of a facultative ant-plant. Biotropica 30: 210-213.

Andersen, A. N. 1991. Sampling communities of ground-foraging ants: pitfall catches compared with quadrat counts in an Australian tropical savanna. Australian Journal of Ecology 16: 273-279.

Andersen, A. N. 1995. A classification of Australian ant communities, based on functional groups which parallel plant life-forms in relation to stress and disturbance. Journal of Biogeography 22: 15-29.

Beach, T. 1998. Soil catenas, tropical deforestation, and ancient and contemporary soil erosion in the Petén, Guatemala. Physical Geography 19: 378-405.

Belshaw, R. y B. Bolton. 1993. The effect of forest disturbance on the leaf litter ant fauna of Ghana. Biodiversity and Conservation 2: 656-666.

Bestelmeyer, B. T. 2000. *A multiscale perspective on ant diversity in semiarid landscapes*. Dissertation. Colorado State University, Fort Collins, CO, USA.

Bestelmeyer, B. T. y J. A. Wiens. 1996. The effects of land use on the structure of ground-foraging ant communities in the Argentine Chaco. Ecological Applications 6:1225-1240.

Bestelmeyer, B.T., D. Agosti, L. Alonso, C. R. Brandão, W. L. Brown Jr., J. H. C. Delabie y R.

Silvestre. *In press*. Field techniques: an overview, description, and evaluation of sampling techniques. In *Measuring and Monitoring Biodiversity: Standard Techniques for Ground-Dwelling Ants* (D. Agosti, J. Majer, T. Schultz, and L. Alonso, eds.). Smithsonian Institution Press, Washington, DC, USA.

Brandão, C. R. F. 1991. Adendos ao catálogo abreviado das formigas da região neotropical Hymenoptera: Formicidae. Revista Brasileira de Entomología 35:319-412.

Colwell, R. K. 1997. *EstimateS 5. Statistical Estimation of Species Richness and Shared Species from Samples*. Web site: http://viceroy.eeb.uconn.edu/estimates.

Colwell, R. K. y J. A. Coddington. 1994. Estimating terrestrial biodiversity through extrapolation. Philosophical Transactions of the Royal Society of London 345:101-118.

Erwin, T. L. y J. C. Scott. 1980. Seasonal and size patterns, trophic structure, and richness of Coleoptera in the tropical arboreal ecosystem: the fauna of the tree *Luehua seemanni* Triana and Planch in the Canal Zone of Panama. Coleopterists Bulletin 34: 305-322.

Feener, D. H. y E. W. Schupp. 1998. Effects of treefall gaps on the patchiness and species richness of neotropical ant assemblages. Oecologia 116:191-201.

Greenslade, P. J. M. y P. Greenslade. 1977. Some effects of vegetation cover and disturbance on a tropical ant fauna. Insectes Sociaux 24: 163-182.

Haila, Y. y C. R. Margules. 1996. Survey research in conservation biology. Ecography 19:323-331.

Hanski, I. y M. Gilpin, eds. 1996. *Metapopulation Dynamics: Ecology, Genetics, and Evolution*. Academic Press, New York, NY, USA.

Harada, A. Y. y A. G. Bandeira. 1994. Estratificacão e densidade de invertebrados em solo arenoso sob floresta primária e plantios arbóreos na Amazonia central durante estacão seca. Acta Amazonica 24: 103-118.

Janzen, D. H. y T. W. Schoener. 1968. Differences in insect abundance and diversity between wetter and drier sites during a tropical dry season. Ecology 49:96-110.

Johnson, R. A. 1992. Soil texture as an influence on the distribution of the desert seed-harvester ants *Pogonomyrmex rugosus* and *Messor pergandei*. Oecologia 89:118-124.

Kempf, W. W. 1972. Catálogo abreviado das formigas da região neotropical Hymenoptera: Formicidae. Studia Entomológica 15:3-344.

Legendre, P. y L. Legendre. 1998. *Numerical Ecology*, 2nd English ed. Elsevier, Amsterdam, the Netherlands.

Levings, S. C. 1983. Seasonal, annual, and among-site variation in the ground ant community of a deciduous tropical forest: some causes of patchy species distributions. Ecological Monographs 53: 435-455.

Levings, S. C. y D. M. Windsor. 1982. Seasonal and annual variation in litter arthropod populations. Pp. 355-387 in *The Ecology of a Tropical Forest: Seasonal Rhythms and Long-term Changes* (E. G. Leigh Jr., A. S. Rand, and D. M. Windsor, eds). Smithsonian Insitution Press, Washington, DC, USA.

Longino, J. T. Online. www.evergreen.edu/user/serv_res/ research/arthropod/ AntsofCostaRica.html).

Longino, J. T. y N. Nadkarni. 1990. A comparison of ground and canopy leaf litter ants (Hymenoptera: Formicidae) in a neotropical montane forest. Psyche 97:81-93.

May, R. M. 1988. How many species are there on Earth? Science 241:1141-1148.

MacKay, W. P. 1991. Impact of slashing and burning of a tropical rainforest on the native ant fauna (Hymenoptera:Formicidae). Sociobiology 3: 257-268.

McCune, B. y M. J. Mefford. 1999. *PC-ORD. Multivariate Analysis of Ecological Data, version 4.0.* MjM Software Design, Gleneden Beach, OR, USA.

Méndez, C. 1999. How old is the Petén tropical forest? Pp. 31-34 in *Thirteen Ways of Looking at a Tropical Forest: Guatemala's Maya Biosphere Reserve* (J. D. Nations, ed.). Conservation

ITINERARIO DE EXPEDICION
DICCIONARIO GEOGRAFICO

Fechas	Actividades
8 de abril	Arribo a la *Estación Biológica Las Guacamayas* * (17°14.75' N, 90°32.80' W)
9 al 14 de abril	Estudios en Area focal 1, *Sitios del Río San Pedro/Sacluc*
15 de abril	Traslado a El Naranjo, muestreo del *sitio Rapids* (17° 16.99' N, 90°42.80' W)
16 al 19 de abril	Muestreo en el Area principal 2, *Río Escondido* (17^0 18.78' N; 90^0 52.15' W)
20 al 22 de abril	Muestreo del Area principal 3, *Flor de Luna* (17°35.97' N, 90°53.84' W)
23 al 25 de abril	Muestreo del Area principal 4, *Río Chocop* (17°36.13' N, 90°24.53' W)
26 de abril	Regreso a la Estación Biológica Las Guacamayas
27 al 29 de abril	Informe preliminar y preparación de muestras
29 de abril	Presentación de los resultados preliminares al CONAP
30 de abril	Regreso a Flores para el retorno de los participantes

* Refiérase al Mapa 1 para la ubicación de los sitios.

Sitios principales de muestreo RAP en el Parque Nacional Laguna del Tigre, Petén, Guatemala

Los siguientes cuatro sitios focales o áreas principales, fueron estudiadas durante la expedición RAP en el PNLT. Se realizó trabajo de campo de un campamento en cada sitio. No todos los grupos taxonómicos fueron investigados en cada sitio. Consulte los Apéndices 1 y 2 para obtener las coordinadas y tipos de macrohábitats del muestreo específico.

Area focal 1: Estación biológica Las Guacamayas/Río San Pedro y Río Sacluc
(17°14.75' N, 90°32.80' W)
La Estación Biológica Las Guacamayas se encuentra localizada en la confluencia de los ríos San Pedro y Sacluc.

Viajamos en pickups hasta el final de los caminos del Río Sacluc y luego nos transportamos por lancha siete kilómetros hasta la estación. La estación se encuentra provista con dormitorios, una cocina y área para alimentación y baños con regaderas.

Los hábitats acuáticos en el área incluyen ríos profundos rodeados por ciénagas o bosque ribereño y alimenta el río San Pedro. Los hábitats terrestres en el área incluyen bosques no inundados grandes sobre las colinas o *bosques altos* (una muy cercana a la estación), o *guamiles*, bosques de tipo varzea cerca de la villa de Paso Caballos hacia el este y un bosque de encinos secos (*encinal*) a lo largo del Río Sacluc. El área inmediatamente al rededor de la estación está comparativamente menos cambiada debido a la presencia de la estación, lo que ha servido para desanimar la población ilegal. Existen varias poblaciones dentro de las fronteras del parque a lo largo del San Pedro hacia el oeste. No observamos población alguna en el Sacluc entre el lugar en donde estacionamos la lancha y la estación.

Los Rápidos
Aunque no es parte de nuestro plan de muestreo original, pasamos una serie de rápidos en nuestro camino hacia El Naranjo. En parte de los rápidos encontramos que el fondo consistía de un arrecife hecho de conchas (moluscos) mezcladas con individuos vivos. Esta área fue algo único y constituye un descubrimiento importante.

Area focal 2: Río Escondido
(17⁰ 18.78' N; 90⁰ 52.15' W)
Acampamos a lo largo del lado oeste del Río Escondido justo al sur del Biotopo de la Laguna del Tigre. El nombre del río, "río escondido", es adecuado; después del norte del río desde su confluencia con el San Padro, parece ser un callejón sin salida en una laguna rodeada por una gran ciénaga. De hecho, el río continua hacia el norte pero como un canal poco profundo, estrecho que pasa por los humedales. Eventualmente, el río se hace más profundo y se hace más ancho por lo que es más fácil navegarlo. Aprovechamos un poblado abandonado para refugiarnos y acampar. Los hábitats acuáticos son diversos, incluyendo los estrechos profundos del río, los canales con pantanos, lagunas e isletas. Los hábitats terrestres que se encuentran aquí, incluyen el bosque alto quemado previamente, guamiles y prados que resultan de la práctica de quema y rosa y un bosque ribereño alto al limite norte de nuestro muestreo en donde el río se hace más estrecho. Ocho familias viven a lo largo del Río Escondido hacia el norte de nuestro campamento y algunas poblaciones probablemente existen dentro de las fronteras del biotopo.

Area focal 3: Laguna Flor de Luna
(17°35.97' N, 90°53.84' W)
Aquí acampamos al final del camino al lado de la Laguna Flor de Luna. Este camino pasa un punto de revisión militar y los campos petroleros de Xan. Los hábitats acuáticos incluyen solamente las lagunas. Las muestras de tejido de pescado y el sedimento examinado en análisis tóxicos de los campos de Xan. Los hábitats terrestres fueron muestreados al rededor de Flor de Luna e incluyeron el bosque alto y bosque inundado (bosque bajo) el que se encuentra dominado por el árbol de tinto (*Haematoxylum campechianum*) en áreas cercanas a la laguna (tintales). No observamos poblaciones cerca de Flor de Luna pero existen varios cerca de las lagunas al sureste.

Area principal 4: Río Chocop
(17°36.13' N, 90°24.53' W)
Este campo se localizó en el puesto de monitoreo del CONAP en donde el río Chocop pasa por el camino. Otro punto de registro militar se encuentra en la ruta hacia la estación; esto es para evitar la colonización ilegal del área. Los hábitats acuáticos encontrados en el área incluyeron dos ríos poco profundos y pequeños—el Río Chocop y el Río Candelaria que se encuentra cerca—así como varias lagunas. Los hábitats terrestres muestreados incluyen el bosque bajo/ tintales y bosques altos. Las áreas a lo largo del camino cerca de Santa Amelia se encuentran ocupadas por varias familias, se han distribuido y se están regenerando o actualmente utilizan las cosechas o pastos.

GLOSARIO

Aguada: Hundimientos pequeños de piedra caliza que forman estanques que algunas veces contiene agua durante las épocas secas.

Akalché: Hundimientos más grandes de piedras calizas que forman lagunas y a menudo contienen agua durante todo el año.

Arroyo: Río pequeño de caudal bajo.

Bosque alto: Bosques de terrenos elevados con árboles altos, dosel denso y abundante basura de hojas.

Bosque bajo: Un bosque pequeño (15-20 m) saturado con un dosel relativo al bosque alto y crecimiento abundante y denso de palmas y hiervas.

Caño: Un cuerpo de agua turbia por medio del cual las ciénagas se conectan a un río.

Comunidad: Un grupo de especies, usualmente delimitados por afiliaciones taxonómicas que ocupan una localidad.

Complemento: Una medida del grado en el cual difiere la composición de especies (es decir, la identidad de especies) entre dos áreas. Un complemento alto indica que pocas especies son compartidas y que la riqueza de especies de los dos sitios juntos es mucho más alto que cualquiera de los dos solo.

Encinales: Bosques dominados por el encino.

Endémico: Nativo de y restringido a una región geográfica en particular.

Fragilidad: Una medida cualitativa del grado al cual las comunidades o ecosistemas se pueden resistir o recuperar después de un cambio.

Guamil: Término vernacular local para un área de vegetación joven regenerada (normalmente de 1 a 20 años) en donde la vegetación del bosque ha cambiado.

Cárstico: Pertenece a un estrato de piedra caliza irregular que está protegida por el subsuelo y los hundimientos.

Laguna: Español para "lagoon."

Mioceno: Una época geológica durante el período Terciario, de ca. 26 hasta 5 millones de años antes del presente.

Morfoespecies: Grupos de especies informales, utilizados cuando los organismos no pueden ser asignados a una especie descrita formalmente.

Río: Español para "river."

Varzea: Bosques inundados bajos.

APPENDICES/
APENDICES

Aquatic sampling points of the RAP expedition to Laguna del Tigre National Park, Petén, Guatemala

Puntos acuáticos de muestreo de la expedición RAP en el Parque Nacional Laguna del Tigre, Petén, Guatemala

SITE CODES/CODIGO DE LUGARES

SP:	Río San Pedro
E:	Río Escondido
F:	Laguna Flor de Luna
C/CA:	Río Chocop

** Macrohabitat Type

R: Rivers; bodies of running water
    RR: Rapids, freshwater reef
    RB: River embayments permanently connected to the river
    RP: Permanent stretches of river, deep portions that are never dry
    RS: Shallow stretches of river that may be dry periodically
        RS-m: RS-type river with a muddy bottom substrate
        RS-r: RS-type river with a rocky botton substrate
T: Tributary rivers or streams that feed into the river channels
C: Locally known as caños; narrow lotic environments with turbid, almost stagnant, waters
O: Oxbow lagoons formed by river bends that exist as pockets nearly detached from a river channel
L: Lagoons, not associated with rivers

** Tipo de Microhábitat

R: Ríos, cuerpos de agua corriente
    RR: Cascadas, saltos de agua fresca
    RB: desembocaduras a ríos permanentemente conectadas
    RP: tramos permanentes de río, porciones profundas que nunca se secan
    RS: tramos de ríos llanos que pueden secarse pemanentemente
        RS-m: tipo de río con subtrata de fondo fangoso
        RS-r: tipo de río con subtrata rocosa
T: Ríos tributarios o riachuelos que alimentan hacia los canales de los ríos
C: Localmente conocidos como caños; ambientes lóticos con aguas turbias estancadas
O: Lagunas Oxbow formadas por curvaturas que salen como projecciones, casi independiente, de los canales del río
L: Lagunas no asociadas con ríos

See Gazetteer for site descriptions.

Coordinates are presented as degrees and minutes. An asterisk indicates an aquatic point located at the focal point that defines the center of each focal area. See text in Chapter 1 for a detailed explanation. Note that some of the place names provided here are informal names used in this expedition only, whereas others are formal place names or local names used by local inhabitants.

Ver descripciones de los lugares en el Diccionario Geográfico

Las coordenadas están en grados y minutos. Un asterisco indica un punto acuático localizado en el punto focal que define el centro de cada área focal. Referirse al texto en el Capítulo 1 para una explicación más detallada. Notar que algunos de los nombres provistos son nombres informales usados solamente en esta expedición, mientras que otros son nombres locales y formales usados por los habitantes locales.

Sample Points Puntos de Muestreo	Latitude Latitud	Longitude Longitud	Macrohabitat Type** Tipo de Macrohabitat
Arroyo Yalá	17°15.34' N	90°14.04' W	RP
Paso Caballos	17°15.53' N	90°14.32' W	T
Arroyo La Pista	17°15.56' N	90°14.51' W	T
El Sibalito	17°15.62' N	90°15.44' W	RP
Murciélago	17°14.76' N	90°17.71' W	RP
Caracol	17°14.59' N	90°18.44' W	T
San Juan	17°13.72' N	90°20.04' W	T
La Caleta	17°13.83' N	90°19.83' W	C
Pato 1*	17°14.22' N	90°19.26' W	RP
Pato 2	17°14.30' N	90°19.16' W	RB
Río Sacluc	17°13.88' N	90°17.87' W	RP
Rápidos	17°16.99' N	90°42.80' W	RR
Est. Guacamayas	17°14.75' N	90°32.80' W	RP
Bahía Brandon*	17°19.91' N	90°51.17' W	RB
Bahía Jorge	17°20.23' N	90°50.89' W	RB
Tasistal	17°20.76' N	90°50.55' W	RP
Laguna Cocodrilo	17°24.01' N	90°48.65' W	O
Laguna Jabirú	17°24.25' N	90°48.42' W	O
Caleta Escondida	17°25.48' N	90°47.33' W	O
Punto Icaco	17°26.35' N	90°47.08' W	RP
Canal Manfredo	17°17.74' N	90°52.85' W	O
Nuevo Amanecer	17°26.12' N	90°47.38' W	RS
Pueblo R. Escondido	17°25.94' N	90°38.65' W	RP
Laguna Buena Vista	17°30.78' N	90°45.09' W	L
Laguna Santa Amelia	17°26.74' N	90°39.98' W	L
Garza1	17° 33.12' N	90° 47.69' W	L
Roman1	17° 33.87' N	90° 51.19' W	L
Laguna Flor de Luna*	17°35.97' N	90°53.84' W	L
Pozo Xan 3	17°32.45' N	90°47.57' W	L
Vista Hermosa	17° 30.81' N	90° 45.11' W	L
Estanque sin nombre	17° 32.87' N	90° 47.35' W	L
Río Chocop*	17°36.13' N	90°24.53' W	RS
Laguna Guayacán	17°36.64' N	90°25.65' W	L
Laguna La Pista	17°39.17' N	90°32.87' W	L
Tintal	17° 33.92' N	90° 20.30' W	L
Establo	17° 36.71' N	90° 26.13' W	L
Laguna La Lámpara	NA		L
Laguna Poza Azul	NA		L
Candelaria I	17°40.33' N	90°31.86' W	RS-m
Candelaria II	17°40.31' N	90°31.99' W	RS-r

NA Información no disponible

Terrestrial sampling points of the RAP expedition to Laguna del Tigre National Park, Petén, Guatemala

Puntos terrestres de muestreo de la expedición RAP en el Parque Nacional Laguna del Tigre, Petén, Guatemala

Brandon T. Bestelmeyer

SITE CODES/ CODIGO DE LUGARES

SP: Río San Pedro
E: Río Escondido
F: Laguna Flor de Luna
C/CA: Río Chocop

See Gazetteer for site descriptions.
Ver Diccionario Geográfico para descripción del lugar.

Sample Point Punto de Muestreo	Latitude Latitud	Longitude Longitud	Macrohabitat Type Tipo de Microhabitat
SPT1: Pato 1*	17°14.22' N	90°19.26' W	Riparian forest with guamil away from river/ Vegetación Riparia con guamil lejos del río
SPT2: Pato 2	17°14.30' N	90°19.16' W	Riparian forest/ Vegetación Riparia
SPT3: Galería	17°14.48' N	90°18.65' W	Riparian forest/ Vegetación Riparia
SPT4: Encinal 1	17°13.86' N	90°17.87' W	Encinal
SPT5: Encinal 2	17°14.12' N	90°17.77' W	Encinal
SPT6: Ramonal	17°15.34' N	90°17.57' W	Upland forest/Bosque alto
ET1: Mono Araña*	17°18.84' N	90° 52.14 W	Upland forest/Bosque alto
ET2: Saraguate	17°18.83' N	90°52.10' W	Upland forest with riparian forest and grassland patches/ Bosque Alto con vegetación riparia y parches de pastizales
ET3: Punto Icaco	17°26.33' N	90°47.08' W	Upland forest and riparian forest/ Bosque alto y vegetación riparia
FT1: Flor de Luna 1*	17°35.97' N	90°53.84' W	Lowland forest/Bosque bajo (Tintal)
FT2: Flor de Luna 2	17°35.99' N	90°53.95' W	Lowland forest/Bosque bajo (Tintal)
FT3: Flor de Luna 3	17°35.99' N	90°53.95' W	Lowland forest/Bosque bajo (Tintal)
CT1: Río Chocop*	17°36.14' N	90°24.53' W	Upland forest/Bosque alto
CT2: Río Chocop 2	17°36.14' N	90°24.53' W	Upland forest and riparian forest/ Bosque alto y vegetación riparia
CT3: Guayacán 1	17°36.74' N	90°25.68' W	Disturbed Upland forest/ Bosque alto perturbado
CT4: Guayacán 2	17°36.74' N	90°25.68' W	Upland forest/Bosque alto

* Terrestrial points that coincide with the focal point of each sampling area.
* Puntos terrestres que coinciden con el punto focal de cada área de muestreo.

APPENDIX 3   Phytoplankton species identified at each aquatic sampling point in Laguna del Tigre National Park, Petén, Guatemala

APENDICE 3   Especies de fitoplancton identificadas en cada punto acuático en el Parque Nacional Laguna del Tigre, Petén, Guatemala

Karin Herrera

See chapter 2 for descriptions of sampling points. Species are listed by class.
Ver Capítulo 2 para descripción de los puntos de muestreo. Especies estan listadas por clase.

Taxa	SP1	SP2	SP3	SP4	SP5	SP6	SP7	SP9	SP10	SP11	SP12	E1	E2	E3	E5	E6	E7	E8	E9	E10	E11	F1	F2	F5	C1	C2	C3	CA1	CA2
CHROOCOCCACEAE																													
Microcystis aeruginosa	X	X	X	X	X	X	X	X	X	X	X	X	X	X	X	X	X	X				X	X	X	X	X		X	
Chroococcus limneticus	X	X	X	X	X	X	X	X	X	X	X	X	X	X	X	X		X	X	X	X	X	X	X	X	X	X	X	X
C. minutus	X	X	X	X	X	X	X			X	X	X	X	X	X	X	X	X				X	X	X	X	X	X	X	X
Merismopedia sp.	X	X	X	X	X	X	X			X	X		X	X	X	X	X					X	X		X	X			
OSCILLATORIACEAE																													
Lyngbya limnetica	X	X	X	X	X	X	X	X	X	X	X	X	X	X	X	X	X	X	X			X	X	X	X	X	X	X	
Oscillatoria sp. 1	X	X	X	X	X	X	X			X	X	X	X	X	X	X	X		X	X		X	X	X	X	X	X	X	X
Oscillatoria sp. 2	X	X	X	X	X	X	X	X	X	X	X	X		X	X							X	X		X				
Spirulina sp.	X	X	X		X		X			X	X	X		X			X					X	X	X	X	X		X	X
NOSTOCACEAE																													
Anabaena flos-aquae	X	X	X	X	X	X	X	X	X	X	X	X	X		X	X		X		X			X		X			X	X
SCYTONEMATACEAE																													
Tolypothrix sp.			X	X			X	X	X	X	X	X	X	X			X		X			X	X	X		X	X	X	
XANTHOPHYCEAE																													
Botryococcus braunii	X	X	X	X	X	X	X				X	X	X	X	X	X	X	X						X	X	X	X	X	X
EUGLENACEAE																													
Euglena sp.	X	X		X	X	X	X			X	X	X	X	X	X	X			X			X	X	X	X				
Phacus sp.			X																	X	X	X	X						
VOLVOCACEAE																													
Volvox sp.	X	X	X			X								X									X						
MICRACTINIACEAE																													
Golenkinia radiata		X	X	X			X		X	X	X	X	X									X	X		X		X		
Micractinium sp.																						X		X			X		
COELASTRACEAE																													
Coelastrum reticulatum							X					X				X						X			X	X	X	X	
HYDRODICTYACEAE																													
Sorastrum sp.																									X	X			
Pediastrum duplex	X	X		X		X			X					X								X	X	X	X		X		X
P. simplex	X	X		X		X	X		X	X	X	X										X	X	X	X	X	X		X

Taxa	SP1	SP2	SP3	SP4	SP5	SP6	SP7	SP9	SP10	SP11	SP12	E1	E2	E3	E5	E6	E7	E8	E9	E10	E11	F1	F2	F5	C1	C2	C3	CA1	CA2
OÖCYSTACEAE																													
Chlorella vulgaris		X							X	X	X	X	X	X					X			X	X	X			X		
Ankistrodesmus falcatus	X	X	X	X		X				X	X	X				X							X					X	X
Closteriopsis sp.			X		X			X	X	X	X								X			X	X						
Tetraëdron limneticum		X	X																										
T. regulare											X												X	X					
Kirchneriella sp.		X	X	X																									
SCENEDESMACEAE																													
Scenedesmus quadriacuada			X	X	X	X	X				X											X	X	X	X	X	X		
Crucigenia sp.			X	X							X											X	X				X		
MESOTAENIACEAE																													
Cylindrocystis sp.									X	X	X			X	X			X				X		X					X
DESMIDIACEAE																													
Closterium sp.	X			X														X				X	X				X		
Micrasterias sp.																						X	X						
Staurastrum sp. 1																						X	X	X			X		
Staurastrum sp. 2																						X	X						
Cosmarium monomazum									X	X	X					X													
Staurodesmus spencerianus	X	X	X	X	X	X																X	X	X	X	X	X	X	
ZYGNEMATACEAE																													
Spirogyra sp.																						X	X				X		
Zygnema sp.		X		X	X		X		X	X	X					X	X			X	X	X	X	X	X				X
CLADOPHORACEAE																													
Cladophora sp.										X		X	X	X	X				X			X	X				X	X	X
CENTRITRACTACEAE																													
Centritractus sp.	X	X	X	X	X	X	X	X	X	X	X			X	X	X	X					X	X				X		
MALLOMONADACEAE																													
Mallomonas sp.												X	X		X		X	X				X	X					X	X
SYNURACEAE																													
Synura uvella	X	X	X			X																							
OCHROMONADACEAE																													
Dinobryon sertularia			X	X	X																								
PERIDINIACEAE																													
Peridinium sp.	X	X	X	X	X	X	X	X	X			X	X	X	X			X							X		X	X	X
CERATIACEAE																													
Ceratium sp.						X							X	X	X		X						X				X		
MELOSIRACEAE																													
Melosira varians	X	X	X	X	X	X	X	X	X	X	X	X	X	X	X	X	X	X				X	X	X	X	X		X	X
M. granulosa	X	X	X			X	X	X	X											X	X	X	X						
THALASSIOSIRACEAE																													
Cyclotella sp.		X			X	X									X	X							X						X
Stephanodiscus sp.	X	X	X	X	X	X	X	X	X	X	X	X			X								X		X	X	X		
DIATOMACEAE																													
Tabellaria sp.	X	X	X	X	X	X	X	X	X	X	X	X	X	X	X	X	X	X	X			X	X	X	X		X	X	X

Phytoplankton species identified at each aquatic sampling point in LTNP, Petén, Guatemala

Taxa	SP1	SP2	SP3	SP4	SP5	SP6	SP7	SP9	SP10	SP11	SP12	E1	E2	E3	E5	E6	E7	E8	E9	E10	E11	F1	F2	F5	C1	C2	C3	CA1	CA2
Meridion circulare	X	X	X		X		X	X	X						X						X				X		X		X
Diatoma vulgare											X	X	X	X	X	X		X	X		X	X	X	X		X	X		X
Fragilaria capucina	X	X	X	X	X	X	X	X	X	X	X	X	X	X	X	X	X	X	X	X	X	X	X	X	X	X	X	X	X
Synedra ulna	X	X	X	X	X	X	X		X	X	X	X	X	X	X	X	X					X	X				X		
S. acus	X	X	X	X	X	X	X	X	X	X	X	X	X	X	X	X	X			X	X	X	X	X			X		
ACHNANTHACEAE																													
Cocconeis sp.	X	X	X	X	X	X	X		X	X	X	X			X							X						X	X
Navicula sp. 1	X	X	X			X	X	X	X	X	X	X	X	X	X	X	X	X		X		X	X	X	X		X	X	X
Navicula sp. 2	X	X	X	X	X				X	X	X	X			X	X	X	X	X	X	X	X	X		X	X	X		X
Navicula sp. 3		X		X		X	X	X	X	X		X	X	X	X			X	X	X		X	X	X		X		X	X
Pinnularia sp.	X	X	X	X			X			X	X	X	X	X	X	X	X	X				X	X	X	X	X	X	X	X
Stauroneis sp.	X	X	X	X			X	X	X	X	X	X	X			X				X		X						X	X
Amphiprora sp.	X	X			X	X	X								X	X	X												
Gyrosigma sp.	X	X	X	X	X	X	X				X			X	X	X	X				X					X	X		X
Achnanthes sp.		X			X		X	X	X	X	X	X	X		X	X	X				X			X				X	X
GOMPHONEMACEAE																													
Gomphoneis sp.	X	X		X	X	X	X	X	X	X	X	X	X		X		X			X		X	X	X	X	X		X	
Cymbelaceae																													
Cymbella sp.	X	X		X	X	X	X	X	X	X	X	X	X		X		X	X		X		X	X	X	X	X	X	X	X
EPITHEMIACEAE																													
Denticula sp.		X	X		X		X	X	X	X	X	X		X	X	X				X		X	X	X	X	X	X	X	X
Epithemia sp.									X	X	X	X								X								X	X
NITSCHIACEAE																													
Nitzschia acircularis			X	X	X				X	X		X								X		X				X		X	X
SURIRELLACEAE																													
Surirella sp. 1					X	X					X				X		X												
Surirella sp. 2			X	X		X	X		X			X		X						X		X	X				X		
TOTAL	37	42	41	39	37	35	41	25	34	37	44	38	34	29	38	30	26	23	18	19	36	42	42	30	28	19	36	27	29

** In SP9 and E9 there were very few phytoplankton, so densities were not quantified.

** En SP9 y E9 había poca cantidad de fitoplancton por lo que no se cuantificaron las densidades.

Density (organisms/liter) of phytoplankton identified at each sample point, organized by class

Densidad (organismos/litro) del fitoplancton identificada en cada punto de muestreo organizado por clase

APPENDIX 4

APENDICE 4

Karin Herrera

See Chapter 2 for descriptions of sampling points.
Ver Capítulo 2 para descripción de los puntos de muestreo.

Taxa	SP 1	SP 2	SP 3	SP 4	SP 5	SP 6	SP 7	SP 10	SP 11	SP 12	E 1	E 2	E 3	E 5	E 6	E 7	E 8	E 10	E 11	F 1	F 2	F 5	C 1	C 2	C 3	CA 1	CA 2
CHROOCOCCACEAE																											
Microcystis aeruginosa	62	1062	387	142	382	319	1718	79	42	34	9	34	3	29	9	16	0	3800	1894	23	216	93	0	0	1969	0	0
Chroococcus limneticus	30	332	126	42	97	254	71	396	183	45	26	39	32	36	0	8	11	380	178	30	11	348	9	2	1353	9	3
C. minutus	2	13	10	2	4	22	29	0	108	11	4	5	5	3	27	4	3	76	41	60	29	13	4	14	34	4	3
Merismopedia sp.	10	13	10	17	4	4	29	0	0	3	4	0	3	3	14	8	5	61	18	0	7	7	0	0	0	0	0
OSCILLATORIACEAE																											
Lyngbya sp.	60	13	10	2	12	4	29	101	50	28	4	10	3	73	55	8	5	152	41	8	90	9	4	2	9	0	0
Oscillatoria limosa	6	53	97	52	8	44	321	18	17	6	9	15	2	31	0	12	3	76	6	8	50	28	18	2	34	4	3
Oscillatoria sp. 2	2	40	10	2	4	17	86	35	17	6	4	0	2	88	0	0	0	91	6	0	18	0	0	0	0	0	0
Spirulina sp.	3	27	10	0	4	0	43	18	25	6	0	0	3	0	0	0	3	8	6	8	25	4	0	0	0	4	6
NOSTOCACEAE																											
Anabaena flos-aquae	6	53	58	8	24	13	143	70	8	3	35	29	0	5	9	0	8	167	0	8	0	11	0	0	0	4	8
Scytonemataceae																											
Tolypothrix sp.	0	0	10	2	0	0	14	132	21	132	21	6	9	0	3	0	0	0	3	15	15	0	8	25	4	0	0
XANTHOPHYCEAE																											
Botryococcus braunii	105	332	300	96	24	13	14	0	0	6	4	69	2	3	27	16	0	0	0	8	0	2	22	7	223	9	0
EUGLENACEAE																											
Euglena sp.	7	13	0	19	16	22	71	44	17	28	9	59	5	16	0	0	8	0	0	0	58	6	13	9	0	0	0
Phacus sp.	0	0	10	0	0	0	0	0	0	0	0	0	0	0	0	0	0	15	18	15	7	0	0	0	0	0	0
Volvocaceae																											
Volvox sp.	1	7	10	0	0	4	0	0	0	0	0	0	2	0	0	0	0	0	0	0	0	2	0	0	0	0	0
MICRACTINIACEAE																											
Golenkinia radiata	0	7	10	2	0	0	7	0	4	3	4	5	2	0	0	0	0	8	6	0	0	4	0	0	197	0	0
Micractinium sp.	0	0	0	0	0	0	0	0	0	0	0	0	0	0	0	0	0	0	6	0	0	7	0	0	9	0	0
COELASTRACEAE																											
Coelastrum reticulatum	0	0	0	0	0	0	14	0	0	0	0	0	2	0	0	4	0	0	0	15	0	47	9	0	137	17	0
HYDRODICTYACEAE																											
Sorastrum sp.	0	0	0	0	0	0	0	0	0	0	0	0	0	0	0	0	0	0	0	0	4	4	0	0	0	0	0
Pediastrum duplex	33	20	0	8	0	0	57	0	0	6	0	0	0	0	9	0	0	0	24	38	4	0	4	0	43	0	3
P. simplex	1	27	0	0	8	0	0	9	0	6	4	5	0	0	9	0	0	0	24	38	4	0	4	2	171	0	3
OÖCYSTACEAE																											
Chlorella vulgaris	0	13	0	0	0	0	0	519	441	34	0	25	162	26	0	0	0	0	0	45	4	61	0	0	17	0	0
Ankistrodesmus falcatus	8	13	10	0	8	0	14	0	58	6	0	39	0	0	9	0	0	0	0	8	0	0	0	0	0	17	3
Closteriopsis sp.	0	0	39	0	8	0	0	176	67	26	0	0	0	0	0	0	0	0	18	15	0	0	0	0	0	0	0
Tetraëdron limneticum	0	13	10	0	0	0	0	0	0	0	0	0	0	0	0	0	0	0	0	0	0	0	0	0	0	0	0
T. regulare	0	0	0	0	0	0	0	0	0	6	0	0	0	0	0	0	0	0	0	0	0	0	9	22	0	0	0
Kirchneriella sp.	0	13	10	2	0	0	0	0	0	0	0	0	0	0	0	0	0	0	0	0	0	0	0	0	0	0	0

Taxa	SP 1	SP 2	SP 3	SP 4	SP 5	SP 6	SP 7	SP 10	SP 11	SP 12	E 1	E 2	E 3	E 5	E 6	E 7	E 8	E 10	E 11	F 1	F 2	F 5	C 1	C 2	C 3	CA 1	CA 2
SCENEDESMACEAE																											
Scenedesmus quadriacuada	0	0	10	10	12	9	14	0	0	17	0	0	0	0	0	0	0	152	71	8	4	2	0	0	51	0	0
Crucigenia sp.	0	0	0	8	16	0	0	0	0	17	0	0	0	0	0	0	0	0	0	0	14	11	0	0	34	0	0
MESOTAENIACEAE																											
Cylindrocystis sp.	0	0	0	0	0	0	0	0	21	40	71	0	0	114	36	0	0	0	0	68	0	15	0	0	0	0	3
DESMIDIACEAE																											
Closterium sp.	2	0	0	2	0	0	0	0	0	0	0	0	0	0	0	0	3	0	0	0	47	19	0	0	17	0	0
Micrasterias sp.	0	0	0	0	0	0	0	0	0	0	0	0	0	0	0	0	0	0	0	15	0	6	0	0	0	0	0
Staurastrum sp. 1	1	0	0	0	0	0	0	0	0	0	0	0	0	0	0	0	0	0	0	75	32	22	0	0	522	0	0
Staurastrum sp. 2	0	0	0	0	0	0	0	0	0	0	0	0	0	0	0	0	0	0	0	8	40	0	0	0	0	0	0
Cosmarium monomazum	0	0	0	0	0	0	0	0	0	6	4	5	0	0	5	0	0	0	0	0	0	0	0	0	0	0	0
Staurodesmus spencerianus	3	13	19	2	12	9	0	0	0	0	0	0	0	0	0	0	0	0	30	233	22	13	13	9	103	0	0
ZYGNEMATACEAE																											
Spirogyra sp.	0	0	0	0	0	0	0	0	0	0	0	0	0	0	0	0	0	0	0	0	18	6	0	0	26	0	0
Zygnema sp.	0	13	0	25	32	0	21	62	8	34	0	0	0	0	0	8	8	8	12	8	29	11	18	0	0	0	6
CLADOPHORACEAE																											
Cladophora sp.	0	0	0	0	0	0	0	0	25	0	4	5	5	41	0	0	0	0	0	8	11	0	9	0	0	9	11
CENTRITRACTACEAE																											
Centritractus sp.	20	13	19	4	20	13	71	70	25	17	9	0	0	10	18	85	3	0	6	68	0	0	0	0	51	0	0
MALLOMONADACEAE																											
Mallomonas sp.	0	0	0	0	0	0	0	0	0	0	4	5	0	88	0	4	3	0	0	391	18	0	0	0	43	4	0
SYNURACEAE																											
Synura uvella	46	93	10	0	0	9	0	0	0	0	0	0	0	0	0	0	0	0	0	0	0	0	0	0	0	0	0
OCHROMONADACEAE																											
Dinobryon sertularia	0	0	10	2	4	0	0	0	0	0	0	0	0	0	0	0	0	0	0	0	0	0	0	0	0	0	0
PERIDINIACEAE																											
Peridinium sp.	21	265	310	10	8	157	1155	44	0	0	4	10	11	16	0	227	0	0	0	0	0	0	9	0	676	4	17
CERATIACEAE																											
Ceratium sp.	0	0	0	0	0	0	100	0	0	0	0	0	5	18	5	0	16	0	0	0	36	0	0	0	9	0	0
MELOSIRACEAE																											
Melosira varians	2	93	10	2	4	70	100	106	50	28	4	29	2	21	73	16	32	0	0	8	47	6	4	2	0	86	6
M. granulosa	5	53	19	0	0	17	207	70	0	0	0	0	0	0	0	0	0	228	47	30	7	0	0	0	0	0	0
THALASSIOSIRACEAE																											
Cyclotella sp.	0	0	10	0	0	166	29	0	0	0	0	0	0	3	73	0	0	0	0	15	0	0	0	0	0	0	14
Stephanodiscus sp.	125	425	135	19	177	87	228	44	83	11	35	49	0	5	0	0	0	0	0	0	25	0	4	5	0	4	0

Density (organisms/liter) of phytoplankton identified at each sample point, organized by class

Taxa	SP 1	SP 2	SP 3	SP 4	SP 5	SP 6	SP 7	SP 10	SP 11	SP 12	E 1	E 2	E 3	E 5	E 6	E 7	E 8	E 10	E 11	F 1	F 2	F 5	C 1	C 2	C 3	CA 1	CA 2
DIATOMACEAE																											
Tabellaria sp.	114	159	77	21	36	52	285	374	141	113	410	368	66	332	555	282	37	0	118	75	108	0	9	0	51	17	6
Meridion circulare	1	7	29	0	4	0	100	9	0	0	0	0	0	21	0	0	0	0	24	0	0	0	31	0	17	0	51
Diatoma vulgare	0	0	0	0	0	0	0	0	0	6	9	5	12	5	18	0	8	0	18	23	29	7	0	11	17	0	14
Fragilaria capucina	10	53	39	8	8	96	86	35	33	40	57	83	116	73	727	137	98	106	24	53	54	6	18	9	103	34	101
Synedra ulna	31	398	126	153	80	385	214	132	42	113	44	147	15	13	82	32	0	0	0	30	87	0	0	0	240	0	0
S. acus	62	465	193	209	121	1141	2138	616	632	340	379	196	291	36	82	121	0	76	24	30	87	0	0	0	86	0	0
ACHNANTHACEAE																											
Cocconeis sp.	4	7	19	8	12	4	14	4	8	9	26	0	0	31	0	0	0	0	0	15	0	0	0	0	0	21	8
Navicula sp. 1	21	27	48	0	0	13	114	264	116	159	132	69	97	104	327	20	11	76	178	8	58	9	0	11	17	21	70
Navicula sp. 2	27	13	19	8	40	0	0	167	166	45	9	0	0	47	118	121	26	38	41	53	32	0	9	25	26	0	3
Navicula sp. 3	0	13	0	8	0	35	185	123	100	0	35	15	8	73	91	0	8	0	30	8	206	0	9	0	0	56	22
Pinnularia sp.	2	7	39	6	0	0	71	0	4	6	154	74	40	78	391	16	8	0	47	38	206	4	22	25	43	17	11
Stauroneis sp.	2	13	10	2	0	0	14	9	58	6	18	39	0	0	9	0	0	0	6	0	0	0	0	0	0	17	3
Amphiprora sp.	6	13	0	0	8	4	86	0	0	0	0	0	5	145	236	0	0	0	0	0	0	0	0	0	0	0	0
Gyrosigma sp.	9	13	10	2	4	4	14	0	8	0	18	5	0	3	9	16	0	0	142	0	0	0	31	5	0	17	0
Achnanthes sp.	0	46	0	0	12	0	29	70	8	34	9	34	0	5	45	32	0	0	6	0	0	2	0	0	0	9	17
GOMPHONEMACEAE																											
Gomphoneis sp.	16	53	0	4	24	4	57	26	17	23	463	132	22	0	95	0	8	0	30	0	263	4	9	9	26	0	17
CYMBELACEAE																											
Cymbella sp.	19	27	0	19	16	13	86	44	890	289	260	132	0	316	0	48	8	0	6	15	72	11	623	11	17	17	17
EPITHEMIACEAE																											
Denticula sp.	0	0	10	10	0	4	0	18	33	62	4	20	0	52	45	16	0	0	47	38	206	4	22	25	43	17	11
Epithemia sp.	0	0	0	0	0	0	0	18	33	28	9	0	0	0	0	0	0	0	0	0	0	0	0	0	0	9	6
NITSCHIACEAE																											
Nitzschia acircularis	0	0	0	4	4	4	0	0	4	6	0	20	0	0	0	0	0	0	6	0	0	0	9	0	17	4	3
SURIRELLACEAE																											
Surirella sp. 1	0	0	0	0	8	4	0	0	0	6	0	0	0	5	0	36	0	0	0	0	0	0	0	0	0	0	0
Surirella sp. 2	0	0	29	2	0	4	43	9	0	0	13	0	2	0	0	0	0	15	12	0	0	0	0	0	0	9	0

APPENDIX 5   **Aquatic insects associated with *Salvinia auriculata*, by sampling point, in Laguna del Tigre National Park, Petén, Guatemala**

APENDICE 5   **Insectos acuáticos asociados con *Salvinia auriculata*, por punto de muestreo en el Parque Nacional Laguna del Tigre, Petén, Guatemala**

Ana Cristina Bailey and Jorge Ordóñez

See Chapter 2 and Appendix 1 for descriptions of sampling points.
Ver Capítulo 2 y Apéndice 1 para descripciones de puntos de muestreo.

ORDER / Morphospecies	SP.1: Yalá	SP.2:Pa. Caballos	SP3:La Pista	SP4:El Sibalito	SP5: Caracol	SP7:San Juan	SP8:La Caleta
HEMIPTERA							
Lipogomphus sp.1			X			X	
Lipogomphus sp.2		X	X		X		
Paraplea sp.1	X	X					
Belostoma sp.1				X	X		
Guerris sp.1						X	
ODONATA							
Neoneura sp.1						X	
Libellula sp.1						X	
Libellula sp.2			X			X	
Coenagrion sp.1					X		
Chromagrion sp.1		X			X		
Amphiagrion sp.1				X	X		
Argia sp.1		X				X	
EPHEMEROPTERA							
Baetodes sp.1				X			
Leptohyphes sp.1		X	X		X	X	
Leptohyphes sp.2					X		
COLEOPTERA							
Hydrochus sp.1		X	X		X	X	X
Hydrochus sp.2		X				X	X
Crenitis sp.1					X		
Paracymus sp.1				X		X	
Derallus sp.1							X
Hydrophilidae desc.1				X		X	
Scirtidae desc. 1		X		X			
Scirtidae desc. 2							
Scirtidae desc. 3		X	X	X			
Scirtidae desc. 4	X			X	X	X	
Elmidae desc. 1		X					
Hydrocanthus sp.1			X		X		
Dytiscidae desc. 1			X				
Hydroscaphidae desc. 1			X		X		
Lutrochus sp.1				X			
Scolitidae desc.1				X			
Ochthebius sp.1					X		
Prasocuris sp.1						X	

ORDER Morphospecies	SP.1: Yalá	SP.2:Pa. Caballos	SP3:La Pista	SP4:El Sibalito	SP5: Caracol	SP7:San Juan	SP8:La Caleta
DIPTERA							
Prionocera sp.1				x			
Probezzia sp.1				x		x	
Ceratopogonidae desc. 1	x	x	x		x		
Chironomidae desc.1	x			x		x	x
Chironomidae desc. 2		x		x	x	x	x
Chironomidae des. 3	x						
Chaoboridae desc. 1	x				x		
Odontomyia sp.1			x				
Stratiomys sp.1			x		x		
Stratiomys sp.2		x	x				

APPENDIX 6   Plant taxa recorded in and around marshes and riparian forests in Laguna del Tigre National Park, Petén, Guatemala

APENDICE 6   Taxones de plantas registrado en y alrededor de los bosques anegados y ribereños en el Parque Nacional Laguna del Tigre, Petén, Guatemala

Blanca León and Julio Morales Can

Life form/Forma de vida:
t: Terrestrial/Terrestre
a: Acuatic/Acuática
am: Amphibian/Anfibio
e: Epifita/Epífita

Family Familia	Genus Género	Species Especies	Author Autor	Local Name Nombre Local	Life Form Forma de Vida
ACANTHACEAE	Aphelandra	scabra	(Vahl) Sm.		t
ACANTHACEAE	Ruellia	cf. brittoniana			t
ALISMATACEAE	Sagittaria	lancifolia subsp. media	(Micheli) Bogin		am
AMARANTHACEAE	Alternanthera	obovata	(Mart. & Gal.) Millsp.		t
ANACARDIACEAE	Bunchosia	sp.			t
ANACARDIACEAE	Metopium	brownei	(Jacq.) Urb.	chechén negro	t
ANACARDIACEAE	Spondias	sp.		jobo	t
ANNONACEAE	Annona	glabra	L.		t
APIACEAE	Centella	asiatica	(L.) Urb.		am
APIACEAE	Hydrocotyle	bonariensis	Lam.		am
APIACEAE	Hydrocotyle	ranunculoides	L.		am
APOCYNACEAE	Rhabdadenia	biflora	(Jacq.) Müll. Arg.		t
ARACEAE	Pistia	stratiotes	L.		a
ARECACEAE	Acoelorraphe	wrightii	(Griseb. & H. Wendl.) H. Wendl. Ex Becc.	taciste	am
ARECACEAE	Bactris	major	Jacq.	güiscoyol	t
ARECACEAE	Desmoncus	orthacantos	Mart.	bayal	t
ARECACEAE	Sabal	sp.		guano	t
ARECACEAE	Thrinax	sp.			t
ASCLEPIADACEAE		sp. 1			t
ASTERACEAE	Eclipta	sp.			t
ASTERACEAE	Fleishmannia	sp.			t
ASTERACEAE	Mikania	micrantha			t
ASTERACEAE	Wedelia	sp.			am
BIGNONIACEAE	Crescentia	cujete			t
BIGNONIACEAE	Clytostoma	binatum	(Thunb) Sandw.		t
BLECHNACEAE	Blechnum	serrulatum	Rich.		am
BOMBACACEAE	Pachira	aquatica			am
BOMBACACEAE	Quararibea	funebris	(La Llave) Vischer	batidor	t
BROMELIACEAE	Aechmea	sp.			e
BROMELIACEAE	Pancratium	sp.			t
BROMELIACEAE	Tillandsia	Streptophylla	Scheidw. & C. Morren		e
CABOMBACEAE	Brasenia	Schreberi	J. F. Gmel.		a
CABOMBACEAE	Cabomba	palaeformis	Fassett.		a
CACTACEAE		sp.			t

Family Familia	Genus Género	Species Especies	Author Autor	Local Name Nombre Local	Life Form Forma de Vida
CAPPARIDACEAE	Capparis	flexuosa	L.		t
CELASTRACEAE				recoma	t
CHRYSOBALANACEAE	Chrysobalanus	icaco	L.	icaco	t
CHRYSOBALANACEAE	Hirtella	americana	L.	aceituno peludo	t
CLUSIACEAE	Calophyllum	cf. brasiliense Camb. var. rekoi	(Standl.) Standl.		t
COMBRETACEAE	Bucida	buceras	L.		am
COMMELINACEAE	Commelina	sp.			t
CONVOLVULACEAE	Ipomoea	indica	(Burm. f) Merr.	quiebracajete	t
CONVOLVULACEAE	Ipomoea	sagittata	Lam.		t
CUCURBITACEAE		sp.			t
CUSCUTACEAE	Cuscuta	sp.			t
CYPERACEAE	Cladium	jamaicense	Crantz		am
CYPERACEAE	Cyperus	lundelli	O'Neill		am
CYPERACEAE	Cyperus	unioloides	(R. Br.) Urb.		am
CYPERACEAE	Eleocharis	geniculata	(L.) Roem. & Schult.		am
CYPERACEAE	Eleocharis	interstincta			am
CYPERACEAE	Eleocharis	rostellata	Torr		am
CYPERACEAE	Fuirena	simplex	Vahl		am
CYPERACEAE	Rynchospora	cephalotes	(L.) Vahl		am
CYPERACEAE	Rynchospora	holoschoenoides	(Rich.) Herter		am
CYPERACEAE	Scleria	sp.			am
DIOSPYRACEAE	Diospyros	digyna		matasano	t
EUPHORBIACEAE	Caperonia	castaneifolia	(L.) A. St.-Hil.		am
FABACEAE	Acacia	sp.		subin	t
FABACEAE	Dalbergia	sp.		xactix	t
FABACEAE	Haemotxylum	campechianum	L.		am
FABACEAE	Inga	sp.			t
FABACEAE	Lysiloma	bahamensis		salam	t
FABACEAE	Lonchocarpus	hondurensis	Benth.	yaxman	t
FABACEAE	Lonchocarpus	sp.		palo de San Juan	t
FABACEAE	Mimosa	pigra			am
FABACEAE	Piscidia	piscipula	(L.) Sarg.		t
FABACEAE	Phitecellobium	sp.		palo Peru	t
FLACOURTIACEAE	Xylosma	flexuosum	(Kunth) Hemsl.	abalche de monte	t
GENTIANACEAE	Eustoma	exaltatum	(L.) Salisb.		t
HYDROCHARITACEAE	Vallisneria	americana	Michx.		a
LAMIACEAE	Teucrium	vesicarium	Mill.		t
LAURACEAE	Nectandra	membranacea		cocche	t
LENTIBULARIACEAE	Utricularia	foliosa	L.		a
LENTIBULARIACEAE	Utricularia	gibba	L.		a
LENTIBULARIACEAE	Utricularia	sp.			a
LOGANIACEAE	Mitreola	petiolata	(J. F. Gmel.) Torr. & A. Gray		am
MALPIGHIACEAE	Heteropterys	lindeniana	A. Juss.	guayabillo de pantano	t
MALVACEAE	Hibiscus	sp.			am
MYRICACEAE	Myrica	sp.			t
MYRTACEAE	Calyptranthes	chytraculia	(L.) Sw.	chilimis hoja ancha	t
NAJADACEAE	Najas	wrightiana	A. Braun		a
NYMPHAEACEAE	Nymphaea	ampla			a

Family Familia	Genus Género	Species Especies	Author Autor	Local Name Nombre Local	Life Form Forma de Vida
OCHNACEAE	*Ouratea*	*lucens*	(Kunth) Engl.		t
ONAGRACEAE	*Ludwigia*	cf. *nervosa*			am
ONAGRACEAE	*Ludwigia*	sp.			am
ORCHIDACEAE	*Eulophia*	*alta*	(L.) Fawc. & Rendle		am
ORCHIDACEAE	*Habenaria*	*bractescens*	Lindl.		am
ORCHIDACEAE	*Habenaria*	*repens*	Nutt.		am
ORCHIDACEAE	*Spiranthes*	sp.			am

Fish species believed to occur in the Río San Pedro and the upper Río Candelaria basins in Petén, Guatemala

Especies de peces que se cree, existen en el Río San Pedro y en la cuenca del Río Candelaria en Petén, Guatemala

Philip W. Willink, Christian Barrientos, Herman A. Kihn, and Barry Chernoff

+ = collected/colectada
- = not collected during RAP survey/no se colectaron durante el estudio del RAP

Taxa	Common Name Nombre Común	Taxa collected during RAP Taxa colectada durante el RAP
SEMIONOTIFORMES		
Lepisosteidae		
Atractosteus tropicus Gill 1863	Peje lagarto	+
ELOPIFORMES		
Megalopidae		
Tarpon atlanticus Valenciennes 1847	Sábalo	-
ANGUILLIFORMES		
Anguillidae		
Anguilla rostrata Lesueur 1817	Anguila	-
CLUPEIFORMES		
Clupeidae		
Dorosoma anale Meek 1904	Sardina de leche	+
Dorosoma petenense Günther 1867	Sardina de leche	+
CYPRINIFORMES		
Catostomidae		
Ictiobus meridionalis Günther 1868	Chopa	-
Cyprinidae		
Ctenopharyngodon idella Valenciennes 1844	Carpa	+
CHARACIFORMES		
Characidae		
Astyanax aeneus Günther 1860	Sardina	+
Brycon guatemalensis Regan 1908	Machaca	+
Hyphessobrycon compressus Meek 1904	Sardinita	+
SILURIFORMES		
Ariidae		
Ariidae sp.	Curruco	+
Cathorops aguadulce Meek 1904	Cabeza de fierro	+
Potamarius nelsoni Evermann & Goldsborough 1902	Curruco	+
Ictaluridae		
Ictalurus furcatus Valenciennes 1840	Jolote	+
Pimelodidae		

Taxa	Common Name Nombre Común	Taxa collected during RAP Taxa colectada durante el RAP
Rhamdia guatemalensis Günther 1864	Filin	+
Rhamdia laticauda Kner 1858	Filin	-
BATRACHOIDIFORMES		
Batrachoididae		
Batrachoides goldmani Evermann & Goldsborough 1902	Pez sapo	+
ATHERINIFORMES		
Atherinidae		
Atherinella alvarezi Diaz-Pardo 1972	Robalito	+
Atherinella cf. *schultzi* Alvarez & Carranza 1952	Robalito	+
CYPRINODONTIFORMES		
Aplocheilidae		
Rivulus tenuis Meek 1904	"none available"	-
Poeciliidae		
Belonesox belizanus Kner 1860	Picudito	+
Carlhubbsia kidderi Hubbs 1936	Pulta	+
Gambusia sexradiata Hubbs 1936	Pulta	+
Gambusia yucatana Regan 1914	Pulta	-
Heterandria bimaculata Heckel 1848	Pupo	+
Phallichthys fairweatheri Rosen & Bailey 1959	Pulta	+
Poecilia mexicana Steindachner 1863	Pupo	+
Poecilia petenensis Günther 1866	Pupo	+
Xiphophorus hellerii Heckel 1848	Pulta	+
Xiphophorus maculatus Günther 1866	Pulta	-
BELONIFORMES		
Belonidae		
Strongylura hubbsi Collette 1974	Agujeta	+
Hemiramphidae		
Hyporhamphus mexicanus Alvarez 1959	Agujeta	-
MUGILIFORMES		
Mugilidae		
Mugil cephalus Linnaeus 1758	Liza	-
Mugil curema Valenciennes 1836	Liza	+
SYNBRANCHIFORMES		
Synbranchidae		
Ophisternon aenigmaticum Rosen & Greenwood 1976	Anguila	-
PERCIFORMES		
Centropomidae		
Centropomus undecimalis Bloch 1792	Robalo	-
Cichlidae		
Cichlasoma bifasciatum Steindachner 1864	Mojarra	+
Cichlasoma friedrichsthalii Heckel 1840	Guapote	+
Cichlasoma helleri Steindachner 1864	Shibal	+
Cichlasoma heterospilum Hubbs 1936	Mojarra negra	+
Cichlasoma intermedium Günther 1862	Mojarra	-
Cichlasoma lentiginosum Steindachner 1864	Corrientera	+
Cichlasoma meeki Brind 1918	Shibal	+
Cichlasoma octofasciatum Regan 1903	Negrita	+

Taxa	Common Name Nombre Común	Taxa collected during RAP Taxa colectada durante el RAP
Cichlasoma pasionis Rivas 1962	Shibal	+
Cichlasoma pearsei Hubbs 1936	Mojarra	+
Cichlasoma robertsoni Regan 1905	Shibal	+
Cichlasoma salvini Günther 1862	Canchay	+
Cichlasoma synspilum Hubbs 1935	Mojarra	+
Cichlasoma urophthalmus Günther 1862	Bul	+
Oreochromis sp.	Tilapia	-
Petenia splendida Günther 1862	Blanco	+
Eleotridae		
Gobiomorus dormitor Lacep de 1800	Pijepadre	-
Gerreidae		
Eugerres mexicanus Steindachner 1863	Mojarra blanca	+
Sciaenidae		
Aplodinotus grunniens Rafinesque 1819	Roncador	+

APPENDIX 8

List of fish species captured during the 1999 RAP expedition to Laguna del Tigre National Park, Petén, Guatemala, and the focal sampling point at which they were collected

APENDICE 8

Lista de especies de peces capturados durante la expedición RAP de 1999 al Parque Nacional Laguna del Tigre, Petén, Guatemala y el punto focal de muestreo en el cual fueron recolectados

Philip W. Willink, Christian Barrientos, Herman A. Khin and Barry Chernoff

See Gazetteer for focal sampling point descriptions.
Ver Diccionario Geográfico para descripciones de los puntos focales de muestreo.

Taxa	San Pedro	Escondido	Flor de Luna	Chocop	Candelaria
SEMIONOTIFORMES					
Lepisosteidae					
Atractosteus tropicus	X	X	X	-	-
CLUPEIFORMES					
Clupeidae					
Dorosoma anale	X	-	-	-	-
Dorosoma petenense	X	X	X	-	-
CYPRINIFORMES					
Cyprinidae					
Ctenopharyngodon idella	X	-	-	-	-
CHARACIFORMES					
Characidae					
Astyanax aeneus	X	X	X	X	X
Brycon guatemalensis	X	-	-	-	-
Hyphessobrycon compressus	X	X	X	X	X
SILURIFORMES					
Ariidae					
Ariidae sp.	X	-	-	-	-
Cathorops aguadulce	X	X	-	-	-
Potamarius nelsoni	X	-	-	-	-
Ictaluridae					
Ictalurus furcatus	X	-	-	-	-
Pimelodidae					
Rhamdia guatemalensis	X	X	X	X	-
BATRACHOIDIFORMES					
Batrachoididae					
Batrachoides goldmani	X	-	-	-	-
ATHERINIFORMES					
Atherinidae					
Atherinella alvarezi	X	-	-	-	-
Atherinella c.f. schultzi	-	-	-	-	X
CYPRINODONTIFORMES					

Taxa	San Pedro	Escondido	Flor de Luna	Chocop	Candelaria
Poeciliidae					
Belonesox belizanus	X	X	X	X	X
Carlhubbsia kidderi	X	X	X	-	-
Gambusia sexradiata	X	X	X	-	-
Heterandria bimaculata	-	X	-	X	X
Phallichthys fairweatheri	X	X	X	-	-
Poecilia mexicana	X	X	X	X	X
Poecilia petenensis	X	X	X	-	-
Xiphophorus hellerii	-	-	-	X	-
BELONIFORMES					
Belonidae					
Strongylura hubbsi	X	X	-	-	-
MUGILIFORMES					
Mugilidae					
Mugil curema	X	-	-	-	-
PERCIFORMES					
Cichlidae					
Cichlasoma bifasciatum	X	X	X	-	X
Cichlasoma friedrichsthalii	X	X	-	X	X
Cichlasoma helleri	X	X	X	X	X
Cichlasoma heterospilum	X	X	X	X	-
Cichlasoma lentiginosum	X	-	-	-	-
Cichlasoma meeki	X	X	X	X	X
Cichlasoma octofasciatum	X	X	-	X	-
Cichlasoma pasionis	X	X	X	X	-
Cichlasoma pearsei	X	X	X	-	-
Cichlasoma robertsoni	X	X	X	X	-
Cichlasoma salvini	X	X	X	-	X
Cichlasoma synspilum	X	X	X	-	X
Cichlasoma urophthalmus	X	X	X	-	-
Petenia splendida	X	X	X	X	X
Gerreidae					
Eugerres mexicanus	X	-	-	-	-
Sciaenidae					
Aplodinotus grunniens	X	-	-	-	-
TOTAL	38	27	22	15	13

APPENDIX 9 List of fish species captured during the 1999 RAP expedition to Laguna del Tigre National Park, Petén, Guatemala, and the macrohabitat in which they were collected

APENDICE 9 Lista de especies de peces capturados durante la expedición RAP de 1999 en el Parque Nacional Laguna del Tigre, Petén, Guatemala y el macrohábitat en donde se recolectaron

Philip W. Willink, Christian Barrientos, Herman A. Kihn, and Barry Chernoff

See Table 1.1 for macrohabitat descriptions.
Ver Tabla 1.1 para descripción de macrohábitat.

Taxa	Lagoons Lagunas	Oxbow Lagoons Lagunas Oxbow	Caños	Tributaries Tributarios	Rivers, seasonal Ríos, estacionales	Rivers, permanent Ríos, permanentes	Rivers, bays Ríos, bahías	Rivers, rapids Ríos, rápidos
SEMIONOTIFORMES								
Lepisosteidae								
Atractosteus tropicus	X	-	-	X	-	X	X	-
CLUPEIFORMES								
Clupeidae								
Dorosoma anale	-	-	-	-	-	X	-	-
Dorosoma petenense	X	X	-	X	X	X	-	X
CYPRINIFORMES								
Cyprinidae								
Ctenopharyngodon idella	-	-	-	-	-	X	-	-
CHARACIFORMES								
Characidae								
Astyanax aeneus	X	X	X	X	X	X	X	X
Brycon guatemalensis	-	-	-	-	-	X	-	-
Hyphessobrycon compressus	X	-	-	-	X	X	X	X
SILURIFORMES								
Ariidae								
Ariidae sp.	-	-	-	-	-	X	-	-
Cathorops aguadulce	-	-	-	X	-	X	-	-
Potamarius nelsoni	-	-	-	-	-	X	-	-
ICTALURIDAE								
Ictalurus furcatus	-	-	-	-	-	X	-	-
PIMELODIDAE								
Rhamdia guatemalensis	X	X	-	X	X	-	-	-
BATRACHOIDIFORMES								
Batrachoididae								
Batrachoides goldmani	-	-	-	-	-	-	-	X
ATHERINIFORMES								

Taxa	Lagoons Lagunas	Oxbow Lagoons Lagunas Oxbow	Caños	Tributaries Tributarios	Rivers, seasonal Ríos estacionales	Rivers, permanent Ríos permanentes	Rivers, bays Ríos, bahías	Rivers, rapids Ríos, rápidos
Atherinidae								
Atherinella alvarezi	-	-	-	X	-	-	-	-
Atherinella c.f. schultzi	-	-	-	-	X	-	-	-
CYPRINODONTIFORMES								
Poeciliidae								
Belonesox belizanus	X	X	-	X	X	X	-	-
Carlhubbsia kidderi	X	X	-	-	X	X	X	-
Gambusia sexradiata	X	X	-	X	X	X	X	-
Heterandria bimaculata	-	-	-	-	X	X	-	-
Phallichthys fairweatheri	X	-	X	X	X	X	-	-
Poecilia mexicana	X	X	-	X	X	X	-	X
Poecilia petenensis	X	X	-	X	X	-	-	-
Xiphophorus helleri	-	-	-	-	X	-	-	-
BELONIFORMES								
Belonidae								
Strongylura hubbsi	X	-	-	-	-	-	-	-
MUGILIFORMES								
Mugilidae								
Mugil curema	-	-	-	-	-	X	-	-
PERCIFORMES								
Cichlidae								
Cichlasoma bifasciatum	X	-	-	-	X	X	-	X
Cichlasoma friedrichsthalii	-	-	-	X	X	X	X	-
Cichlasoma helleri	X	X	X	X	X	X	-	X
Cichlasoma heterospilum	X	X	-	X	-	X	X	X
Cichlasoma lentiginosum	-	-	-	-	-	-	-	X
Cichlasoma meeki	X	X	-	X	X	X	X	X
Cichlasoma octofasciatum	-	-	-	X	X	X	-	-
Cichlasoma pasionis	X	X	X	X	X	X	X	-
Cichlasoma pearsei	X	X	-	-	-	X	-	X
Cichlasoma robertsoni	X	-	-	X	X	X	X	-
Cichlasoma salvini	X	X	X	X	X	X	-	X
Cichlasoma synspilum	X	X	X	X	X	X	X	X
Cichlasoma urophthalmus	X	X	X	X	X	X	X	-
Petenia splendida	X	X	-	X	X	X	X	X
Gerreidae								
Eugerres mexicanus	-	-	-	-	-	X	-	-
Sciaenidae								
Aplodinotus grunniens	-	-	-	-	-	X	-	-
TOTAL	23	17	7	22	24	33	13	14

APPENDIX 10 Detailed methodology for DNA strand breakage and flow cytometry assays

APENDICE 10 Metodología detallada para ensayos de separación de filamentos de ADN y citometría de flujo.

Chris W. Theodorakis and John W. Bickham

For flow cytometric analysis, nuclei were isolated from blood cells and stained with propidium iodide according to a modification of the methods of Vindelov et al. (1983) and Vindelov and Christianson (1990; see Chapter 5 for references). An increase in the amount of chromosomal damage is reflected by an increase in cell-to-cell variation in DNA content (Bickham, 1990). This variation (measured as half-peak coefficient of variation [CV]) was measured using an Epics Profile II Flow Cytometer (Coulter Corp., Hialeah, CA), which measures the cell-to-cell half-peak coefficient of variation (gated CV) in DNA content for cells in the G_1/G_1 phase of the cell cycle. Alignment, focus, and instrumental gain were set prior to analysis using 0.097 mm fluorescent microspheres (Coulter Corp.) and all samples were run in one day. Differences between sites were tested using the Kruskal-Wallis test.

For DNA extractions, 20 ml whole blood was suspended in 500 ml TEN (50 mM Tris, 10 mM EDTA, 100mM NaCl, pH 8.0). DNA was then extracted and purified according to Theodorakis et al. (1996), and dissolved in TE (10 mM Tris, 1 mM EDTA, pH 8.0). DNA was quantified spectrophotometrically at 260 nm (1 AU = 50 mg DNA/ml).

The DNA was subjected to electrophoresis with alkaline or neutral running buffer (Theodorakis et al., 1996). Alkaline electrophoresis was performed with 30 mM NaOH, 2 mM EDTA (pH 12.5) as the running buffer. A total of 0.5 mg of DNA was loaded into each well of a 0.8% agarose gel and subjected to electrophoresis at 5 V/cm for 5 hr. The buffer was constantly recirculated and cooled in an icebath. For neutral gel electrophoresis, TBE (45 mM Tris, 45 mM borate, 0.5 mM EDTA, pH 8.0) was used as the running buffer. A total of 0.05 mg DNA was loaded into a 0.3% agarose gel and subjected to electrophoresis at 0.75 V/cm for 18 hr. Because such a low percentage gel is very fragile, the gels were cast on a basement of 3% agarose. After electrophoresis, gels were stained in ethidium bromide and photographed under UV light. The average molecular length (Ln) of the DNA in each sample was calculated as in Theodorakis et al. (1993).

The Ln calculated under alkaline conditions is affected by both single- and double-strand breaks, while the Ln of neutral gels is affected by double strand breaks only. Because DNA is a long and fragile molecule, double-strand breaks can be caused not only by genotoxicant exposure, but also by physical shearing which may take place during extraction, purification and analysis. In order to take this into account, the number of single-strand breaks (per 10^5 bases) was determined by subtracting the double-strand Ln from the single-strand Ln according to the following formula, as modified from Freeman et al. (1990):

$$\#SSB = \left(\frac{1}{Ln\,(single)} - \frac{1}{Ln\,(double)} \right) \times 100$$

Para los análisis de citometría de flujo, núcleos de células de sangre fueron aislados y coloreados con iodo de propidio de acuerdo a la modificación del método de Vindelov et al. (1983) y de Vindelov y Christianson (1990; ver el Capítulo 5 para referencias). Un incremento en la cantidad de daños a los cromosomas es reflejado en un incremento en la variación del contenido de ADN de célula a célula (Bickham, 1990). Esta variación (medida con la mitad del máximo coeficiente de variación [CV]) fue medida usando un citómetro de flujo Epics Profile II (Coulter Corp., Hialeah, CA), el cual mide la mitad del máximo del coeficiente de variación de célula a célula [*gated* CV] en el contenido de ADN para células en la fase G/G del ciclo celular. (Se determino el alineamiento, enfoque, e incremento instrumental) antes del análisis usando microesferas flourecentes de 0.097 mm (Coulter Corp.) Todas las muestras se analizaron en un día. Las diferencias entre las localidades fueron determinadas usando la prueba Kruskal-Wallis.

Para extracciones de ADN, se suspendieron 20 mL de sangre en 500 mL TEN (50 mM Tris, 10 mM EDTA, 100 mM NaCl, pH 8.0). El ADN fue extraído y purificado de acuerdo con Theodorakis et al. (1996), y fue disuelto en TE (10mM Tris, 1 mM EDTA, pH 8.0). El ADN fue cuantificado por medio espectrofotométrico a 260 mm (1 AU = 50 mg ADN/mL).

El ADN fue sometido a electroforesis con un amortiguador (*buffer*) alcalino o neutral (Theodorakis et al., 1996). Se llevo a cabo una electroforesis alcalina con 30 mM NaOH, 2mM EDTA (pH 12.5) como buffer. Cada pocito de 0.8% gel agaroso se llenó con un total de 0.5 mg de ADN; este fue sometido a electroforesis a 5 V/cm por 5 hr. El buffer fue recirculado constantemente y enfriado en hielo. Para electroforesis con gel neutral, se usó TBE (45 mM Tris, 45 mM borato, 0.5 mM EDTA, pH 8.0) como buffer. Se llenaron un total de 0.05 mg de ADN en gel agaroso de 0.3% y sometido a electroforesis a 0.75 V/cm por 18 hr. Ya que un porcentaje bajo de gel es muy frágil, los gels se formaron en usando un fondo de 3% de agarosa. Después de la electroforesis, los gels se colorearon con bromidio etídico y se fotografiaron bajo la luz Ultra Violeta. La longitud molecular promedio (Ln) de cada muestra de ADN fue calculada como en Theodorakis et al. (1993).

La calculada bajo condiciones alcalinas es afectada por rompimientos simples y dobles de hebras de ADN, mientras que el La de gels neutrales es afectado por rompimientos dobles solamente. Ya que el ADN es una molécula larga y frágil, rompimientos doble puede ocurrir no solo por exposición a genotoxicantes, sino también por un corte físico, el cual puede ocurrir durante la extracción, purificación y análisis. Para tomar esto en cuenta, el número de rompimientos simples (por cada 10^5 bases) fue determinado al sustraer la Ln de dobles hebras del Ln de hebras simples de acuerdo con la fórmula siguiente: electroforesis modificada de Freeman et al. (1990):

$$\#SSB = \left(\frac{1}{Ln\,(single)} - \frac{1}{Ln\,(double)}\right) X\,100$$

Birds species found in Laguna del Tigre National Park, Petén, Guatemala during the RAP survey

APENDICE 11 Especies de aves que se encontraron en el Parque Nacional Laguna del Tigre, Petén, Guatemala durante la expedición RAP de 1999

Edgar Selvin Pérez and Miriam Lorena Castillo Villeda

Scientific Name Nombre Científico	Common English Name Nombre común en inglés	Common Spanish Name Nombre común en español
TINAMIDAE		
Crypturellus boucardi	Slaty-breasted Tinamou	
Crypturellus cinamomeus	Thicket Tinamou	Mancolola
Crypturellus soui	Little Tinamou	Mancolola Enana
Tinamus major	Great Tinamou	Mancolola Grande
PODICIPEDIDAE		
Tachybatus dominicus	Least Grebe	Zambullidor Enano
PELECANIDAE		
Pelecanus occidentalis	Brown Pelican	Pelícano Pardo
PHALACROCORACIDAE		
Phalacrocorax brasilianus	Olivaceous Cormorant	Malache, Pato Coche
ANHINGIDAE		
Anhinga anhinga	Anhinga	Pato Aguja, Anhinga.
ARDEIDAE		
Agamia agami	Agami Heron	Garcita Vientre Castaño
Ardea herodias	Great Blue Heron	Garzón Azulado
Botaurus pinnatus	Pinnated Bittern	Garza Tigre
Bubulcus ibis	Cattle Egret	Garcita Garrapatera
Butorides striatus	Striated Heron	Garcita Verde
Casmerodius albus	Great Egret	Garza Real
Cochlearius cochlearius	Boat-billed Heron	Garza Pico Zapato, Cucharón
Egretta caerulea	Little Blue Heron	Garza Gris
Egretta thula	Snowy Egret	Garcita Blanca
Egretta tricolor	Tricolored Heron	Garcita Tricolor
Nycticorax violaceus	Yellow-crowned Night-Heron	Garza Nocturna Corona Amarilla
Tigrisoma mexicanum	Bare-throated Tiger-Heron	Garza Tigre
CICONIDAE		
Jabiru mycteria	Jabiru	Ciguieña Jabirú
Mycteria americana	Wood Stork	Garzón Pulido
ANATIDAE		
Anas discors	Blue-winged Teal	Cerceta ala Azul
Cairina moschata	Muscovy Duck	Pato Real
Dendrocygna autumnalis	Black-bellied Whistling-Duck	Pijije
CATHARTIDAE		
Cathartes aura	Turkey Vulture	Viuda, Mayón
Coragyps atratus	Black Vulture	Zopilote Negro, Zope
Sarcoramphus papa	King Vulture	Rey Zope
ACCIPITRIDAE		
Accipiter bicolor	Bicolored Hawk	Gavilán Bicolor
Accipiter cooperii	Cooper's Hawk	Gavilán
Busarellus nigricollis	Black-collared Hawk	Gavilán Pescador
Buteo brachyurus	Short-tailed Hawk	Gavilán
Buteo magnirostris	Roadside Hawk	Gavilán de los Caminos
Buteo nitidus	Grey Hawk	Gavilán Gris
Buteo platypterus	Broad-winged Hawk	Gavilán Aludo

Scientific Name Nombre Científico	Common English Name Nombre común en inglés	Common Spanish Name Nombre común en español
Buteogallus urubitinga	Great Black Hawk	Aguililla Negra
Elanoides forficatus	Swallow-tailed Kite	Gavilán Tijereta
Geranospiza caerulescens	Crane Hawk	Gavilán Ranero
Ictinia mississippiensi	Mississippi Kite	Gavilán Grisillo
Ictinia plumbea	Plumbeous Kite	Gavilán Plomizo
Leptodon cayenensis	Grey-headed Kite	
Leucopternis albicollis	White Hawk	Gavilán Blanco
Morphnus guianensis	Crested Eagle	Aguila Monera*
Pandion haliaetus	Osprey	Aguila Pescadora
Parabuteo unicinctus	Harris' Hawk	Gavilán Mixto
Rostrhamus sociabilis	Snail Kite	Gavilán Caracolero
Spizaetus ornatus	Ornate Hawk-Eagle	Aguilucho de Penacho*
Spizatur melanoleucus	Black-and-White Hawk-Eagle	Aguilucho Blanco y Negro
FALCONIDAE		
Falco deiroleucus	Orange-breasted Falcon	Halcón Pecho Naranja
Falco rufigularis	Bat Falcon	Halcón Murcielaguero
Herpetotheres cachinans	Laughing Falcon	Halcón Guaco
Micraster ruficollis	Barred Forest-Falcon	Halcón de Monte Rayado
M. semitorquatus	Collared Forest-Falcon	Halcón de Collar
CRACIDAE		
Crax rubra	Great Curassow	Pajuil, Faisán.
Ortalis vetula	Plain Chachalaca	Chachalaca, Chacha
Penelope purpurascens	Crested Guan	Cojolita
PHASIANIDAE		
Agriocharis ocellata	Ocellated Turkey	Pavo ocelado, Pavo Petenero
*Dactylortyx thoraxicus**	Singing Quail	Bolonchaco*
*Odontophorus guttatus**	Spotted Wood-Quail	Bolonchaco, Cobanitos*
RALLIDAE		
Aramides cajanea	Grey-necked Wood-Rail	Gallinola
Gallinula chloropus	Common Moorhen	Gallinola Frente Colorada
Laterallus ruber	Ruddy Crake	Gallineta Colorada
Porphyrula martinica	Purple Gallinule	Gallinola Morada
HELIORNITHIDAE		
Heliornis fulica	Sungrebe	Pájaro Cantil
ARAMIDAE		
Aramus guarauna	Limpkin	Totolaca,Caraú
CHARADRIDAE		
Charadrius collaris	Collared Plover	Chorlito de Collar
Charadrius vociferus	Killdeer	Collarejo
JACANIDAE		
Jacana spinosa	Northern Jacana	Gallito de Pantano
RECURVIROSTRIDAE		
Himantopus mexicanus	Black-necked Stilt	Candelero Americano
SCOLOPACIDAE		
Actitis macularia	Spotted Sandpiper	Alzaculito
Calidris melanotus	Pectoral Sandpiper	Playerito
Calidris minimus	Least Sandpiper	Playerito
Calidris fuscicola	White-rumped Sandpiper	Playerito
Tringa flavipe	Lesser Yellowlegs	idem
COLUMBIDAE		
Claravis pretiosa	Blue Ground-Dove	Tortolita Celeste
Columba flavirostris	Red-billed Pigeon	Paloma piquirojo
Columba nigrirostris	Short-billed Pigeon	Paloma pico corto
Columba speciosa	Scaled Pigeon	Paloma Escamosa
Columbina minuta	Plain-breasted Ground-Dove	Tortolita café
Columbina talpacoti	Ruddy Ground-Dove	Tortolita
Geotrygon montana	Ruddy Quail-Dove	Paloma de Montaña, Paloma Perdiz

Scientific Name Nombre Científico	Common English Name Nombre común en inglés	Common Spanish Name Nombre común en español
Leptotila rufaxilla	Grey-fronted Dove	Paloma Cabeza gris
L. verreauxi	White-tipped Dove	Susuy
Zenaida asiatica	White-winged Dove	Torcaza
PSITTACIDAE		
Amazona albifrons	White-fronted Parrot	Loro Frente blanca
Amazona autumnalis	Red-lored Parrot	Loro Cara amarilla, Chelec
Amazona farinosa	Mealy Parrot	Loro Real, Loro Cabeza azul, Cochá
Ara macao	Scarlet Macaw	Guacamaya Roja
Aratinga nana	Aztec Parakeet	Perico Grande
Pionopsitta haematoti	Brown-hooded Parrot	Loro Cabeza Sucia
CUCULIDAE		
Crotophaga sulcirostris	Groove-billed Ani	Pijuy
Dromoccoccyx phasianellus	Pheasant Cuckoo	
Piaya cayana	Squirrel Cuckoo	Piscoy, Bop-pic
Tapera naevia	Striped Cuckoo	Cuclillo Listado
STRIGIDAE		
Ciccaba virgata	Mottled Owl	Buho Café
Glaucidium brasilianum	Ferruginous Pygmy-Owl	
Glaucidium minutissimum	Least Pygmy-Owl	Lechucita, Tecolotito
Lophostrix cristata	Crested Owl	Tecolote Cuernos blancos
Otus guatemalae	Vermiculated Screech-Owl	Tecolotito de Guatemala
Pulsatrix perspicillata	Spectacled Owl	Tecolote de Anteojos
CAPRIMULGIDAE		
Caprimulgus vociferus	Northern Whip-poor-will	
Nyctidromus albicollis	Pauraque	Tapacaminos, Pucuyo
Nyctiphrinus yucatanicus	Yucatan Poorwill	Tapacaminos
NYCTIBIDAE		
Nictibius griseus	Grey Potoo	
Nictibius grandis	Great Potoo	Gran tapacamino
Nictibius jamaicensis	Northern Potoo	
APODIDAE		
Chaetura vauxi	Vaux's Swift	Vencejo Común
TROCHILIDAE		
Amazilia beryllina	Berylline Hummingbird	Colibrí Cola roja
Amazilia candida	White-bellied Emerald	Chupaflor, Colibrí, Gorrioncito Panza blanca
Amazilia tzacatl	Rufous-tailed Hummingbird	Chupaflor Cola roja
Campylopterus brevipennis	Wedge-tailed Saberwing	Colibrí Grande cabeza azul
Chlorostilbon canivetii	Canivet's Emerald	Chupaflor, colibrí, Gorrioncito Esmeralda
Heliothryx barroti	Purple-crowned fairy	Chupaflor, Colibrí
Phaeochroa cuvierii	Scaly-breasted Hummingbird	Colibrí Escamado
Phaethornis longuemareus	Little Hermit	Hermitaño Pequeño
Phaethornis superciliosus	Long-tailed Hermit	Hermitaño Grande
TROGONIDAE		
Trogon collaris	Collared Trogon	Aurora cola rayada
Trogon massena	Slaty-tailed Trogon	Aurora pecho rojo
Trogon melanocephalus	Black-headed Trogon	Cocochana
Trogon violaceus	Violaceous Trogon	Cocochana violeta
MOMOTIDAE		
Eumomota superciliosa	Turquoise-browed Motmot	Tolobojo ceja azul
Hylomanes momotula	Tody Motmot	Tolobojito, motmot enano
Momotus momota	Blue-crowned Motmot	Tolobojo garganta azul
ALCEDINIDAE		
Ceryle alcyon	Belted Kingfisher	Martín pescador
Ceryle torquata	Ringed Kingfisher	Martín pesacador grande
Chloroceryle aenea	Pygmy Kingfisher	Martín pescador enano
Ceryle americana	Green Kingfisher	Martincito pescador

Scientific Name Nombre Científico	Common English Name Nombre común en inglés	Common Spanish Name Nombre común en español
BUCCONIDAE		
Bucco macrorhynchos	White-necked Puffbird	Pájaro bobito collarejo
Malacoptila panamensis	White-whiskered Puffbird	Pájaro Bobito Barbon, Viejito Barbudo
GALBULIDAE		
Galbula ruficauda	Rufus-tailed Jacamar	Jacamar, Guardabarrancos
RAMPHASTIDAE		
Pteroglossus torquatus	Collared Aracari	Pichit. Tucancito Negro.
Ramphastos sulfuratus	Keel-billed Toucan	Tucán Real, Tucán Pico Verde
PICIDAE		
Campephilus guatemalensis	Pale-billed Woodpecker	Carpintero Real, Cheje Grande
Celeus castaneus	Chestnut-colored Woodpecker	Cheje Café
Dryocopus lineatus	Lineated Woodpecker	Carpintero Grande
Melanerpes aurifrons	Golden-fronted Woodpecker	Cheje Común
Melanerpes pucherani	Black-cheecked Woodpecker	Cheje de Monte
Melanerpes pygmaeus		Carpintero atacabado
Melanerpes uropygialis		
Veniliornis fumigatus	Smoky-brown Woodpecker	Carpintero Atacabado
Automolus ochrolaemus	Buff-throated Foliage-Gleaner	Furnárido Pardo
Sclerurus guatemalensis	Scaly-throated Leaftosser	Excarvador
Synallaxis erythrothorax	Rufous-breasted Spinetail	Guito Pechirojo.
Xenops minutus	Plain Xenops	Picolezna
Dendrocinchla homochroa	Ruddy Woodcreeper	Trepador Anaranjado, Chiclero
D. anabatina	Tawny-winged Woodcreeper	Trepador Café, chiclero
Dendrocolaptes certhia	Barred Woodcreeeper	Trepador Grande Rayado, Chiclero Grande
Glyphorynchus spirurus	Wedge-billed Woodcreeper	Trepadorcito
Lepidocolaptes souleyetii	Streak-headed Woodcreeper	Trepador
Sittasomus griseicapillus	Olivaceous Woodcreeper	Trepadorcito Aceituno, Chiclerito Gris
Xiphocolaptes turdinus		Trepadorcito
Xiphocolaptes promerhophirinchus	Strong-billed Woodcreeper	
Xiphorhyinchus flavigaster	Ivory-billed Woodcreeper	Trepador Grande Pico Blanco, Chiclero Picudo
Taraba major	Great Antshrike	Batara Mayor
Cercomacra tyranina	Dusky Antbird	Hormiguero
Dysithamnus mentalis	Plain Antvireo	Hormiguero
Formicarius analis	Black-faced Antthrush	Hormiguero Gallito
Microrhopias quixensis	Dot-winged Antwren	Hormiguero ala Punteada
Thamnophilus doliatus	Barred Antshrike	Pavito (macho), Batara Rayado
Attila spadiceus	Bright-rumped Attila	Bigotón
Camptostoma imberbe	Northern Beardless Tyrannulet	
Empidonax difficilis	Pacific Slope Flycatcher	Mosquerito
Empidonax flaviventris	Yellow-bellied Flycatcher	Mosquerito
Empidonax minimus	Least Flycatcher	Mosquerito
Empidonax trailli	Willow Flycatcher	Mosquerito
Empidonax virescens	Acadian Flycatcher	Mosquerito
Leptogon amauricephalus	Sepia-capped Flycatcher	Mosquerito Espalda Amarilla
Megarynchus pitanga	Boat-billed Flycatcher	Taquilla
Myarchus sp	Myiarchus species	Mosquero
M. tyrannulus	Brown-crested Flycatcher	Bigotón
M. yucatanensis	Yucatan Flycatcher	Mosquero
Myobius sulphureipygius	Sulphur-rumped Flycatcher	Mosquerito Culo Amarillo
Myodinastes luteiventris	Sulphur-bellied Flycatcher	Mosquerito
Myopagis viridicata	Greenish Elaenia	Elaenia verde
Myozetetes similis	Social Flycatcher	Mosquero Corona blanca
Oncostoma cienreigulare	Northern Bentbill	Mosquerito Oliva, Pico ganchudo
Onychorhynchus coronatus	Royal Flycatcher	Mosquero Real
Ornithion semifalvum	Yellow-bellied Tyrannulet	Mosquerito
Pitangus sulphuratus	Great Kiskadee	Chepillo, Taquilla

Scientific Name Nombre Científico	Common English Name Nombre común en inglés	Common Spanish Name Nombre común en español
Platyrinchus mystaceus	White-throated Spadebill	Mosquerito Pico Zapato
Pyrocephalus rubinus	Vermilion Flycatcher	Mosquerito Rojo
Rhytipterna holerythra	Rufous Mourner	Planidera Rojiza
Terenotriccus erythrurus	Ruddy-tailed Flycatcher	Mosquerito Cola roja
Todirostrum sylva	Slate-headed Tody-Flycatcher	Titiriji Común
Tolmomyias sulphurescens	Yellow-olive Flycatcher	Mosquerito
Tyrannus melancholicus	Tropical Kingbird	Chatilla
Tyrannus savana moncahus	Fork-tailed Flycatcher	Mosquero Tijereta
Tytira inquisitor	Black-crowned Tityra	Torrejo Rechinador
Tytira semifasciata	Masked Tityra	Puerquito
Xenotriccus mexicanus	Pileated Flycatcher	Mosquerito
COTINGIDAE		
Cotinga amabilis	Lovely Cotinga	Cotinga Linda
Lipaugus unirufus	Rufous Piha	Pija
PIPRIDAE		
Manacus candei	White-collared Manakin	Señorita
Pipra mentalis	Red-capped Manakin	Sargento
Schiffornis turdinus	Thrush-like Manakin	Tontillo
HIRUNDINIDAE		
Hirundo rustica	Barn Swallow	Golondrina
Steigidopteryx sp.	Rough-winged Swallow species	Golondrina
Steigidopteryx serripennis	Northern Rough-winged Swallow	Golondrina ala de Sierra
Tachycineta albilinea	Mangrove Swallow	Golondrina
CORVIDAE		
Cyanocorax beechei	Purplish-backed Jay	Shara
Cyanocorax yncas	Green Jay	Chara Verde
Cyanocorax yucatanicus	Yucatan Jay	Shara
Psilorhinus morio	Brown Jay	Pea, Urraca
TROGLODITIDAE		
Thryothorus maculipectus	Spot-breasted Wren	Chinchivirín Pinto
Thryothorus ludovicianus	Carolina Wren	Chinchivirín
Uropsila leucogastra	White-bellied Wren	Chinchivirín Panza Blanca
Henicorhina leucostica	White-breasted Wood-Wren	Chinchivirín Pècho Blanco
MUSCICAPIDAE (Sylviinae)		
Poliptila plumbia	Tropical Gnatcatcher	Perlita Tropical
Ramphocaenus melanurus	Long-billed Gnatwren	Pajarito Picudo
MUSCICAPIDAE (Turdinae)		
Catharus ustulatus	Swainson's Thrush	Zorzal
Hylocichla mustelina	Wood Thrush	Zorsal Pinto
Turdus grayi	Clay-colored Thrush	Calandria
MIMIDAE		
Dumetella carolinensis	Grey Catbird	Pájaro Come Chile, Pajaro Gato.
VIREONIDAE		
Hylophilus decuratus	Lesser Greenlet	Vireo Verde
H. ochraceiceps	Tawny-crowned Greenlet	Vireo Leonado
Vireo olivaceus	Red-eyed Vireo	Vireo
Vireo griseus	White-eyed Vireo	Vireo Ojo Blanco
Vireo pallens	Mangrove Vireo	Vireo
Vireo philadelphicus	Philadelphia Vireo	Vireo
Vireolanius puchellus	Green Shrike-Vireo	
EMBERIZIDAE (Parulinae)		
Basileuterus culicivorus	Golden-crowned Warbler	Chip
Dendroica fusca	Blackburnian Warbler	Chipe
Dendroica magnolia	Magnolia Warbler	Chipe Rayado
Dendroica pensylvanica	Chestnut-sided Warbler	Chipe Anaranjado
Dendroica petechia	Yellow Warbler	Chip Amarillo
Dendroica pinus	Pine Warbler	Chipe
Geothylypis nelsoni	Hooded Yellowthroat	Chipe de Antifaz

Scientific Name Nombre Científico	Common English Name Nombre común en inglés	Common Spanish Name Nombre común en español
Geothlypis poliocephala	Grey-crowned Yellowthroat	idem
Geothlypis trichas	Common Yellowthroat	Chip Antifacito
Granatellus sallaei	Grey-throated Chat	Granatello
Helmitheros swaisonii	Swainson's Warbler	Chipe
H. vermivorus	Worm-eating Warbler	Chipe come Gusanos
Ictera virens	Yellow-breasted Chat	Buscabrena
Mniotilita varia	Black-and-White Warbler	Chiclerito Rayado
Oporornis formosus	Kentucky Warbler	Chipe Patudito
Oporornis philadelphia	Mourning Warbler	
Oporornis tolmei	MacGillivray's Warbler	Chipe
Parula americana	Northern Parula	Chip
Protonotaria citrea	Prothonotary Warbler	Chip Anaranjado
Seiurus aurocapillus	Ovenbird	Pizpita
S. novebaracensis	Northern Waterthrush	Chip
Setophaga ruticilla	American Redstart	Rey Chipe
Vermivora chrysoptera	Golden-winged Warbler	Chip
Vermivora peregrina	Tennessee Warbler	Chipe
Wilsonia citrina	Hooded Warbler	Chipe
Wilsonia pusilla	Wilson's Warbler	Chipe Gorra Negra
(Coerebinae)		
Coereba faveola*	Bananaquit	Siquita*
(Thraupinae)		
Cyanerpes cyaneus	Red-legged Honeycreeper	Mielero Patas Rojas
E. affinis	Scrub Euphonia	Calandrita
Eucometis penicillata	Grey-headed Tanager	Hormiguero Cabeza Gris
Euphonia gouldi	Olive-backed Euphonia	Calandrita Verde
Euphonia hirundinaceae	Yellow-throated Euphonia	Calandrita
Habia fuscicauda	Red-throated Ant-Tanager	Hormiguero Rojo
Habia rubica	Red-crowned Ant-Tanager	Hormiguero Rojo
Pheucticus ludovicianus*	Western Tanager	Tangara
Piranga rubra	Summer Tanager	Quitrique Colorado
Rhampocelus passerinii*	Scarlet-rumped Tanager	Terciopelo*
Tangara larvata	Golden-hooded Tanager	Quitrique Careto
Thraupis abbas	Yellow-winged Tanager	Carbonero
T. episcopus	Blue-grey Tanager	Pito
(Cardinalinae)		
Cardinalis cardinalis	Northern Cardinal	Cardenal
Caryothraustes poliogaster	Black-faced Grosbeak	Piquigrueso Enmascarado
Cyanocompsa cyanoides	Blue-black Grosbeak	Pico grueso Azul
Cyanocompsa parellina	Blue Bunting	Ruiz Azul
Guiraca caerulea	Blue Grosbeak	Pico Grueso Azul
Passerina ciris	Painted Bunting	Colorín Siete Colores
Passerina cyanea	Indigo Bunting	Colorín Azul
Saltator atriceps	Black-headed Saltator	Chinchigorrión Cabeza Negra
Saltator coerulescens	Greyish Saltator	Chinchigorrión
Saltator maximus	Buff-throated Saltator	Chinchigorrión
(Emberizinae)		
Arremonops aurantirostris	Orange-billed Sparrow	Pico de Oro
Arremonops chloronotus	Green-backed Sparrow	Talero
Arremonops rufivigatus	Olive Sparrow	Gorrion Aceitunado
Sporophila minuta	Ruddy-breasted Seedeater	Jaulin
S. torqueola	White-collared Seedeater	Jaulín de Collar
Volantina jacarina	Blue-black Grassquit	Jaulín Azulado
(Icterinae)		
Agelaius phoeniceus	Red-winged Blackbird	Tordo Sargento
Amblycercus holosericeus	Yellow-billed Cacique	Zanate Pico Amarillo
Dives dives	Melodious Blackbird	Zanate Cantor
Icterus chrysater	Yellow-backed Oriole	Chorcha

Scientific Name Nombre Científico	Common English Name Nombre común en inglés	Common Spanish Name Nombre común en español
Icterus cucullatus	Hooded Oriole	Chorcha
Icterus dominicensis	Black-cowled Oriole	Chorcha, Bolsero Capa Negra
Icterus galbula	Baltimore Oriole	Chorcha
Icterus gradacauda	Audubon's Oriole	Chorcha
Icterus mesomelas	Yellow-tailed Oriole	Chorcha
Icterus spurius	Orchard Oriole	Chorcha
Molothrus aeneus	Bronzed Cowbird	Tordo Ojo Rojo
Oryziborus funereus	Thick-billed Seedfinch	
Psarocolius montezuma	Montezuma's Oropendola	Oropéndola
Quiscalus mexicanus	Great-tailed Grackle	Clarinero

Programa de Evaluación Rápida

Amphibian and reptile species of Laguna del Tigre National Park, Petén, Guatemala

APPENDIX 12

Especies de anfibios y reptiles del Parque Nacional Laguna del Tigre, Petén, Guatemala

APENDICE 12

Francisco Castañeda Moya, Oscar Lara, and Alejandro Queral-Regil

See Appendices 1 and 2 for descriptions of sampling points.
Ver Apéndices 1 y 2 para descripciones del lugar de muestreo.

FAMILY/FAMILIA Species/Especies	SPT1: Pato 1	SPT5: Río Sacluc	ET2: Saraguate	ET3: Icaco	Laguna Flor de Luna*	Río Chocop*	Lagoons along Xan road
BUFONIDAE							
Bufo marinus	X+						
Bufo valliceps	X*	X		X		X*	
MICROHYLIDAE							
Gastrophryne elegans			X	X		X*	
HYLIDAE							
Agalichnys callidryas	X*	X					
Hyla ebracatta						X*	
Hyla microcephala	X	X	X		X*	X*	
Hyla picta	X	X	X	X	X*	X*	
Phrynohyas venulosa	X	X				X*	
Scinax staufferi	X	X	X			X*	
Smilisca baudinii	X		X			X*	
LEPTODACTYLIDAE							
Leptodactylus melanonotus	X	X	X	X	X*	X*	
Leptodactylus labialis		X				X*	
RANIDAE							
Rana berlandieri	X	X	X	X	X*	X*	
Rana vaillanti		X	X	X		X*	
GEKKONIDAE							
Coleonyx elegans	X+						
Sphaerodactylus glaucus	X+					X+	
TEIIDAE							
Ameiva undulata	X						
CORYTOPHANIDAE							
Basiliscus vittatus	X	X	X				
SCINCIDAE							
Mabuya brachiopoda	X						
POLYCHROTIDAE							
Norops bourgeaei	X	X	X				
Norops uniformis	X						
Norops sp3	X						
BOIDAE							
Boa constrictor	X+						

FAMILY/FAMILIA *Species/Especies*	SPT1: Pato 1	SPT5: Río Sacluc	ET2: Saraguate	ET3: Icaco	Laguna Flor de Luna*	Río Chocop*	Lagoons along Xan road
COLUBRIDAE							
Coniophanes bipunctatus	X⁺						
Coniophanes quinquevittatus	X⁺						
Drymarchon corais			X⁺				
Drymobius margaritiferus	X⁺					X⁺	
Imantodes cenchoa	X	X					
Leptodeira polysticta	X			X	X⁺		
Tretanorhinus nigroluteus	X⁺						
VIPERIDAE							
Bothrops asper				X			
KINOSTERNIDAE							
Kinosternon leucostomum	X			X	X⁺		X⁺
Staurotypus triporcatus	X				X⁺		
DERMATEMYDIDAE							
Dermatemys mawii							X⁺
EMYDIDAE							
Trachemys scripta	X⁺		X				
Totals from transects	17	13	11	9	--	--	0

* Localities sampled and species found by Méndez et al. (1998).
+ Species found during the RAP expedition but not in the study transects (anecdotal observations).

Mammal species recorded within each focal area in Laguna del Tigre National Park, Petén, Guatemala

Especies de mamíferos registrados dentro de cada área focal en el Parque Nacional Laguna del Tigre, Petén, Guatemala

Heliot Zarza and Sergio G. Pérez

See Appendix 2 and Gazetteer for sampling site descriptions.
Ver Apéndice 2 y Diccionario Geográfico para descripciónes del lugar de muestreo

Sampling Sites/ Lugar de muestreo:
I.   Río San Pedro
II.   Río Escondido
III. Laguna Flor de Luna
IV.   Río Chocop

Order/Orden	Species/Especies	Sampling Sites/Lugar de Muestreo			
		I	II	III	IV
CHIROPTERA	Rhynchonycteris naso		X		X
	Saccopteryx bilineata	X			
	Noctilio leporinus	X			
	Mormoops megalophylla	X			
	Pteronotus parnellii	X	X	X	X
	Micronycteris schmidtorum		X		
	Desmodus rotundus	X			
	Chrotopterus auritus	X			
	Artibeus intermedius	X	X	X	X
	Artibeus jamaicensis	X	X		X
	Artibeus lituratus	X	X		X
	Dermanura phaeotis	X	X	X	X
	Dermanura watsoni	X	X	X	X
	Carollia brevicauda	X	X		
	Centurio senex	X			
	Glossophaga commissarisi	X			
	Mimon bennettii	X		X	
	Mimon crenulatum		X		
	Sturnira lilium	X	X	X	
	Tonatia brasiliense				X
	Tonatia evotis	X	X	X	
	Uroderma bilobatum		X	X	
	Molossus ater	X			
PRIMATES	Alouatta pigra	X	X	X	X
	Ateles geoffroyi	X	X	X	X
RODENTIA	Sciurus deppei				X
	Heteromys desmarestianus	X			
	Heteromys gaumeri				X
	Oryzomys couesi	X			
	Ototylomys phyllotis	X			
	Peromyscus yucatanicus			X	
	Sigmodon hispidus		X		
	Agouti paca	X	X		

Order/Orden	Species/Especies	Sampling Sites/Lugar de Muestreo			
		I	II	III	IV
CARNIVORA	*Potos flavus*	X	X		
	Nasua narica	X			X
	Procyon lotor			X	
PERISSODACTYLA	*Tapirus bairdii*	X			
ARTIODACTYLA	*Tayassu tajacu*	X		X	
	Mazama americana	X			
	Odocoileus virginianus			X	X

CONSERVATION INTERNATIONAL   **Programa de Evaluación Rápida**

Ant (Hymenoptera: Formicidae) species recorded in Laguna de Tigre National Park, Petén, Guatemala

APPENDIX 14

Especies de hormigas (Hymenoptera:Formicidae) registradas en el Parque Nacional Laguna del Tigre, Petén, Guatemala

APENDICE 14

Brandon T. Bestelmeyer and Roy R. Snelling

Biogeographic range type:
C= Central America
Cn= northern Central America
Cw= Central America and into United States
M= Maya forest
N= Neotropical
Nw= Neotropical and into United States

Functional group:
At= Attini, Cr= cryptic species
Dd= dominant dolichoderines
Gm= generalized myrmicines
Lg= legionary species
Op= opportunists
Sc= subordinate camponotini
Sp= specialist predators

Habitat:
A= Arboreal
T= Terrestrial

Tipos de áreas biográficas:
C= América Central
Cn= América Central septentrional
Cw= América Central hacia Estados Unidos
M= Bosque Maya
N= Neotrópico
Nw= Neotrópico hacia Estados Unidos

Grupo funcional:
At= Attini, Cr= especies crípticas
Dd= Dolicoderinos dominantes
Gm= Myrmisinos generalizados
Lg= Especies legionarias
Op= Oportunistas
Sc= Camponotini subordinados
Sp= Depredadores especializados

Hábitat:
A= Arbóreo
T= Terrestre

SUBFAMILY SUB FAMILIA	Genus Género	RangeType Tipo de Rango	Functional Group/ Habitat Grupo Funcional/Hábitat
PONERINAE	*Belonopelta* cf. *deletrix* Mann	Cn	Cr/T
	Discothyrea cf. *horni* Menozzi	?	Cr/Tor A
	Ectatomma ruidum Roger	N	Op/T
	Ectatomma tuberculatum (Olivier)	N	Op/T, A
	Gnamptogenys tornata (Roger)	C	Cr/T
	Hypoponera nitidula (Emery)	Cn	Cr/T, A
	Hypoponera sp. A	--	Cr/T, A
	Hypoponera sp. B	--	Cr/T, A
	Odontomachus bauri Emery	N	Sp/T
	Odontomachus brunneus Patton	?	Sp/T
	Odontomachus minutus Emery	N	Sp/T
	Odontomachus yucatecus Brown	M	Sp/T
	Pachycondyla apicalis (Latrielle)	N	Sp/T, A
	Pachycondyla harpax (Fabricius)	Nw	Sp/T, A
	Pachycondyla lineaticeps (Mayr)	C	Sp/T, A
	Pachycondyla stigma (Fabricius)	N	Sp/T, A
	Pachycondyla unidentata (Mayr)	N	Sp/T, A
	Pachycondyla villosa (Fabricius)	Nw	Sp/T, A
	Thaumatomyrmex ferox Mann	C	Cr/T

Ant (Hymenoptera: Formicidae) species recorded in LTNP, Petén, Guatemala

SUBFAMILY SUB FAMILIA	Genus Género	RangeType Tipo de Rango	Functional Group/ Habitat Grupo Funcional/Hábitat
PSEUDOMYRMECINAE	*Pseudomyrmex boopis* (Roger)	N	Sp?/A
	Pseudomyrmex elongatus (Mayr)	N	Sp?/A
	Pseudomyrmex ejectus (F. Smith)	?	Sp?/A
	Pseudomyrmex ferrugineus (F. Smith)	Cn	Sp?/A
	Pseudomyrmex gracilis	N	Sp?/A
	Pseudomyrmex oculatus (F. Smith)	?	Sp?/A
	Pseudomyrmex pepperi (Forel)	Cn	Sp?/A
	Pseudomyrmex PSW-13	?	Sp?/A
	Pseudomyrmex simplex (F. Smith)	Nw	Sp?/A
ECITONINAE	*Eciton burchelli* (Westwood)	N	Lg/T
	Eciton hamatum (Fabricius)	N	Lg/T
	Labidus c.f. *coecus* (Latrielle)	N	Lg/T
	Labidus praedator (F. Smith)	N	Lg/T
MYRMICINAE	*Acromyrmex octospinosus* (Reich)	N	At/T
	Apterostigma pilosum complex	--	At/T
	Atta cephalotes (Linné)	N	At/T
	Cephalotes basalis (F. Smith)	Cn	?/A
	Cephalotes biguttatus (Emery)	Cn	?/A
	Cephalotes cristatus (Emery)	N	?/A
	Cephalotes kukulcan (Snelling)	?	?/A
	Cephalotes minutus (Fabricius)	N	?/A
	Cephalotes scutulatus (F. Smith)	C	?/A
	Cephalotes umbraculatus (Fabricius)	N	?/A
	Crematogaster brevispinosus complex	--	Gm/A
	Crematogaster sp. B	--	Gm/A
	Crematogaster sp. C	--	Gm/A
	Cyphomyrmex minutus Mayr	Nw	At/A
	Cyphomyrmex sp. A (new sp.?)	--	At/?
	Hylomyrma dentiloba (Sanstchi)	C	?
	Leptothorax sp.	--	Cr/A
	Monomorium ebeninum Forel	C	Gm/T
	Monomorium floricola (Jerdon)	N	Gm/T
	Neostruma sp.	--	Cr/T
	Octostruma balzani (Emery)	N	Cr/T
	Pheidole fimbrata Roger	N	Gm, Cr/T, A
	Pheidole punctatissima Mayr	C	Gm, Cr/T, A
	Pheidole sp. A	--	Gm, Cr/T, A
	Pheidole sp. B	--	Gm, Cr/T, A
	Pheidole sp. C	--	Gm, Cr/T, A
	Pheidole sp. D	--	Gm, Cr/T, A
	Pheidole sp. E	--	Gm, Cr/T, A
	Pheidole sp. F	--	Gm, Cr/T, A
	Pheidole sp. H	--	Gm, Cr/T, A
	Pheidole sp. I	--	Gm, Cr/T, A
	Pheidole sp. J	--	Gm, Cr/T, A
	Pheidole sp. K	--	Gm, Cr/T, A
	Pheidole sp. L	--	Gm, Cr/T, A
	Pheidole sp. M	--	Gm, Cr/T, A
	Pheidole sp. N	--	Gm, Cr/T, A
	Pheidole sp. O	--	Gm, Cr/T, A
	Pheidole sp. P	--	Gm, Cr/T, A
	Pheidole sp. Q	--	Gm, Cr/T, A
	Pheidole sp. R	--	Gm, Cr/T, A
	Pheidole sp. S	--	Gm, Cr/T, A
	Rogeria belti Mann	C	Cr/T

Programa de Evaluación Rápida

SUBFAMILY SUB FAMILIA	Genus Género	RangeType Tipo de Rango	Functional Group/ Habitat Grupo Funcional/ Hábitat
	Rogeria cornuta Kugler	Cn	Cr/T
	Sericomyrmex aztecus (Forel)	M?	At/T
	Smithistruma margaritae (Forel)	Nw	Cr/T
	Solenopsis geminata (Fabricius)	Nw	Gm/T
	Solenopsis succinea Emery	C?	Gm, Cr/T, A
	Solenopsis sp. A	--	Gm, Cr/T, A
	Solenopsis sp. B	--	Gm, Cr/T, A
	Solenopsis sp. C	--	Gm, Cr/T, A
	Solenopsis sp. D	--	Gm, Cr/T, A
	Strumigenys elongata Roger	N	Cr/T
	Trachymyrmex saussurei (Forel)	Cn	At/T
	Trachymyrmex sp. A	--	At/T
	Trachymyrmex sp. B	--	At/T
	Wasmannia auropunctata (Roger)	Nw	Cr/T,A
	Xenomyrmex skwarrae Wheeler	M	?
DOLICHODERINAE	*Azteca forelii* Emery	Cn	Dd/A
	Azteca instabilis complex	--	Dd/A
	Azteca sp. A	--	Dd/A
	Azteca sp. B	--	Dd/A
	Azteca sp. C	--	Dd/A
	Dolichoderus bispinosus (Olivier)	N	Dd/A
	Dolichoderus diversus Emery	C?	Dd/A
	Dorymyrmex sp.	--	Op/T
	Forelius sp.	--	Hc/T
	Tapinoma inrectum Forel	C	Op?/T
	Tapinoma ramulorum Emery	C?	Op?/T
FORMICINAE	*Brachymyrmex* sp.	--	?
	Camponotus atriceps (F. Smith)	N	?/T
	Camponotus auricomus Roger	C	Sc/T,A
	Camponotus novogranadensis Mayr	N	Sc/T,A
	Camponotus planatus Roger	Cw	Sc/T,A
	Camponotus rectangularis Emery	N	Sc/T,A
	Camponotus sericeiventris (Guerin)	N	Sc/T,A
	Camponotus sp. A	--	Sc/T,A
	Paratrechina terricola (Buckley)?	Nw	Op/T
	Paratrechina sp. A	--	Op/T
	Paratrechina sp. B	--	Op/T